Big Data Analytics and Intelligent Techniques for Smart Cities

Big Data Analytics and Intelligent Techniques for Smart Cities

Edited by
Kolla Bhanu Prakash, Janmenjoy Nayak,
B. T. P. Madhav, Sanjeevikumar Padmanaban,
and Valentina Emilia Balas

CRC Press
Taylor & Francis Group
Boca Raton London New York

CRC Press is an imprint of the
Taylor & Francis Group, an **informa** business

First edition published 2022
by CRC Press
6000 Broken Sound Parkway NW, Suite 300, Boca Raton, FL 33487-2742

and by CRC Press
2 Park Square, Milton Park, Abingdon, Oxon, OX14 4RN

Library of Congress Cataloging-in-Publication Data
Names: Prakash, Kolla Bhanu, editor.
Title: Big data analytics and intelligent techniques for smart cities /
edited by Kolla Bhanu Prakash, Janmenjoy Nayak, B.T.P. Madhav,
Sanjeevikumar Padmanaban, and Valentina E. Balas.
Description: First edition. | Boca Raton, FL: CRC Press, 2021. |
Includes bibliographical references and index.
Identifiers: LCCN 2021019343 (print) | LCCN 2021019344 (ebook) |
ISBN 9780367753559 (hbk) | ISBN 9781032034522 (pbk) |
ISBN 9781003187356 (ebk)
Subjects: LCSH: Smart cities. | Big data.
Classification: LCC TD159.4 .B54 2021 (print) | LCC TD159.4 (ebook) |
DDC 307.1/16028557—dc23
LC record available at https://lccn.loc.gov/2021019343
LC ebook record available at https://lccn.loc.gov/2021019344

ISBN: 9780367753559 (hbk)
ISBN: 9781032034522 (pbk)
ISBN: 9781003187356 (ebk)

Typeset in Times
by codeMantra

Dedicated to

Parents, family members, students and Almighty

Contents

Preface

In modern days, cities are viewed as the reflection of the face of a nation. A smart city is a multifaceted and modernistic urban area that addresses and serves the needs of inhabitants. The latest projections indicate that in developing regions by 2030, the world is expected to have 43 megacities with over 10 million inhabitants. For any successful development of a country, sustainable urbanization is also a key factor. Artificial Intelligence (AI) and Machine Learning (ML) approaches have progressively become an integral part of numerous industries. They are now finding their way to smart city projects to simplify and advance operational processes at large. The smart city analyzes through the lens of mobile crowd-sensing and computing through emphasizing a wider aspect with hybrid sensing, participatory, opportunistic, and mobile social networking with a focus on the integration of machines and human intelligence in this platform as an emerging yet promising solution. A synergistic activity between machines and humans is generated by this approach, although machines may process the bulk of raw data and enhance the decision-making process. Intelligent techniques such as Deep Learning provide promising future directions in the implementation of several aspects of the smart city including smart transportation, intelligent infrastructures, smart governance, smart urban modeling, sustainability, smart health solutions, smart education, security, and privacy. The concept behind a smart city is to use advanced technology and data analytics to efficiently provide services to smart city residents on data collected by sensors. Deep Learning plays a vital role in intelligent computer vision for effective decision-making that can be used significantly to obtain data insights, understand data patterns for classification, and/or predict data. Smart cities prefer the direction of transfer learning for the distribution of training and testing, transferred from one platform to another. Deep Learning approaches with semantic technologies make smart city applications to enable better interaction of smart devices with users. The use of Deep Reinforcement learning algorithms combined with virtual objects will help construct virtual representations of physical objects so that the objects would work automatically. These techniques derive a future interest for smart cities for the incorporation of speech recognition technologies that allow comprehension of natural language in smart devices. The potential area of intelligent learning technologies such as wearables and mobile devices in smart cities allows space for senior citizens and lesser technically savvy users

Intelligent learning techniques have transformed the concept of a smart city into existence with the evolution of IoT alongside Big Data analytics. The concept behind a smart city is to use advanced technology and data analytics to efficiently provide services to the inhabitants of the smart city on data collected by sensors. As we know, smart city-oriented anticipatory platforms collect data from various sources (e.g., sensors) to distinguish the context and apply intelligent approaches to envisage the future outcome. The context of Big Data is a recent development study in data analytics for smart cities. As Big Data true power comes in the form of data analytics which derives qualitative and quantitative information to provide a compact framework toward assisting the citizens of smart cities with effective decision-making. They examine the domain, priorities, resources needed, and available (cloud-based and centralized) frameworks for data-analytical techniques implementation with advanced management tools (database) to provide insight into decision-making for smart cities.

This book is a selective collection of basics to advance approaches of Big Data analytics for smart cities, and it explores the possible future applications and challenges of this technology. It also simulates the theory and applications of Big Data modeling in the context of smart cities and illustrates the case studies of some of the existing smart cities across the globe. With the help of present technological innovations and digital practices, it shows the ways to develop sustainable smart cities, which includes several important aspects such as system design, system verification, real-time control and adaptation, Internet of Things, and test beds. Moreover, a detailed applicative and analytical viewpoint is described in the context of Smart Transportation/Connected Vehicle and Intelligent Transportation Systems for improved mobility, safety, and environmental protection. This book addresses a few important subjects regarding smart cities, such as smart education, smart culture, and smart transformation management for social and societal changes that are brought up by the implementation and institutionalization of smart cities.

In Chapter 1, Ayesha et al. discuss the description of a smart education system and present a conceptual framework. The framework of smart pedagogies and key features of a smart learning environment is planned for adoptive smart learners who require mastery of knowledge and skill in the 21stcentury era of learning. A smart pedagogics framework embraces class-based discriminated instructions, assemblage-based personalized learning, and mass-based procreative learning. A smart education system advances from current educational information maintained by technologies such as cloud computing, IoT, and mobile internets; builds a pervasive network atmosphere and cloud computing data center; and creates a sensing system of multi-dimensional IoT. Further, this chapter carries out research on the application of Big Data in a smart education system, endorsing transformation of the information portal of universities into the services portal. Additionally, technological architecture of a smart education, which accentuates the role of smart computing, has also been discussed.

Chapter 2 sheds light on analytics for voltage prediction of various buses in an IEEE standard power system, using R programming. Prasad and Sudhakar have proposed a simplified approach for the prediction of voltage using R STUDIO for IEEE 6 bus systems. The result of the prediction when compared with the conventional method seems to be more accurate, reducing the computation time with good convergence. The difficulties of the conventional techniques such as accuracy, complexity in matrix formation, speed, efficiency, and limited Machine Learning are overcome by this methodology. The output of the system seems to be more efficient in comparison with that of the conventional method such as load flow analysis using Newton–Raphson method. Moreover, the proposed approach on account of its effectiveness, simplified algorithms, and lesser execution time will be befitting for online practical implementations.

In Chapter 3, a robust regularized coding model and an iteratively reweighted regularized robust coding algorithm for robust face recognition (FR) has been proposed by Sandhya and Anitha. One significant advantage of RRC is its robustness to a variety of outlier pixels by seeking an approximate maximum a *posteriori* estimation solution to the coding problem. By assigning the weights to the pixels, adaptively and iteratively, according to their coding residuals, the IR^3C algorithm is able to robustly identify the outliers and thereby diminish their effects on the coding process. Also, it was shown that the l2-norm regularization entails a much lower computational cost when compared with l1-norm regularization, without compromising on the performance in RRC. The proposed RRC methods were thoroughly assessed on face recognition under a variety of conditions, including nonuniform illumination, expression variation, occlusion, and corruption. The experimental results altogether suggest that RRC performs remarkably better than various state-of-the-art techniques. More specifically, RRC with l2-norm regularization could realize very high recognition rates, while offering the benefit of low computational costs, thus proving it to be a good candidate model for practical robust face recognition systems.

Chapter 4 provides introduction to Big Data, analysis, interpretation, and management of secured smart healthcare. Further, Sucharita et al. have discussed challenges of Big Data analytics such as the high volume of data collected from various healthcare centers across multiple platforms. The main intention is to propose a secured smart healthcare framework using Big Data. In this chapter, security and privacy have been proposed by using a data security and privacy layer. It provides additional security features including monitoring activity, masking data, and homomorphic encryption. The proposed framework provides uniqueness in maintaining the security of the patient's data.

Chapter 5 addresses the involvement of Big Data analysis for smart healthcare systems. Arnaja et al. have explored the various methods of artificial intelligence used in the health system. In this portion, several selective diseases were also used to address the role of ML techniques to support the health sector. The IoT-enabled healthcare systems and the significance of different sensors as IoT devices are highlighted. Further, the chapter also deliberates the storage and protection issues for patients' private medical data. Besides, the cloud computing aspects are also considered in the sense of intelligent healthcare systems involving massive data storage. Finally, this chapter also outlines some future research directions in the coming years.

Chapter 6 provides a comprehensive overview of smart energy systems in Big Data context. Initially, Tanya et al. discussed the different methods for Big Data analysis involved in the load and prediction of prices in the smart grid environment. Moreover, the authors also discuss a range of other issues relating to smart networking, such as creating a cloud-based smart grid platform, linking smart grids together with the Internet of Things, automated demand response, real-time smart grid pricing, etc. Further, the chapter offers Big Data analysis for the management of energy. Additionally, it addresses Big Data analysis' participation in smart cities and advanced metering schemes. The role of large data in various industrial applications is highlighted. Last but not the least, many different applications of Big Data such as energy internet, maintenance systems, and social and environmental sustainability are also considered.

In Chapter 7, the best possible placement of multiple Distribution Generators in the distributed system has been suggested by Kavitha et al. The main aim of the chapter is to reduce the power losses in the distribution side. This can be achieved by Distribution Generators being operated at power factors nearer to the power factor of the combined load in the considered system. Moreover, the Voltage Stability Index in the distribution network was improved by placing multiple Distribution Generators. In this work, the ability of the Distribution Generators to withstand the load growth for years is also discussed. This chapter suggests shuffled frog leaping algorithm for optimal placement of multiple Distribution Generators. This proposed methodology is tested on IEEE radial distribution system having 33 buses. The base case analysis is done along with single, double, and triple Distribution Generator placement. The results were compared with a few optimizations methods such as improved analytical approach, mixed integer nonlinear programming, and particle swarm optimization. The comparison encourages the use of shuffled frog leaping algorithm for Distribution Generator placement. The load growth analysis helps us to keep track of future expansion planning requirement.

Chapter 8 is mainly focused on Big Data for smart energy. The chapter addresses how Big Data analytics is useful in smart grids. Smart energy systems, smart appliances, smart meters, and synchrophasors are discussed. The benefits of smart grids and importance of IoT in smart grids research are also outlined.

Chapter 9 proposes an intelligent ML-based framework approach for distinguishing and classifying anomalies from normal behavior based on the type of attack. Further, Karun et al. have estimated the complete experimentation performance and evaluations of ML algorithms for recognition of categorical attacks such as data probing, DoS attack, malicious control, malicious operation, scan, spying, and wrong setup found in the DS2OS data set. The experimental results of the simulation model report that the Gradient Boosting algorithm performs well in categorizing the attacks.

Chapter 10 focuses on problems faced by the education sector due to the outbreak of the global pandemic disease COVID-19. Further, Saumyadip and Souvik describe the process of switching over to smart education, which included the delivery of lectures, webinars, and conferences online. The organization of virtual laboratories is also described. The chapter also discusses the various problems that may be faced by attendees during the processes. The students may face some difficulties due to switching over from conventional methods. Virtual laboratories or simulation software are proving helpful in making the students understand their theoretical subjects in a better way. The quality of laboratories may improve in the future. Webinars have quite nicely replaced the seminars, and this may prove very helpful in the coming years as people from different states or even different countries will be able to attend them. There are some areas of improvement for organizing the e-conferences. Some lack from the management side has been observed, which may be due to the many aspects managed at the same time and this is getting better day by day.

Chapter 11 focuses on the concept of smart building and summarizes the various Big Data research projects occurring in this area for energy conservation and occupants' comfort. Further, Manimala and Sivanthi have performed a detailed review of the application of IoT-based sensors and the processing techniques for occupant detection and human action recognition using ML algorithm for reducing energy consumption. Even though the advancements in recent technologies make the concept of smart buildings realistic, still there are various issues and challenges that limit full-scale implementation of real-world smart buildings. Addressing these challenges is a powerful driving force for technical advancements in both industrial and academic areas of smart building research.

Chapter 12 classifies and reviews previous trolley models for shopping. In addition, Raju et al. have presented a comparative analysis of different existing models along with their strengths and weaknesses. Moreover, this chapter also introduces a proposed idea of a futuristic trolley by focusing on some limitations of current research activities. In the future, this idea has the potential to become one of the frameworks which will make life simpler for the users in stores.

Chapter 13 provides details and relationship between Big Data and IoT appliances related to a smart healthcare system. In addition, Chetana and Satyanarayana discuss several challenges for actually implementing the Big Data concept using IoT-based secure applications in the healthcare industry along with several opportunities for future research directions.

Editors
Kolla Bhanu Prakash
Janmenjoy Nayak
B. T. P. Madhav
Sanjeevikumar Padmanaban.
Valentina E. Balas

MATLAB® is a registered trademark of The MathWorks, Inc. For product information, please contact:
The MathWorks, Inc.
3 Apple Hill Drive
Natick, MA 01760-2098 USA
Tel: 508-647-7000
Fax: 508-647-7001
E-mail: info@mathworks.com
Web: www.mathworks.com

Acknowledgments

We would like to say thank you to the Almighty and our parents for the endless support, guidance, and love through all our life stages. We are thankful to our beloved family members for standing beside us throughout our careers, helping us to move our careers forward, and through the process of editing this book. We dedicate this book to our family members.

We would like to specially thank Sri. Koneru Satyanarayana, President, K. L. University, India, for his continuous support and encouragement throughout the preparation of this book.

Our great thanks to our students and family members who have put in their time and effort to support and contribute in some manner. We would like to express our gratitude toward all who supported, shared, talked things over, read, wrote, offered comments, allowed us to quote their remarks and assisted in editing, proofreading and designing throughout this book's journey. We pay our sincere thanks to the open data set providers.

We believe that the team of authors provides the perfect blend of knowledge and skills that went into authoring this book. We thank each of the authors for devoting their time, patience, perseverance and effort toward this book; we think that it will be a great asset to the all researchers in this field!

We are grateful to the CRC Press team, who showed us the ropes for creating this book. Without that knowledge, we would not have ventured into starting this book, which ultimately led to this. Their trusting in us, their guidance, and their provision of the necessary time and resources gave us the freedom to manage this book.

Last, but definitely not least, we'd like to thank our readers who gave us their trust, and we hope our work inspires and guides them.

Editors
Kolla Bhanu Prakash
Janmenjoy Nayak
B. T. P. Madhav
Sanjeevikumar Padmanaban
Valentina Emilia Balas

Editors

Dr. Kolla Bhanu Prakash is a Professor and Research Group Head in CSE Department, K. L. University, Vijayawada, Andhra Pradesh, India. He earned his M.Sc. and M.Phil. in Physics from Acharya Nagarjuna University, Guntur, India, M.E. and Ph.D. in Computer Science Engineering from Sathyabama University, Chennai, India. Dr. Kolla Bhanu Prakash has 15+ years of experience working in academia, research, teaching, and academic administration. His current research interests include AI, Deep Learning, Data Science, Smart Grids, Cyber-Physical Systems, Cryptocurrency, Blockchain Technology and Image Processing. Dr. Prakash is an IEEE Senior Member. He is a Fellow-ISRD, Treasurer – ACM Amaravathi Chapter, India, LMISTE, MIAENG, SMIRED. He has reviewed more than 130 peer-reviewed journals that are indexed in Publons. He is the editor of six books published by Elsevier, CRC Press, Springer, Wiley, and Degryuter. He has published 75 research papers, has six patents, and authored seven books, four of which are accepted. His scopus H-index is 14. He is a frequent editorial board member and TPC member in flagship conferences and refereed journals. He is reviewer for *IEEE Access Journal*, Springer Nature, Inderscience Publishers, *Applied Soft Computing Journal* – Elsevier, *Wireless Networks Journal, IET Journals, KSII Journal*, and IEEE Computer Society journals. He is series editor for "Next Generation Computing & Communication Engineering", Wiley publishers; under this series, at present, a 5-book agreement is signed. He is series editor for "Industry 5.0: Artificial Intelligence, Cyber-Physical Systems, Mechatronics and Smart Grids", CRC Press.

Dr. Janmenjoy Nayak is an Associate Professor, Aditya Institute of Technology and Management (AITAM) (An Autonomous Institution), Tekkali, K Kotturu, AP, India. He has published more than 120 research papers in various reputed peer-reviewed, refereed journals, presented at international conferences, and written book chapters. Being a two time Gold Medalist in Computer Science in his career, he has been awarded with INSPIRE Research Fellowship from Department of Science & Technology, Govt. of India (both at JRF and SRF levels) and the best researcher award from Jawaharlal Nehru University of Technology, Kakinada, Andhra Pradesh for the AY: 2018–2019; he also has many more awards to his credit. He has edited 12 books and 8 special issues on various topics including Data Science, Machine Learning, and Soft Computing with reputed international publishers including Springer, Elsevier, Inderscience, etc. His area of interest includes data mining, nature inspired algorithms, and soft computing.

Dr. B. T. P. Madhav was born in India, A.P., in 1981. He earned his B.Sc., M.Sc., MBA, and M.Tech. degrees from Nagarjuna University, A.P., India in 2001, 2003, 2007, and 2009, respectively. He earned his Ph.D. in the field of antennas from K. L. University. Currently, he is working as Professor and Associate Dean at K. L. University. He has published more than 486 papers in international and national journals and presented at many conferences. Scopus and SCI publications of 321 with H-Index of 31 and total citations are more than 3207. He is a reviewer for several international journals including IEEE, Elsevier, Springer, Wiley, and Taylor & Francis and has served as a reviewer for several international conferences. Research interests include antennas, liquid crystals applications, and wireless communications. He is a member of IEEE, a life member of ISTE, IACSIT, IRACST, IAENG, and UACEE, and a fellow of IAEME. He has received several awards including Indian book of records, Asian book of records, outstanding reviewer from Elsevier, best researcher, and distinguished researcher from K. L. University. He received the best teacher award from K. L. University for 2011, 2012, 2013, 2014, 2015, 2016, 2017, 2018, and 2019. He is an editorial board member for 36 journals, has authored 15 books, and has 10 patents to his credit. He guided three Ph.D. scholars who won awards; three Ph.D. scholars have submitted their thesis and six scholars are pursuing their Ph.D. under his guidance.

Dr. Sanjeevikumar Padmanaban (Member 2012–Senior Member 2015, IEEE) earned his Ph.D. in electrical engineering from the University of Bologna, Bologna, Italy in 2012. He was an Associate Professor with VIT University from 2012 to 2013. In 2013, he joined the National Institute of Technology, India, as a Faculty Member. In 2014, he was invited as a Visiting Researcher at the Department of Electrical Engineering, Qatar University, Doha, Qatar, funded by the Qatar National Research Foundation (Government of Qatar). He continued his research activities with the Dublin Institute of Technology, Dublin, Ireland, in 2014. Further, he served as an Associate Professor in the Department of Electrical and Electronics Engineering, University of Johannesburg, Johannesburg, South Africa, from 2016 to 2018. From March 2018 to February 2021, he has been a Faculty Member in the Department of Energy Technology, Aalborg University, Esbjerg, Denmark. Since March 2021, he has been with the CTIF Global Capsule (CGC) Laboratory, Department of Business Development and Technology, Aarhus University, Herning, Denmark. Dr. S. Padmanaban has authored more than 300 scientific papers and was the recipient of the Best Paper cum Most Excellence Research Paper Award from IET-SEISCON'13, IET-CEAT'16, IEEE-EECSI'19, IEEE-CENCON'19 and five best paper awards from ETAEERE'16 sponsored Lecture Notes in Electrical Engineering, Springer book series. He is a Fellow of the Institution of Engineers, India, the Institution of Electronics and Telecommunication Engineers, India, and the Institution of Engineering and Technology, U.K. He is an Editor/Associate Editor/Editorial Board for refereed journals, in particular the *IEEE SYSTEMS JOURNAL, IEEE Transaction on Industry Applications, IEEE ACCESS, IET Power Electronics, IET Electronics Letters*, and Wiley-International *Transactions on Electrical Energy Systems*, Subject Editorial Board Member—Energy Sources—*Energies Journal*, MDPI, and the Subject Editor for the *IET Renewable Power Generation, IET Generation, Transmission and Distribution*, and *FACETS* journal (Canada).

Dr. Valentina Emilia Balas is a Full Professor in the Department of Automatics and Applied Software at the Faculty of Engineering, Aurel Vlaicu University of Arad, Romania. She earned a Ph.D. cum laude in Applied Electronics and Telecommunications from Polytechnic University of Timisoara. Dr. Balas is the author of more than 350 research papers in refereed journals and has presented at international conferences. Her research interests are in Intelligent Systems, Fuzzy Control, Soft Computing, Smart Sensors, Information Fusion, Modeling, and Simulation. She is the Editor-in Chief for *International Journal of Advanced Intelligence Paradigms (IJAIP)* and *International Journal of Computational Systems Engineering (IJCSysE)*, Editorial Board member of several national and international journals, and is an evaluator expert for national and international projects and Ph.D. thesis. Dr. Balas is the director of Intelligent Systems Research Centre in Aurel Vlaicu University of Arad and Director of the Department of International Relations, Programs and Projects in the same university. She served as General Chair of the International Workshop Soft Computing and Applications (SOFA) in nine editions organized in the period of 2005–2020 and held in Romania and Hungary. Dr. Balas participated in many international conferences as Organizer, Honorary Chair, Session Chair, member in Steering, Advisory or International Program Committees, and Keynote Speaker.

Currently, she is working on a national project with EU funding support: BioCell-NanoART = Novel Bio-inspired Cellular Nano-Architectures – For Digital Integrated Circuits, 3M Euro from National Authority for Scientific Research and Innovation. She is a member of European Society for Fuzzy Logic and Technology (EUSFLAT), member of Society for Industrial and Applied Mathematics (SIAM), and a Senior Member IEEE, member in Technical Committee – Fuzzy Systems (IEEE Computational Intelligence Society), chair of the Task Force 14 in Technical Committee – Emergent Technologies (IEEE CIS), and member in Technical Committee – Soft Computing (IEEE SMCS). Dr. Balas was former Vice-President (responsible with Awards) of IFSA – International Fuzzy Systems Association Council (2013–2015), is a Joint Secretary of the Governing Council of Forum for Interdisciplinary Mathematics (FIM) – A Multidisciplinary Academic Body, India, and recipient of the "Tudor Tanasescu" Prize from the Romanian Academy for contributions in the field of soft computing methods (2019).

Contributors

Chetan S. Arage
Department of CSE
School of Technology
Sanjay Ghodawat University
Kolhapur, India

Banerjee Arnaja
Department of Computer Science and
   Engineering
Thapar Institute of Engineering and
   Technology
Patiala, India

Dr. Anil Badarla
University of Technology
Jaipur, India

H.S. Behera
Departmemt of Information Technology
Veer Surendra Sai University of
   Technology
Burla, India

Swadhin Chakrabarty
Department of Electrical Engineering
Regent Education and Research
   Foundation
Barrackpore, India

R. Divya
Department of Electrical and
   Electronics Engineering
Thiagarajar College of Engineering
Madurai, India

Dr. Ahmed A. Elngar
Faculty of Computers and Information
Beni-Suef University
Beni Suef, Egypt

Souvik Ganguli
Department of Electrical and
   Instrumentation Engineering
Thapar Institute of Engineering and
   Technology
Patiala, India

Deepika Ghai
School of Electronics and Electrical
   Engineering
Lovely Professional University
Phagwara, India

Saumyadip Hazra
Department of Electrical and
   Instrumentation Engineering
Thapar Institute of Engineering and
   Technology
Patiala, India

Nikhil Karande
Department of Computer Engineering
GH Raisoni Institute of Engineering
   and Technology
Pune, India

D. Kavitha
Department of Electrical and
   Electronics Engineering
Thiagarajar College of Engineering
Madurai, India

Abhimanyu Kumar
Department of Electrical and
   Instrumentation Engineering
Thapar Institute of Engineering and
   Technology
Patiala, India

B. Ashok Kumar
Department of Electrical and
    Electronics Engineering
Thiagarajar College of Engineering
Madurai, India

K. Manimala
Department of Electrical and Electronic
    Engineering
Dr. Sivanthi Aditanar College of
    Engineering
Tiruchendur, India

Bighnaraj Naik
Department of Computer Application
Veer Surendra Sai University of
    Technology
Burla, India

Ayesha Naureen
University of Technology
Jaipur, India
and
B.V. Raju Institute of Technology
Tuljaraopet, India

Yash Negi
Institute of Advanced Computing
SAGE University
Indore, India

Vidyullatha Pellakuri
Department of Computer Science and
    Engineering
Koneru Lakshmaiah Education
    Foundation
Vaddeswaram, India

Anitha Perla
Electronics and Communication
    Engineering
Malla Reddy College of
    Engineering
Hyderabad, India

H. Prasad
Department of Electrical and
    Electronics Engineering
St. Joseph's College of Engineering
Chennai, India

Balla Adi Narayana Raju
School of Electronics and Electrical
    Engineering
Lovely Professional University
Phagwara, India

P. Venkateswara Rao
Department of Computer Science and
    Engineering
Narayana Engineering College
Gudur, India

Kirti Rawal
School of Electronics and Electrical
    Engineering
Lovely Professional University
Phagwara, India

Dukka Karun Kumar Reddy
Departmemt of Information
    Technology
Veer Surendra Sai University of
    Technology
Burla, India

Hare Ram Sah
Institute of Advanced Computing
SAGE University
Indore, India

K.V.V. Satyanarayana
Department of CSE
Koneru Lakshmaiah Education
    Foundation
Guntur, India

S. Senthilrani
Department of Electrical and
    Electronics Engineering
Velammal College of Engineering and
    Technology
Madurai, India

Tanya Srivastava
Department of Computer Science and
    Engineering
Thapar Institute of Engineering and
    Technology
Patiala, India

Yashonidhi Srivastava
Department of Electrical and
    Instrumentation Engineering
Thapar Institute of Engineering and
    Technology
Patiala, India

V. Sucharita
Department of Computer Science and
    Engineering
Narayana Engineering College
Gudur, India

T.D. Sudhakar
Department of Electrical and
    Electronics Engineering
St. Joseph's College of Engineering
Chennai, India

Sandhya Swaminathan
Electronics and Communication
    Engineering
Malla Reddy College of Engineering
Hyderabad, India

1 Big Data for Smart Education

Ayesha Naureen
University of Technology
B.V. Raju Institute of Technology

Anil Badarla
University of Technology

Ahmed A. Elngar
Beni-Suef University

CONTENTS

1.1   INTRODUCTION

Consistent intelligent design is applied in the field of education as a result of rapid technological developments in terms of what can be instrumented. Smart education, which has recently gained prominence, will be discussed here. Smart education-focused educational programmers must conduct a global study of the current time span. Malaysia decided to take part in a smart education initiative back in 1997. As a result, Malaysia has adopted the smart school strategy. Smart schools, which are backed by the government, aim to expand educational programmers and achieve nationwide education while also ensuring that job force is prepared to face the challenges of the 21st century. Intellectual theory has also been used as a criterion in Singapore, and technology-assisted education was found to be important in 2006. As a result, both schools will place a greater emphasis on the variety of learning environments in the future (traditional as well as smart school strategies). The smart interdisciplinary student central education system was designed in collaboration with IBM in Australia. The system will connect schools and tertiary institutions, and also participate in staff training. In South Korea, the smart education project is the most important challenge for reforming the country's educational system and improving educational infrastructure. A smart school initiative in New York emphasizes the importance of technology in the classroom. Global smart education was given focus in Finland, based on a news release from 2011, and in the UAE, from 2012 onward, and much progress has been made in this novel trend. The related research topics of smart education growth are reviewed in the following sections, and perceptions of smart education and the intangible context aimed at research are proposed, as well as research frameworks for smart education. Furthermore, smart computing is portrayed as a form of smart education in technical architecture. Universities efficient data had reached a stage of smart education, strategy, but rapid commercialization of technologies such as mass storage, cloud computing, and IoT in universities, application of Big Data (BD) has become a specific core application of a smart education system (SES). The BD applications link physical data storage systems to computation-supporting platforms for data collection and sorting, while the management framework is for data analysis and processing. Data is used to link the various sections of a SES. Large amounts of data on the current status and behavior of each component mutually together will be organized through Big Data Analytics (BDA) to existing development designs and placed into smart application [1]. In conclusion, a provocation of easing smart education is provided to stimulate researchers and educators who are interested in smart education project and expansion.

1.1.1 Big Data (BD)

Big Data is a collection of data [2] that is so large and complex that traditional systems are unable to process it. The word also refers to the methods and software used to switch "BD" and sample BD from the large amount of data exchanged on the internet on a daily basis, such as YouTube video views, Twitter feeds, and mobile data location tracking. The current data would create an environment conducive to learning.

1.1.1.1 Big Data Analytics (BDA)

BD has recently been applied to datasets that have become so large that performing tasks with a conventional database management system (DBMS) has become challenging. To collect, stockpile, actively use, and release the data at the required time, extremely large datasets necessitate the use of software resources and storage systems [3]. The size of BDs is continuously growing, ranging from a few hundred TB to several PBs of data in a single dataset. As a result, analytics, storage, discovery, data capture, data exchange, and visualization, as well as exploration of the volume of a massive, broad dataset to discover previously unknown truths, are all complications associated with BD [4].

BDA is a sophisticated analytic method that can be used for BD sets. The analytic foundation of a large data sample allows for the disclosure and exploiting of commercial variance in large datasets. In terms of control, there is an additional issue [4].

Features of BD

Data of a size, distribution, multiplicity, or otherwise timeless nature necessitates the use of newly technical architecture, analytic, and tool orderly allowing insight such as the discovery of a new source of commercial cost. BD's main characteristics are as follows: volume, variety, and velocity, or any three V's volume data, its scope, and how huge it is. The pace at which data changes, or how often it is produced, is defined by velocity. To sum it up, diversity encompasses a wide range of data formats, categories, and uses, as well as various types of data analysis [5]. Furthermore, BD offers the benefit of velocity, or speed, which is largely determined by the density of a data peer group or the density of data dispatch. Flowing data collected in real time from websites [4] is the most significant advantage of BD. Few researchers and organizations have considered the inclusion of a fourth V, veracity, which is concerned with data accuracy. Data inconsistency, deficiency, uncertainty, inactivity, deceit, and estimates are used to classify BD output as good, mediocre, or unspecified (Figure 1.1).

1.1.2 Big Data Architecture Aimed at Learning Analytics

It is essential to plan an architecture for learning analytic framework. The aim of the system is to use a phase structure to seamlessly combine generation, addition,

FIGURE 1.1 Big Data V's.

cleaning, and other types of preprocessing, storage, and management, as well as analytic, visual, and other alert systems. Furthermore, the optimized architecture should
be generic. However, in the following sections, an additional optimized outline aimed
at the real-world implementation is given in Figure 1.2.

- *Data Gathering Device*: The framework is made up entirely of artifacts
  and devices that are responsible for gathering raw data at each stage of
  the advanced education process. Many data collection devices have been
  blamed for amassing unanalyzed data, particularly at every stage of the
  advanced education process. Various data collection devices, such as student ID cards, communal networks, and learning management system
  (LMS), sense student data and can be used as data sources. Since each student is passed through Data Management Systems (DMS) for review, both
  structured and unstructured data are produced.
- *Data Storage and Management Systems*: The DS framework includes
  a massive DBMS as well as features such as buffering and Return Time
  Query Objective (RTQO). This stage is also responsible for—and focuses
  on—data preprocessing rather than data cleaning. Another important feature of a DS and management system is to process and change raw data
  using the hooked-on approach so that data can be efficiently processed by
  the analytic engine [7].
- *Data Analytics Systems*: This is the system's nerve center. A smart processing algorithm is used in the DA framework, and it is designed to extract
  evocative and useful information from raw streams of otherwise static data.
- *Data Visualization*: An impartial data visualization framework forms the
  pictorial representation of an analysis outcome, allowing for a fast decision.
- *Action*: A learning analytics system's goal is to provide learners, administrators,
  and lecturers with information through alarm and warning systems in order to
  direct systemic progress, course design, and participation in instruction.

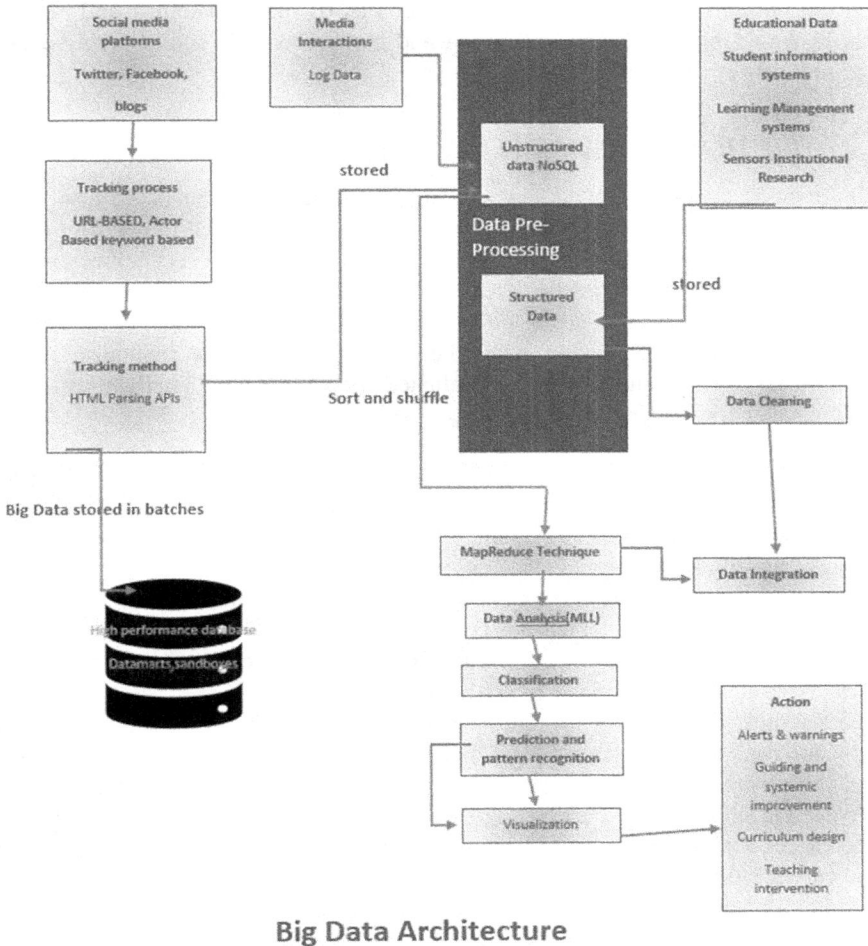

Big Data Architecture

FIGURE 1.2 Big Data architecture.

1.1.3 ROLE OF BIG DATA IN SMART EDUCATION

We may imagine that BD and education both have one or else two to learn from one another. BD systems are used by businesses to capture, store, and analyze large volumes of data as well as gain an invaluable insight into their operations. Innovation is changing the way businesses use data, improve sectors, and transform education. BD systems assist people in cramming information and improving how organizations analyze it, as well as offering opportunities to view data and use the data to make sound decisions. BD systems has assembled a team of experts who can better understand and interpret the data than ever before. Following segmentation and exploring three approaches that BD technology has that are different from the current study climate, BD technology has incredible potential aimed at the future of teaching and learning. Modified syllabi strive for higher-quality learning outcomes, and BD aids this by broadening student presentation and comprehension, thus creating a new learning pathway [8].

1.2   FRUITION OF SMART LEARNING

While smart learning is a groundbreaking academic pattern, it is built on smart devices and intellectual technology (Lee et al. [9]; Kim et al. [10]), despite the fact that technology has been implemented and manipulated within serving learners for decades. It is described as the use of technology to improve studying because technology improves learning and makes educational methods more workable. Technology may take the form of media or some other type of tool aimed at increasing educational access. Bruce and Levin [11] evaluated speech, while Daniel (2012) looked at contact and collaborative construction; Meyer and Latham (2008) investigated assessment in the area of teaching to improve learning [12].

Learning on cell phones has become the dominant TEL (technology-enhanced learning) model, thanks to the proliferation of mobile phones and their related technologies and mobile channels. It emphasizes mobile learning as learning by mobile devices, as well as the learner's ability, as there is no longer a disparity in content, and the out-of-date educational type is now obsolete due to additional support offered by universal technology, which has resulted in further variations in the moveable learning style, that emphasizes that learning can take place at any time and in any place, regardless of time, place, or setting (Hwang et al. [13]).

Numerous scholars have recently begun to emphasize the value and necessity of authentic activity, in which students focus on real-world problems [13]. There is a trend to design learning with the inclusion of a virtual eLearning system in order to place students in a secure learning environment. Smart learning combines a few elements of ubiquitous learning and is extended as 1 to 1 in the TEL model, in which the learner studies through time and place, as well as having the ability to switch learning scenarios through their smart private device (Chan et al. [14]).

Other intelligent technologies that facilitate the advent of smart education include cloud computing, learning analytics, BD, Internet of things (IoT), and wearable technology. Cloud computing, learning analytics, and BD, which concentrate on how learning data can be collected, analyzed, and guided toward enhancing learning and teaching, aid in the creation of personalized and adaptive learning and teaching (Lias and Elias [15]; Mayer-Schonberger, Cukier 2013, Piocciano 2012) [1]. With these adaptive learning technologies, a learning platform can respond to individual learner data and adjust instructional resources accordingly using cloud storage and learning analytics, and it can also use aggregated data from a large number of learners to gain insights into the design and adaptation of curricula using BD (NMC 2015).

IoT and wearable technology are stifling the creation of effective research and unified learning in this area. This IoT will connect people, things, and devices. Learners use smart devices, which have benefits due to the abundance of related knowledge available in the environment (NMC 2015). Wearable technology integrates location data and workout records into learning, as well as a social media interface and a casual realism tool.

1.2.1 INSIGHT OF SMART LEARNING

There is no clear and consistent definition of smart learning. Scholars from various disciplines and educational experts are constantly debating the concept of smart learning. Despite this, only a few key elements have been discussed in the literature. Deliberate smart learning is described by Hwang [16] and Scott and Benlamri [22] as a situation of conscious pervasive education. Gwak [17] imagines how people would perceive the world in the future.

1. It is attentive to learners, in addition to bringing additional content on devices;
2. It is reliable, intellectual, and delivers personalized learning based on advanced IT setup.

Because technology is so important in the support of smart eLearning, it's vital to focus on more than just the use of smart devices [17]. Smart learning, according to Kim et al., would combine the advantages of a social network with the advantages of a traditional classroom. Others also attempted to identify the characteristics of a smart eLearning. Self-reliant, driven, agile, resource-enriching, and technology-embedded are MEST available features of smart learning [20]. Smart learning, according to Lee et al. [9], includes proper and appropriate learning, and social and mixed learnings as well-located learning, implementation, and material value have all changed.

1.2.2 SMART LEARNING ENVIRONMENT

Smart learning environments should, in general, be accurate, efficient, and appealing. The nucleus of a smart eLearning community is often thought to be the learners. The aim of the smart eLearning environment is to provide self-education, self-mutation, and customized services that include learning content rendering to suggested alteration (Kim et al. [18]). For the purpose of facilitating better and faster learning, Koper suggested a smart learning environment that is distinct from the corporeal environment and is enriched with interactive, contextual, aware, and adaptive devices [21]. According to Hwang, potential expectations of a smart learning environment include contextual awareness, the ability to suggest ideas instantly, the ability to adjust to the learner's needs, as well as the ability to acclimate the learner interface and subject material [16]. SLE not only allows learners to access omnipresent resources and interact with learning systems whenever and wherever they want, but also provides them with important learning guides, suggestions, or other helpful tools in the most accurate form, at the most accurate time, and in the most accurate place.

Smart devices can be used to learn anywhere and at any time. In SLE, the aware facet of circumstances plays a significant role. It makes provision for learners to have access to adequate learning resources. SLE is designed by Kim et al. [10]. SLE should be focused on cloud computing. Smart learning facilities provide context-aware support for smart learning graded against learners through data collection and analysis, with the goal of providing updated and customized learning services to learners. Scott and Benlamri developed a SLE that is both learner-centric and service-oriented, as well as a ubiquitous computing scenario, based on a semantic web plume [22]. A universal interactive learning space dominates the educational landscape, which it transforms into a conventional learning space or an intellectual ambient learning environment focused on context awareness and real-time learning. Consider SLE to be the best digital environment for learning context awareness, identifying learner types, providing flexible learning services and inventing communication tools, recording learning procedures frequently, and assessing learning outcomes. Its mission is to make learning more enjoyable, engaging, and active for students.

SLE is learner-driven and collaborative, with a focus on sharing resources and services. Spector (2014) believed that SLE should be productive, reliable, engaging, versatile, adaptive, and reflexive, and that it should help learner and teacher preparation as well as creative alternatives. These features may help with teamwork, stress management, and motivation. SLE promotes learner-centric, customized, and integrated learning services; serves as a connecting and collaborative tool; allows context-aware improvements; and offers omnipresent access, according to a literature review. SLE aims to assist in the realization of reliable, efficient, and meaningful learning for learners.

1.2.3 DENOTATION OF SMART IN SMART LEARNING

Desire for a smart education is for workers to master 21st-century skills as well as ability to meet the needs and challenges of a society's intellectual technology plays an important role in the construction of a smart educational environment. In a smart educational environment, learning can occur whenever and wherever it encompasses a variety of learning styles, such as proper and impromtu learning, as well as social learning plus aims to maintain a learner's eLearning experience by offering personalized learning services and adaptive content tailored to their background, personal abilities, and needs. Hence, "smart" in a smart education typically refers to intellectual, personalized, and adaptive learning. Then, there are different meanings of smart for different entities and/or educational circumstances. For learners, smart means the ability to allow people to think quickly and creatively in a variety of situations.

Smart refers to an educational technology's ability to achieve its purpose both efficiently and competently (Spector 2014); technology refers to both hardware and software. Smart is a hardware term that refers to a smart computer that is both portable and affordable. It is simply intended to provide support for the learner, since they can access the learning at any time and from any location using a smart device. Smartphones and laptops may also identify and collect learning data to keep learners engaged in a specific context and provide coherent learning. Aimed at software, smart refers to its adaptive and modular nature, as well as its ability to personalize eLearning for each learner, according to their specific needs, using adaptive learning technologies such as BD, cloud computing, learning analytics, and adaptive engines.

Smart refers to engaging, bright, and scalable smart educational environments that can provide custom-made and personalized learning services to engage learners in an effective, accessible, and expressive learning system. To support convergence, increase ease, and encourage smart device and learning, open system architectures must be improved.

1.2.4 SMART LEARNER

Learning is now broadly defined as the process of acquiring capability and understanding. It leads to new abilities to do and comprehend things that were previously unknown. Capability is often defined as the possession of a precise skill, or as the possession of precise information.

People in the 21st century are expected to have certain skills and capabilities in order to survive and live efficiently through work and leisure. Education is needed to create a workforce that is prepared to take on the challenges of working in the 21st century.

An abundance of new organizations are springing up to teach 21st-century skills to individuals. In the 21st century, organizations aiming at monetary and educational advancement as well as development have been grouped into four categories: ways of intelligence, resources for work, ways of work, and ways to live. Firm for the 21st century (P21 2015) The following skills suggest a structure for learning and show that students must master knowledge and skills: key subject plus 21st-century theme; learning and innovation skills; information, media, and technology ability; life and career skill; here, Lab proposes that digital age literateness, inventive thinking, active communication, and hi-tech communication are key skills for the digital era.

Based on these observations, we propose four levels of smart education capabilities that students can master in order to meet the needs of today's society. Basic knowledge and essential skills, such as capabilities, personalizing proficiency, and enhancing communal intelligence, are included in these capabilities. These capabilities are divided into four categories: awareness, ability, attitude, and meaning. The four levels of capabilities are presented in depth in the subsequent pages.

i. *Elementary Knowledge and Essential Skills*: Elementary knowledge and skills, such as those for information and skills in important subjects like STEM, reading, writing, and painting, are all important. The mastery of basic subjects is critical to a student's success (P21 2015). Reading, writing, and mathematics are the critical skills for the 21st century.

ii. *Comprehensive Capabilities*: Skills for identifying and addressing current global circumstances are included in comprehensive capabilities. The majority of 21st-century competence systems place greater emphasis on people's thoughtful behaviors. Students may use proper logic and all-inclusive thinking to find dissimilar nuanced solutions with these skills. Student should resolve dissimilar difficulties and come up with improved solutions.

iii. *Personalized Expertise*: This level of capability necessitates the student's mastery of knowledge as well as technological literacy, ingenuity, and innovation abilities. Information and technology literacy necessitates that students master Information Communication Technology (ICT) skills, which include using a variety of Intensive Training Capabilities (ITC) applications, combining cognitive abilities, and developing other thoughtful learning skills. Creativity and creative skills require students to think and work in new ways to access their brain's abilities where invention can take place.

The importance of communal intelligence as a mode of activity for communication and collaboration cannot be overstated. Communal intelligence refers to information gained through contact and interaction by a group of people. Students find it necessary to share and relay findings or outputs to others after previous work on information and knowledge. As a result, the student wishes to communicate in a variety of ways that are both simple and effective. In addition, affiliation allows students to work effectively and deferentially in a variety of teams.

1.3   FRAMEWORK OF SMART EDUCATION

A smart education, as well as the denotation framework of smart, as a perception of a smart education, is presented based on generalizations of different countries. Harvey (2012) [30] described the essence of smart education as the creation of an intellectual environment through the use of smart technologies. As a result, smart pedagogy could make it easier to provide customized learning services while also allowing students to explore their talent and wisdom with better, more advanced thought skills and more stable behavior (Figure 1.3). And, as seen in the diagram below, a research framework based on this concept of smart education is being created. The framework identifies the three critical elements in smart education and highlights ideologies for better education, so it's probably best to rename it smarter education. Whatever is being done to address the demand for smart pedagogy as a methodology subject, a

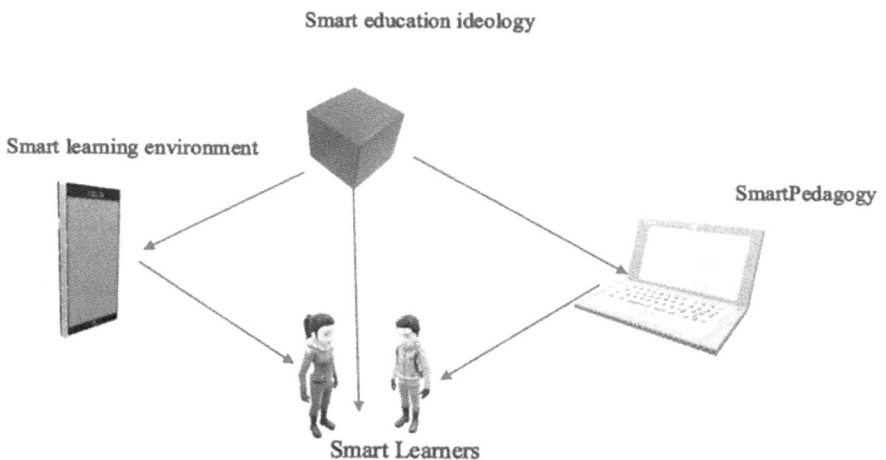

Smart education ideology

Smart learning environment

SmartPedagogy

Smart Learners

FIGURE 1.3 Framework of smart education.

smart learning environment as a technology issue, and an advanced education goal to foster smart learners as a result. A smart pedagogy can have a major impact on a smart environment; thus, smart pedagogies and smart environments both contribute to the development of smart learners.

1.3.1 SMART PEDAGOGY

With the rapid advancement in technology, students have learned using increasingly versatile and effective learning methods. Knowledge and skills are intimately linked, according to cognitive science research. To achieve the level of understanding that the learner requires, context knowledge and procedural ability must be combined. The learners then put their newfound knowledge into practice in order to improve their results. Serious thinking and learning skills are extremely important because they cannot be acquired solely by training; some appropriate, practical experience is required in a particular domain and context. A smart learner is one who uses thoughtful instructional or learning plans in a connected manner to cultivate information and ability; thus, we searched the literature regarding related pedagogy or learning plans. On analyzing the literature, we summarize the assumed applicable practical method.

Students also embrace common knowledge as well as core skills in schoolroom learning target lines and procedures, which are consistently the same for each student in a typical classroom. Then, there are students with a variety of needs due to their varied backgrounds. Each student is entitled to a rigorous education that is aligned with content as well as standards that promote thoughtfulness. Different learners, willingness levels, interests, and learning outlines will be accommodated in different classrooms. Exceptional teaching focuses on the unique needs of each student and cultivating fundamental knowledge as well as core skills in students.

Students who have different abilities, whether in the classroom or online, often need to learn in groups or squads to complete their shared assignments or achieve common goals; with concentrated effort, learners can nurture inclusive skills alongside serious thinking and enhance their problem-solving abilities. Students are expected to take responsibility for their own learning by disseminating knowledge and participating in discussions at an advanced stage.

The learning protocol should be tailored to the student's specific learning needs, which include the student's requirements, context, interests, preferences, and so on.

Intelligence is the ability to complete tasks. Sternberg describes effective intelligence as having the three basic aspects: analytics, thoughtful, creative intelligence, and applied applications, as previously stated. For the learner, we foster skills such as problem-solving and decision-making, creative thoughtfulness, and inter-driven learning. Intelligence is born from the need to assimilate skills. It's akin to transmission, in which a person learns a specific situation and then applies it to other situations that are unrelated. Learning is a multiplicative method in which the eLearner is an active addressee of information who works to construct evocative, thoughtful information derived from a situation. Learners may use propagative learning to become adaptable, apply what they've learned, and generate new ideas.

We propose four instructional plans to better describe learners' requirements, as seen in Figure 1.4. Class-based differentiated directions, group-based collaborative learning, independent personalized learning, and mass-based generative learning are all included in these plans. These plans cover both correct and appropriate learning in the physical and digital worlds. The following are the four levels of a smart plan in detail.

i. *Class-Based Distinguished Instruction*: Distinguished teaching is a means of delivering instruction and learning techniques to students with varying abilities in the same class. It coexists in the classroom with standard-based education and is seen as a community in which students are regarded as unique learners. Teachers use differentiated instruction to set different standards of expectation for learning and job completion within lessons or units, allowing students to develop their own learning preferences and learn more effectively.

ii. *Group-Based Collaborative eLearning*: Collaborative learning is a term used to describe a situation in which two or more people study together in some way. Teachers develop a blended learning system to facilitate student thinking through problem-solving and to create an expressive learning environment. Computer support for learning has emerged as a result of technological advancements, with the use of computers and information technology (IT) to enhance learning. Koschmann (2002) describes computer-supported collaborative learning as an arena of education concerned with meaning as well as the process of making meaning in the form of combined action, and the way in which these practices are mediated by design artifacts. Computer-supported collaborative learning would involve students in combined problem-solving through design software to provide meaning formation. In small groups of students, pay attention to their problem-solving methods and encourage discussion (Figure 1.4).

iii. *Discrete-Based Personalized Learning*: Personalized learning is described as fine-tuning speed, fine-tuning approach, and connecting learner interest and experience to meet student needs, and then providing support for further improving learning capacity among individual students to achieve goals or discover interests based on inspiration. However, when a student engages with an individual learning environment, it is not enough for material to be versatile to engage the student's interest. Their knowledge literacy as well as their technological literacy will be improved. They will participate in educational activities, and their creativity will be encouraged in the learning process. Here are five main issues to consider in order to personalize learning through information technologies: help students make informed learning decisions, improve and broaden diverse knowledge and skills, build a variety of learning environments, and emphasize assessment knowledge and ability, and concentrate on evaluation and input from students.

iv. *Mass-Based Generative Learning*: This is the core concept of productive eLearning, which entails both the development and alteration of individual

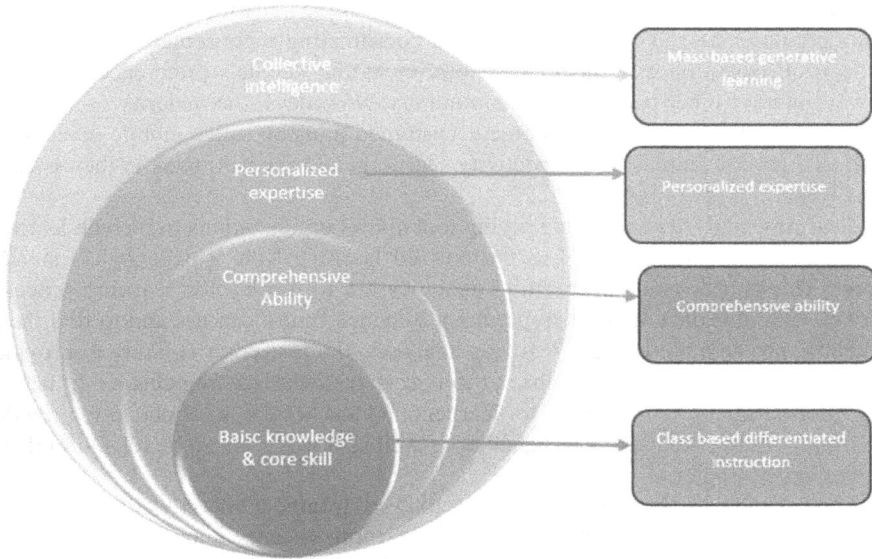

FIGURE 1.4 Four-tier architecture of smart pedagogy.

mental constructions about the world. Engle suggested a content-and-contextual-analysis-based theoretical paradigm for productive learning. The goal is for students to participate in the creation and transmission of knowledge, as well as learning and transmission of contextual information, to create Interco textuality while students are learning online, able to relate new information to old, obtain meaningful information, and use their metacognitive abilities. Interactivity, teamwork, and ingenuity are all strong. These features will aid in the accumulation of knowledge as well as participation in the creation of intercontextuality in the learning process skills, especially communication and cooperation.

1.3.2 Smart Learning Environments

The standard lecture podium has been criticized for being too artificial, rigid, and insensitive to current cultural needs [23]. In the digital age, with the advancement of novel technology and the growth of novel pedagogy, technology is being used to facilitate learning and to engage learners in becoming global marvels. Piccolo et al. [24] define and extend the dimensions of learning environments, which include space, time, technology, power, and collaboration; therefore, it is possible to create new learning environments that are both technically and pedagogically innovative.

Ambient intelligence is rapidly evolving from a technological standpoint as new research models are developed (Shadbolt [25]). In AI/ML environments, devices support people in carrying out their daily activities and tasks in a simple and natural manner by using knowledge, and information from the network system can connect and communicate self-sufficiently without coordinating with people, and it makes a decision based on a number of factors, including people's preferences and the involvement of other people in the community. Nowadays, the majority of students are digital natives, who have become accustomed to using smart mobile devices as well as digital resources for connectivity, learning, and entertainment in their everyday lives.

Learning analytics as the underlying tool allows organizations to include learner assembly development as well as customized learning from a pedagogical standpoint. The overall goals of learning analytics are to display the learning process as well as to use data analysis to predict a student's future success and to find their potential problem. During the learning analytics, the instructor is likely to provide the learner with instructive feedback through virtualized learning classes. In learning information by visualization for learners and teachers, it is beneficial to provide an overall view of the learners' behavior as well as how these can be related to their nobles or extra actors.

The use of technology in a smart eLearning environment not only allows learners to access digital resources and interact with learning systems at any time and in any place, but also actively provides them with critical learning guidance, helpful tools, or learning suggestions at the right time, in the right place, and in the right procedure. Many different types of technology, including both hardware and software, are used to help and boost education. Touchable objects such as communicating whiteboards, smart tables, electronic pockets, cell phones, wearable devices, smart devices, and sensors that use omnipresent computing, cloud computing, context intelligence technology, and so on are included in the hardware category. Learning systems, learning tools, online resources, educational games that use social networking, learning analytics, visualization, virtual reality, and other types of software are all examples of software.

The goal of a smart learning environment, which is based on the provision of various technologies, is to deliver engaging, customized, and unified learning practice aimed at the learner, as well as a smart environment that includes both proper and unstructured learning to enjoy personalized learning practice. By using learning analytics, a smart learning environment will provide precise as well as entertaining learning services. We suggested ten core features of a smart learning environment based on the smart education mandate:

1. *Location Concern*: Intuitive learner's current location in real time.
2. *Context Concern*: Discover various setups as well as action-related material.
3. *Social Concern*: Intellectual social connection.
4. *Interpretability*: A common norm in the face of varying sources, services, and stages.
5. *Unified Connection*: Deliver continuous facility connecting devices.
6. *Adaptability*: Learning opportunities that are aligned to learning accesses, preferences, and mandates.
7. *Pervasive*: Predict learner appeal before it is clear; provide a pictorial and transparent way for learners to access learning sources and facilities.

8. *Entire Record*: Record's learning journey to mine as well as to analyze in-depth, then include rationale evaluation, proposal, and push on-demand services.
9. *Usual Communication*: Senses of many modes of communication are transferred, as well as location and face recognition.
10. *High Rendezvous*: Immerse yourself in a variety of directional communication learning scenarios in a technologically rich setting.

1.3.3 TECHNICAL ARCHITECTURE OF A SMART EDUCATION ENVIRONMENT

Smart computing is the most recent series of technology discovery and progress, which began in 2008, and is a critical technology in a smart learning environment since it combines elements of hardware, software, and network, as well as digital sensor, smart computer, net technology, BDA, computational intelligence, and intelligent machine to understand a wide range of advanced applications. Overall, these innovations effectively allow learning to occur in difficult situations; moreover, advancements in computing technologies lead to new dimensions in smart computing as well as improved learning methods.

We proposed ten core features of a smart learning environment in smart education in the previous portion. The current technology architecture of smart education environment is focused on smart computing to well comprehend the featured variety of learning environments.

1.3.4 3-TIER ARCHITECTURE OF SMART COMPUTING

As mobile devices become smaller, smarter, and more reasonable, the world is rapidly moving toward an era of a single network. The omnipresence of such a system is critical for location-based services, eLearning, and data transmission. In addition, computation is rapidly shifting away from the conventional computer and toward this. A smart learning environment's three-tier architecture is important because it includes cloud computing, fog computing, and swarm computing. Three-tier design, cloud, fog, and swarm are companies that are all important right now. Cloud, fog, and swarm components can all be present in an educational application. Cloud and fog may help manage and control the group's resources. Learning content as well as analyses can be exchanged through this three-tier architecture.

   i. *Cloud Computing*: Cloud computing, which provides SAS, is the innermost layer. It sets up a network of remote servers and applications that allow for central DS as well as online access to a computer service or resource. Smart eLearning environments are a critical tool for justifying resource management. Its infrastructure supports a smart eLearning environment's point, virtualization, and centralization of DS plus education services within education. It understands smart vision, smart material, and smart thrust in cloud computing and smart learning environments (Kim et al. [10]).

ii. *Fog Computing*: Fog computing, which is currently based on IoTs, is at the heart of tri-tier architecture; in reality, everything can be a part of it, allowing for a wide range of services. This necessitates both enhanced infrastructure and urbane devices in abundance. Fog technology is a highly virtualized stage that offers processing, storing, and networking services between end devices and traditional cloud computing, but it is not entirely located at the network's edge. The smart learning environment understands actual communication thanks to the fog computing functionality. Location awareness, a vast sensor network, and support for versatility are just a few examples.

iii. *Swarm Computing*: The top layer is swarm computing. As computing technology becomes more persistent and widespread, we envision the creation of environments that can understand what we're doing and provide for our daily activities. Due to the omnipresence of sensors, swarm computing, also known as environment-aware computing, is based on swarms of smart devices as well as a sensor network. This sensor data will be examined by the DMS.

1.3.5   KEY FUNCTION OF SMART COMPUTING

Furthermore, smart computing allows computing technology to perform five main functions: consciousness, analysis, alternative, action, and auditability. Provision for consciousness for-computing provision review replaces cloud computing provision operation as well as audibility in the tri-tier architecture of swarm computing. When smart computing is used to create smart learning environments or other systems, they are capable of providing and step of bright behavior.

1. *Consciousness*: Learning happens everywhere and at any moment. It will use technologies such as swarm computing, pattern recognition, data mining, learning analytics, and other resources to detect data on student recognition, location, condition, and place, and it will use the network to move data from the learner computer back to the smart eLearning system's main server for review.

2. *Analysis*: Analysis is used to take stock of learning data, as well as to propose learning patterns and resources to electronic learners, whenever system servers collect actual data when learner devices engage with analytical tools like learning analytic, data mining, and BD.

3. *Alternative*: Using eLearning in flow or workflow engineering allows you to distinguish anything you see at a glance or otherwise with human interpretation as a replacement course of action in response to a learning pattern. As soon as a decision is made, a learning act is triggered.

4. *Action*: In a single connection, the system can send an action to the appropriate procedure app; this procedure app can be customized for a variety of situations, with exact app components pushed to our smart device so we can perform actions; as a result, the learner can access a related learning source in the gallery or get position information outside.

5. *Auditability*: If the correct learning action is taken only, evaluate the auditability within this smart education which is important to show the learning procedure and make it more efficient. A smart eLearning system is required to collect, track, and analyze data from eLearning actions at each stage in order to evaluate and enhance learning.

1.4   BIG DATA SUBSEQUENT REVOLUTION IN EDUCATION

1.4.1   BIG DATA IS MAKING EDUCATION SMARTER

Every 26 seconds, a student drops out of school; across 1,700 high schools, dropout rates are about 40%; by 2020, there will be 120 million jobs available, and American students will be able to be hired for them. BD is more sophisticated method of obtaining employment in American universities. In today's educational technology, BDA is being used to teach and nurture careers to success. Consider a secret to locate a problem, present this data, and then measure the item in question. This personalized data also provides students with answers. Data analytics is rising at a 10% pace in American universities, while the dropout rate is −50%. These systems allow evaluation and immediate feedback simpler for students. A 10% increase in reading ability and math performance; an extra 10,000 schools are thus executing data analytics. Many instructors will use technology in the coming decade to reduce the amount of extra work they have to do by unknown approaches, resulting in innovative systems for instructor data analytics, which is an instructional tool targeted at classroom teachers. BD is a 21st-century measurement instrument [26].

1.5   TECHNIQUE EDUCATORS ARE IMPROVING IN LEARNING PROCESS

Determine how data analytics can help teachers and students make better decisions by providing them with new opportunities to advance the educational process. In the field of education, a professor's pedagogical decision to increase student sympathetic of content or arrange the layout of a course can have the greatest impact on a student's learning as well as graduation rate. High excellence intrusion can reduce the time it takes a student to learn a particular material, allows the student to acquire additional knowledge in the same time frame, and assists the student in making good decisions about what to study. This learning effectiveness not only enhances the student experience and can help professors meet some of their demands, but is also very useful for supporting decision-making; as a result, instructors must increase the consistency of their data analytics learning skills. It provides educators and students with an advantage in thoughtful behavior such as when and how questions can be answered.

Agile software decision-making has long been hailed as a game-changer in product development, and it has now arrived to transform the education sector. By providing teachers and learners with better decisions and data science to push procedure invention, BDA opens up new opportunities to enhance education procedures. We see new technologies and applications being developed on a regular basis to help students and teachers make better use of their time.

Technology will still play an important role in practice. It is, however, how educators use the impact of technology on decision-making, or what is commonly referred to as "soft skills," which is essential. Soft skills are the teacher's social and emotional intelligence, and they are what really drives the eLearning process. In order to ask the right questions and make the best use of BD as a tool to support our decision-making, it's critical for educators and eLearners to understand how data analytics can help the learning process. Three ways to benefit from BD are to take advantage of its ability to assess thoughtfulness, personalize learning skills, and assist us in creating more exciting courses.

1.5.1   Measure, Monitor, and Respond

BDA helps instructors to gather, track, and respond in real time to students in order to help instructors familiarize their teaching styles as well as answer student's needs until it strengthens our ability to address any latent biases we might have against interaction or otherwise presentation of student work.

1.5.2   Epitomize Learning Experience

Different levels of knowledge should be used to keep the course interesting for students. Students in overview courses often progress through different levels of essential information. By using DA to determine where each student is overwhelmed or excelling, it would be possible to have different first materials for each student in the same course. This will pique students' interest in a topic while also indicating to whom and when specific learning material will be delivered.

1.5.3   Designing New Courses

The main challenge for business schools is to quickly grasp what businesses need and to develop curriculum to meet those needs. BDA is used to understand business and job patterns, as well as to coordinate the first course and basic learning concepts.

One of the most important duties of a teacher is to evaluate the thoughtfulness of his or her students. Determine how quickly you can respond to a subject, the amount of information you provide, and the number of related concepts you can deliver in a given amount of time. Frequently, during a lecture at a university level, a professor will ask students to explain or elaborate on a key concept. Give a quiz or use a mid-term test to assess the student's sympathetic and upcoming presentation.

We can now speed up the knowledge process by simply adapting our pedagogical method and responding to the unique needs of each student, thanks to the growing amount of data. We could forecast graduation results to help us determine where, when, and to whom we should devote additional time and resources.

The use of large datasets to inform business decisions is important in education, and it is reasonable for any sector to benefit from the technologies and trends that are driving economic and social change [27].

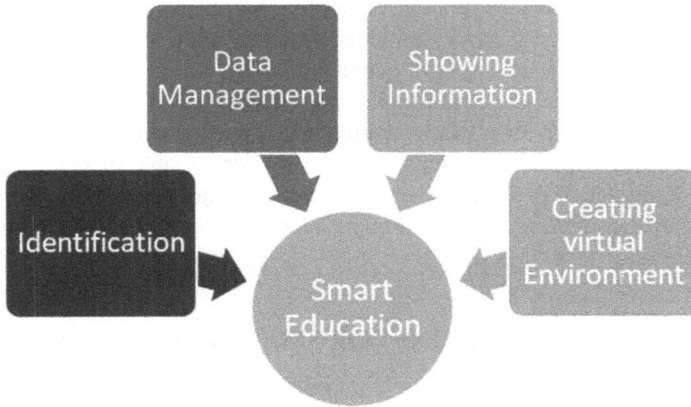

FIGURE 1.5   Four main objectives to develop a system to sustain smart education [28].

1.6   BIG DATA COMPONENTS IN SMART EDUCATION

To sustain smart education using big data, the following components play an important role (Figure 1.5):

- *Identification*: Understand the different scenarios to deliver the smart education. Identify the issues in the specific category of delivery of content in smart education.
- *Data Management*: Huge amount of data needs to be gathered and analyzed to find the specific category of smart education.
- *Showing Information*: The information needs to be revealed to developers to develop big data applications to deal with those issues in specific categories of smart education.
- *Creating Virtual Environment*: Finally, create an environment to deliver the content to reach end users for understanding the issues of smart education.

1.7   BIG DATA TOOLS IN SMART EDUCATION

BD analysis can be aided by using open source software [29], and some of the most useful tools are mentioned below.

MongoDB is a multistage document-oriented database management system. In its position, it uses JSON-like documents. Its architecture is based on tables.

- *Hadoop*: By using a simple programming model, a system for distributed processing of a large dataset and a cross-wise cluster of network computers can be developed.
- *MapReduce* is a programming model and system that is used by Hadoop. It allows massive amounts of data to be processed in small or large clusters of compute nodes.

- *Orange*: Orange is a Python-based application that can be used to process and mine BD files. It has a simple user interface with dragging and falling features, as well as a variety of addenda.
- *Weka*: Weka is a Java-based tool for processing large amounts of data; it contains a large number of algorithms that can be used in data mining.
- *SAP HANA*: In-memory-exclusive RDBMS are capable of handling massive quantities of data. To solve the problem of dealing with BD, it employs a parallel in-memory relational query technique, column-like storage, and compression technology.

1.8   BIG DATA APPLICATIONS FOR SMART EDUCATION

BD application technology enables comprehensive data collection for student affairs administration. In the opinion of a data mining task and system, the primary goal in SMS applications of BD is data visualization, which allows users to immediately detect data processing results. The second goal is to provide predictive analysis and to make possible judgments at a high level of precision in order to visualize data.

BD technology mines knowledge, including information embedded in vast volumes of data, and provides a foundation for human social and economic behaviors, enhancing operational performance. Operating efficiency recovery can also be aided by BDA and mining technologies. BDA and mining technology can also help with operating efficiency recovery. BDA and mining technologies are also used in universities to recover teaching assistance, analyze user activity on campus networks, and shape public attitudes, both of which effectively facilitate in-depth IT and teaching integration.

Endorsing the improvement of a teaching outcome, universities' data resources provide a wealth of knowledge on teaching that can be effectively used. The focus on and effort for curricula can be found by analyzing and mining data such as click rate, download, and recurrence rate of content, and personalized learning can direct students' future career growth. Asset management that works administrators can obtain information about individuals, finances, and properties in the management system according to their access level by completing registration and labeling of assets information such as building information, state of living, and property information by using fixed keyword investigations plus user-defined fuzzy investigations, resulting in actual asset management. Analyze user activity and monitor behaviors in the campus network, then organize and archive the data. Until activity trajectory information is generated and stored, all types of activity information are collected and organized according to user type and time. In addition, administrators may make inquiries on individuals, affairs, and services by analyzing a BD initial alert for irregular activities of a specific object. Information about early warnings and emergencies are required.

Decision-making assists in focusing on data capture and smart decision analysis, resulting in the creation of a service application. The authority and consistency of data used by all systems are guaranteed by creating shared databases using captured data, ensuring the consistency, fairness, and accuracy of data across all business systems in universities with initiative-level information standards. accessibility, intelligibility, consistency, and obtainability.

Using BD, statistical summaries may be generated from a variety of data, such as population statistics, event statistics, and department statistics, and so on, and are shown in the form of reports and charts. BD aids decision-making by allowing administrators to easily access data on individuals and events through several departments under their control. Public opinion polls and university public opinion polls are commonly shared on campus bulletin boards, microblogs, WeChat, and other social media platforms. In the context of BD technology, a standard application of BD is to precisely and quickly understand the trend of online group opinion and direct students in properly expressing their views.

Application of a BD in Education
- Advanced education analytic
- Student engagement
- Bookstore effectiveness

1.8.1 HIGHER EDUCATION ANALYSIS

BD allows students to get the most out of their learning and ensures success in higher education. Students, for example, are often asked, "What is the best major for me?" Similarly, teachers are fascinated by how to personalize learning paths so that no student falls behind. The development of a complete profile map of a student based on their grades, extracurricular interactions, and social activity allows for performance improvement and learning path customization.

1.8.2 STUDENT ENGAGEMENT

Data mining assists universities in gaining a holistic view of students; BD addresses questions such as which learning behaviors are associated with better comprehension and higher grades, or which potential courses are best for such students. This often takes into account the interests of students in social and extracurricular activities. Institutions will construct engaging learning experiences for all students using this holistic approach.

1.8.3 BOOKSTORE EFFECTIVENESS

For applications such as merchandising productivity and textbook inventory optimization, BD is used to increase bookstore effectiveness through analytics.

Today, large stores search for book varieties, buy them all together, and then sell them as a kit, giving consumers a hassle-free experience. They often use social media analytic to strategize their acquisition, and they understand topics in trend. They often use social media analytic to consider trending topics in order to plan their acquisition and inventory.

1.9 HOW BD AND EDUCATION COULD WORK TOGETHER TO BENEFIT STUDENT SUCCESS

Between 2015 and 2017, customers and businesses created more data than the total amount of data produced previously. Technology experts predict that the total amount of data produced will increase from 130 to 40,000 EBs in the first half of 2020, with researchers estimating that a community will produce 1.7 MB of data per second. Growing tenfold in today's digital world (4.4–44 ZBs), you might imagine BD and education having two things to change. BD systems are used by businesses to collect, store, and analyze large volumes of data as well as gain insight into their daily operations. Innovation is changing the way knowledge is leveraged, the market is refined, and the face of education is changing. People can use BD systems to research and develop how organizations analyze and infer knowledge, as well as use those interpretations to make better decisions. As a result, the position of a professional who can combine, understand, and analyze data is more important than ever. In the academic setting, innovation envisions a classroom with a continuous video feed that captures complex details of the learner, such as facial expression, distinct physical activity, and social interaction. This definition may encompass each object that students interact with as well as each statement that the learner and instructor complete. Furthermore, the grading system may include biometric data collected through a wearable device, allowing researchers to assess what motivates students to learn.

For the future of teaching and learning, BD technologies have enormous potential. Three forms that BD technology is changing today's learning situation are described in the following section [28].

1.9.1 CUSTOMIZED CURRICULA AIMED AT IMPROVED LEARNING OUTCOME

BD technology helps new educational models develop, such as digital learning, and then customizes curricula. The majority of modern consumers depend heavily on technology. Today's smartphone cars and applications assist customers in navigating the environment and dealing with essential necessities of existence. Although certain devices entertain, others have substantial life-enhancing functionality, and a few innovations boost educational excellence.

Students also struggle with inscription research papers as well as essays on BD topics, and they seek expert assistance from an academic writer. There are several professional paper inscription services as well as custom writing designs available to assist college and university students all over the world with their academic projects on a variety of topics, including BD. Educators in interactive classrooms use BD technology to collect data on students and create customized learning plans. This is a particularly useful resource for a teacher who is in control of the class's learning course. While school districts incorporate merged learning systems that combine traditional classrooms and online learning, old-style classroom curricula contain static learning plans that are reliably implemented for all learners in the class regardless of academic presentation using an advanced BD method. It enables students to achieve their desired learning outcomes while working at their own pace.

1.9.2   Big Data to Expand Student's Performance

BD is changing education across the income spectrum, from elementary school to colleges, as technology and educational standards advance. BD systems assist teachers in improving their understanding of human actions, formulating new conclusions, and allowing them to better consider current trends in education and the importance of BDA for educators. BD technology is used by modern educators to identify student trouble areas. Students will spend extra time working through exciting topic matter, while continuing to maintain phase with respite from class, by relying on clear tests to reveal matters, as well as by adapting learning technology.

The BD system enables educators to accurately evaluate students while constantly tracking their development and chances of advancement. Technology may also assist educators in creating learning plans that will help students advance in their chosen area.

1.9.3   New Paths of Learning Potentials

Traditional learning paths dictate that students progress solely on the basis of their phase; exceptionally gifted students can join advanced learning tracks but stay in the same class as gifted peers because educators collect presentation results infrequently. With the BD system, educators may scale student presentations more frequently and advance students as a result. In the United States, BD systems can help students learn at a higher level and expand educational models. Technology is transforming humanity into a world that requires constant research and updating. As BD systems become more widely used in classrooms, these systems will become increasingly important for teachers to understand how to interact with technology and use it to improve student learning outcomes. Although it took some time, BD technology eventually established learning site support [28].

1.10   BIG DATA ANALYTICS CONSEQUENCES IN ADVANCED EDUCATION

BD is a knowledge system that has previously changed the object of information as well as the social concept in a variety of fields by transmuting and organizing the decision-making concept [31]. The testing arena of a learning analytic is included in BD (Long & Siemen [32,33]), which had previously been a growing part of education, though learning analytic research has largely been insufficient to explore points of individual student and class presentation. BD opens up new possibilities while still posing a challenge in terms of arranging advanced education. Wagner and Ice [34] also demonstrate how BD can be used in advanced education.

1.11   OPPORTUNITIES ALONG WITH CHALLENGES OF BIG DATA IN SMART EDUCATION

The institution of a higher education dataset is attractive to benefit from targeted analytics because of the large amounts of student information, including enrollment, academic, and corrective records. In a higher education environment, BDA is transformative, transforming the current administrative, instructional, learning, and academic work processes, leading to policy and practice outcomes, and addressing new challenges in higher education.

BD makes an innovative education analytical method accessible to institutions that are important for progressing learning outcomes targeted at discrete students as well as verifying instructional programmers with high excellence standards; by creating a program that collects data on each level of the student learning process, the university can address student needs by customizing module, task, response, and learning trees in the syllabus to promote enhanced and better-off learning.

The application of analytic techniques for BD in higher education is expected to present a range of challenges. Some obstacles include getting users to embrace BD as a conduit for embracing new processes and change management. There would be tremendous cost savings associated with data collection, storage, and developing algorithms to mine the data. Procedures have a tendency to be time-consuming and complex. Furthermore, an important feature of an institutional data system is that it is interoperable; therefore, integrating managerial and classroom data, as well as online data, can present additional challenges.

1.12 CONCLUSION: CHALLENGE OF SIMPLIFYING SMART EDUCATION

A smart education is a novel concept of global education that focuses on advancing the learner's dedication to lifelong learning. Focusing on relative, personalized, and cohesive learning to support learners' intelligence, emerging and simplify their problem-solving capacity in a smart education can face a number of challenges, including pedagogical theory, educational technology, leadership, teacher learning leadership, educational structure, and educational ideology.

Our expectations for smart education and a smart learning environment can reduce learners' cognitive load, allowing them to focus on sense-making and ontology construction. Similarly, a student's learning experience can be influenced as well as extended and aided in their overall growth. In a smart learning environment, students can research on their own time and collaborate with others, effectively substituting the extension of a learner's private as well as collective intelligence.

As a result, the concept of a smart city has paid more attention to the state of a smart education based on a smart city in order to encourage an inclusive goal of a smart education under smart city architecture in order to provide each student with a cohesive learning practice. Learning can take place anywhere and at any time, and it generates a large amount of behavioral data about the learner. In order to afford a seamless learning experience as well as tailor updated service for learners, educators must figure out how to assimilate data from various scenarios in smart cities and create data-centric smart education. Future research will focus on interconnected and interoperable learning services, as well as experience with SMS and other smart city systems.

Consolidating data from various sources offers a strong basis for improving decisions related to key business and technological needs, eliminating redundancies and costly time spent salvaging data from various sources. However, combining a dataset through several variations of unrelated systems can be extremely difficult.

REFERENCES

1. Jiang, D. et al. *Journal of East China Normal University* No S1, pp. 119–125, (2015), (In Chinese).
2. Wikipedia, Big data --- Wikipedia, The Free Encyclopedia, Accessed 2015.
3. Kubick, W.R. Big data, information and meaning. In: *Clinical Trial Insights*, pp. 26–28, (2012).
4. Russom, P. Big data analytics. In: *TDWI Best Practices Report*, pp. 1–40, (2011).
5. EMC. Data science and big data analytics. In: *EMC Education Services*, pp. 1–508, (2012).
6. TechAmerica. Demystifying big data: A practical guide to transforming the business of government. In: *TechAmerica Reports*, pp. 1–40, (2012).
7. Biswas, S., & Sen, J. A proposed framework of next generation supply chain management using big data analytics. *Proceedings of National Conference on Emerging Trends in Business and Management: Issues and Challenges*, Kolkota, India, (2016).
8. How Big Data And Education Can Work Together To Help Students Thrive. https://www.smartdatacollective.com/how-big-data-and-education-can-work-together-help-students-thrive
9. Lee, J., Zo, H., & Lee, H. Smart learning adoption in employees and HRD managers. *Br. J. Educ. Technol* 45(6), 1082–1096, (2014).
10. Kim, S., Song, S.M., & Yoon, Y.I. Smart learning services based on smart cloud computing. *Sensors* 11(8), 7835–7850, (2011).
11. Bruce, B.C., & Levin, J.A. Educational technology: media for inquiry, communication, construction, and expression. *J. Educ. Comput. Res.* 17(1), 79–102, (1997).
12. Meyer, B.B., & Latham, N. Implementing electronic portfolios: benefits, challenges, and suggestions. *Educause Q.* 31(1), 34–41, (2008).
13. Hwang, G.J., Tsai, C.C., & Yang, S.J.H. Criteria, strategies and research issues of context-aware ubiquitous learning. *J. Educ. Technol. Soc.* 11(2), 81–91, (2008).
14. Chan, T.W., Roschelle, J., Hsi, S., Sharples, M., Brown, T., Patton C. et al., One-to-one technology-enhanced learning: an opportunity for global research collaboration. *Res. Pract. Technol. Enhanc. Learn.* 1(01), 3–29, (2006).
15. Elias, T. Learning analytics: the definitions, the processes, and the potential, (2011).
16. Hwang, G.J. Definition, framework and research issues of smart learning environments-a context-aware ubiquitous learning perspective. *Smart Learn. Environ.* 1(1), 1–14, (2014).
17. Gwak, D. The meaning and predict of smart learning, *Smart Learning Korea Proceeding, Korean e-Learning Industry Association*, (2010).
18. Kim, T., Cho, J.Y. J.Y., & Lee, B.G. Evolution to smart learning in public education: a case study of Korean public education, In *Open and Social Technologies for Networked Learning*, ed. by Tobias, L., Mikko, R., Mart, L., & Arthur T. (Berlin Heidelberg, Springer), pp. 170–178, (2013).
19. Middleton, A. *Smart Learning: Teaching and Learning with Smartphones and Tablets in Post Compulsory Education* (Media-Enhanced Learning Special Interest Group and Sheffield Hallam University, 2015).
20. MEST: Ministry of Education, Science and Technology of the Republic of Korea, Smart education promotion strategy, President's Council on National ICT Strategies, (2011).
21. Koper, R. Conditions for effective smart learning environments. *Smart Learn. Environ.* 1(1), 1–17, (2014).
22. Scott, K., & Benlamri, R. Context-aware services for smart learning spaces. *IEEE Trans. Learn. Technol*, 3(3), 214–227, (2010).
23. Kinshuk, S. *Graf, Ubiquitous Learning* (Berlin Heidelberg New York, Springer Press, 2012).

24. Piccoli, G., Ahmad, R., & Ives, B. Web-based virtual learning environments: A research framework and a preliminary assessment of effectiveness in basic IT skills training. MIS quarterly, pp. 401–426, (2001).
25. Shadbolt, N. From the editor in chief: ambient intelligence. *IEEE Intell. Syst.* 18(4), 2–3 (2003).
26. Big Data's Making Education Smarter
27. https://master.edhec.edu/news/three-ways-educators-are-using-big-data-analytics-improve-learning-process.
28. Nsunza, WW, et al. Smart Education: A review and future research directions.
29. https://www.smartdatacollective.com/how-big-data-and-education-can-work-together-help-students-thrive/.
30. Cynthia Harvey, "50 Top Open Source Tools for Big Data", http://www.datamation.com/data-center/50-topopen-source-tools-for-big-data-1.html. Accessed 2012.
31. Zhu, Z.T., & He, B. Smart Education: new frontier of educational informatization. *E-Edu Res.* 12, 1–13, (2012).
32. Boyd, D., & Crawford, K. (2012). Critical questions for Big Data. *Commun. Soc.*, 15(5), 662–679. Doi: 10.1080/1369118X.2012.678878.
33. Long, P., & Siemen, G. Penetrating the fog: analytics in learning and education. *EDUCAUSE Rev.* 46(5), 30–40, (2011).
34. Wagner, E., & Ice, P. Data changes everything: delivering on the promise of learning analytics in higher education. *EDUCAUSE Rev.*, 33–42, (2012, July/August).

2 Big Data Analytics Using R for Offline Voltage Prediction in an Electric Power System

H. Prasad and T. D. Sudhakar
St. Joseph's College of Engineering

CONTENTS

2.1 INTRODUCTION

Energy Management Systems or EMS have found widespread application with the advent of distributed generation, variable loading, and the complexity of power grid growing day by day. EMS includes control of a variety of multidimensional process with more number of variables with varying dependency that spikes up the nonlinearity of the problem. This has led to the evolution of conventional power Grid into Smart Grid, which involves continuous monitoring of such complex power system variables, thus leading to the generation of extremely large amount of data almost in an intermittent manner. Data of vital parameters, namely, voltage at the generation and distribution sides, power flow, current, power factor, frequency, etc, are acquired from remote locations like generating stations, transmission networks, distribution utilities, consumers, etc. Analysis of these data is essential for addressing various problems like load forecasting, fault detection, load sharing, voltage control, reactive power compensation, power quality improvement, electric energy trade, assessment of peak demand, etc.

Of these, voltage is a vital parameter directly connected to maintaining the stability of the grid. Sudden drops in voltage magnitude on major parts of the system may cause brownouts or even blackouts. Thus, prediction of probable voltage sags and swells in the system, using data collected from different buses and loads as an offline study, becomes an indispensable requisite for system stability. Voltage collapse occurs mostly in tap changing transformers and other devices when they fail to prevent the collapse and hence defer the quick estimation of voltage profile with continuous monitoring of the system. Better identification tools for voltage prediction are needed. Several offline approaches like the P-Q iteration methodology [1], load flow analysis, distribution factor method [2], the bounding method, pattern recognition method and concentric relaxation technique-based methods to estimate bus voltages for a real-time system have been proposed by various researchers. Hence, prediction of voltage magnitude on the power system is necessary to prevent the system instability. Voltage prediction also helps to improve the preventive control capability of the system by giving a premonition of transient instabilities that would probably occur in the system. Effective voltage prediction requires a deep learning of loading patterns, load changes, identification of vulnerable loads and load center, reviewing energy pricing strategies for critical loads or customers, assessing grid stability, studying reactive power compensation, etc. Though these may look like loosely coupled concepts, Big Data analytics effectively identifies the nexus point. With such voluminous amount of various data generated at high velocity and operational significance, voltage prediction is a perfect candidate for Big Data Analytics.

But before moving into Big Data analytics, let us take a look at conventional techniques used in this regard so far. A new ANN (Artificial Neural Network)-based algorithm with minimally predetermined inputs proposed to estimate the voltage magnitude of each of the buses due to a fault or modifications in the load is discussed [3,11]. Cellular Simultaneous Recurrent Neural Network (CSRN) is a better tool for fast bus voltages prediction of highly interconnected power systems that is more accurate than neural networks and does not have the computational complexity and increased runtime, owing to subtraction of the measured voltages from steady-state values at each bus [4]. A new approach to improve the accuracy and reduce computation time called parallel self-organizing hierarchical neural network (PSHNN) minimized learning and recall times and rendered optimized system complexity. In this regard, two PSHNN were used to reduce errors [5]. Neural networks including a wide range of methods starting from multilayer perceptron (MLP) to radial basis function have been implemented to obtain solutions for various power system problems [6]. Many algorithms that put to use the synchro phasor technology for online voltage stability monitoring using extrapolation technique have been synthesized. In Security assessment of Voltage profile, an Artificial Intelligence-based algorithm is incorporated as a test methodology, where conventional Neural Network techniques and Evolutionary Algorithms are used [7]. A widely used method for load flow analysis of power systems is the Newton–Raphson (N-R) power flow method based on solving a set of nonlinear equations that estimates the voltage magnitude, real power, and reactive power controls using changes that are derived from the derivative-based Jacobian matrix [8].

In comparison with the above methods, Big Data Analytics involves acquiring of data, cleaning or filtering of data to get a meaningful subset, analysis of data in accordance with the problem, and then providing prospective solutions. There are various Big Data tools and technologies being put to use like Hadoop, DataWrapper, MongoDB, Cassandra, etc., to name a few. Of them all, R is a powerful programming language along with its R Studio environment and can serve as a framework for statistical computing, analytics, and graphical visualization. It is an open source framework with vast libraries, enabling exploratory analytics with in-built Machine Learning (ML) and modelling algorithms. It is also simple to use for beginners. This discussion sheds light on analytics for voltage prediction of various buses in an IEEE standard power system, using R programming. The analysis is done as an academic research study with data obtained from power flow analysis of the system under study subjected to various load conditions. The data acquired is subjected to analytical algorithms in R programming framework to predict the voltages of various buses of the power system under study. This new approach based on data analytics, that utilizes the application of analytical software called R-STUDIO,is proposed. Prediction of voltage magnitude is done through limited lines of coding in R-STUDIO. The results obtained seems to be more accurate compared to the conventional methods. All the bottlenecks of the conventional methods such as accuracy, complexity in matrix formation, speed, efficiency, and Machine Learning are reduced in the proposed methodology. The proposed approach owing to its simplified algorithms and reduced execution time will be appropriate for online practical implementations by connecting the system with RTU (Remote Terminal Unit).

2.2   DATA ANALYTICS

Analytics can be described as the finding, refining, interpreting, and indicating meaningful relatable patterns in data in areas where the acquired information is abundant. It is dependent on the combination of logical programming and optimization to express the performance of the system which can also be implemented in business and scientific organizations. Analytics has become increasingly usable in all fields such as media, health care, sports, finance, and many others. All this relies heavily on data taken for analysis. Data can be defined as a set of obtained values attributed to be as qualitative, logical, or quantitative variables in the process. This information is coded in some form in a logically decipherable way suitable for better usage or processing. Data transforms businesses and allows social interactions in the future society. Data Analytics is the science of inspecting data in the form of either organized or raw sets to arrive at any inferences about the information they contain, with the help of upgraded systems and new software technologies. It is the process of utilizing these data to construct inference models that lead to better decisions for institutions in any process they take up. The main ingredients are data, model, decisions and values. IBM, which has been running its business successfully for the past 100 years, has changed its business focus to analytics and investments of about $20 billion right from the year 2005 and has grown its analytics businesses across various places. More than $120 billion has been invested by companies by 2015 on analytics, hardware, software, and services.

Some of the application frameworks of analytics are as follows:

- IBM Watson
- eHarmony
- The Framingham Heart study

Data analytics is akin to Business Analytics that is used for business, but data analytics has a broad focus which helps to increase the operational efficiency, optimize market campaign, respond promptly to the rising markets, and attain a competitive upper hand, all with the motive of improving business efficiency. Based on the application, the statistical data can be obtained from the historical records or even new data can be created from real-time analytics. The data can also be a mixed form of raw and organized data from internal as well as external data sources.

2.3　THE PROCESS FLOW

Data analytics encompasses data exploration that is used to find the patterns and data relationships which is compared to the detective work and Confirmatory analytics (CDA) and implements techniques to estimate whether the hypotheses that belongs to a data set is true or false. It also includes both quantitative and qualitative analysis. In quantitative analytics, analysis of data with the variables that can be measured and analyzed numerically is done. Qualitative method is mostly logical interpretation that deals with verbal and nonverbal data, like images, text, audio, and video.

Modern methods of data analytics involve the art of data mining that sorts and segregates huge data chunk to identify trends, hidden patterns, and inherent relationships following predictive analytics for customer and Machine Learning that uses contemporary algorithms to search data sets more swiftly than data scientists would be able to do by using conventional methods. The pedagogy of Big Data applies all three of these analytics into groups that often contain mixed form of a combination of unstructured, structured, and even semi-structured data, often called as raw data.

A data scientist's job is to collect the data and identify and filter by truncating the subset they require for an analytics application and then analyze it either individually or as a team. Data from variable sources may require augmentation by usage of data integration routines and then be transformed into a basic form and inserted into the chosen analytics platform, namely Hadoop clusters, NoSQL DBMS, or any other data warehousing and mining framework.

Once the data required are procured, next on the agenda would be to find and fix problems related to data quality that could possibly affect the precision of the process. This is done by means of data profiling and cleansing by regression and correlation methodologies to ensure that the data set is logically connected with the problem and that the redundant entries are effectively filtered off. Initially, the model is run with the algorithm against a truncated partial data set to ensure its accuracy; it is then updated and tested repeatedly, just like it is done in case of neural networks. This is achieved by a process known as "training" the model. Lastly, the model is run in deployment-ready production mode against the full data set, which can be done once to address a specific information need or on a routine basis as the data are updated.

In the event of some complex cases, analytics applications can be trained to implement actions as a corrective measure at the end user side. Generally, the concluding step in the data analytics process is communication of the results obtained from analytical models to concerned executives and users to catalyze their decision-making process (Figure 2.1).

2.4 R - AS A PROGRAMMING LANGUAGE

"R," which has gained popularity in recent times, is an easy-to-learn programming language and environment implemented for computation based on statistics, data interpretation, and graphical representation. It can be described as a public-domain, open-source project. R is similar to the commercial S language developed at Bell Laboratories by the efforts of John Chambers and associates. R initially originated from S in the mid-1990s due to the work done by two members, namely, Ross Ihaka and his colleague Robert Gentleman from University of Auckland, New Zealand. R can be considered as a differentiated implementation of S and is used more in the educational field as a research tool. The main advantages of R are as follows:

- It is a freeware, and there is a good amount of support available online.
- It is quite similar to other programming packages such as MATLAB®, which is a paid software, but more user-friendly than programming languages such as C++ or FORTRAN.
- R can be used standalone, but for educational purposes R can be used in combination with the R-Studio free-to-use interface, which has an organized layout and handy options for amateur developers.

Many other choices for data analytics have been found in addition to R such as SPSS, SAS, MATLAB, Minitab, pandas, and Excel. Among all these R is considered to be the most advanced technology as it is an open-source software available for Mac, Windows, and Linux. R is widely used by almost 2 million people around the world. R is just a command line interface that can be done with popular choices such as R-Studio and Rattle. The most suitable way to learn R is by trial-and-error method. R should be downloaded and installed on the computer before working on it from the download link.

2.4.1 R-STUDIO DESCRIPTION

R-Studio, much to the delight of amateur programmers, is an open-source comprehensive development environment for R, used for graphical analysis, computing, and representation of data. R-Studio was formulated and initially brought to use by JJ Allaire, the Chief Scientist at R-Studio is Hadley Wickham.

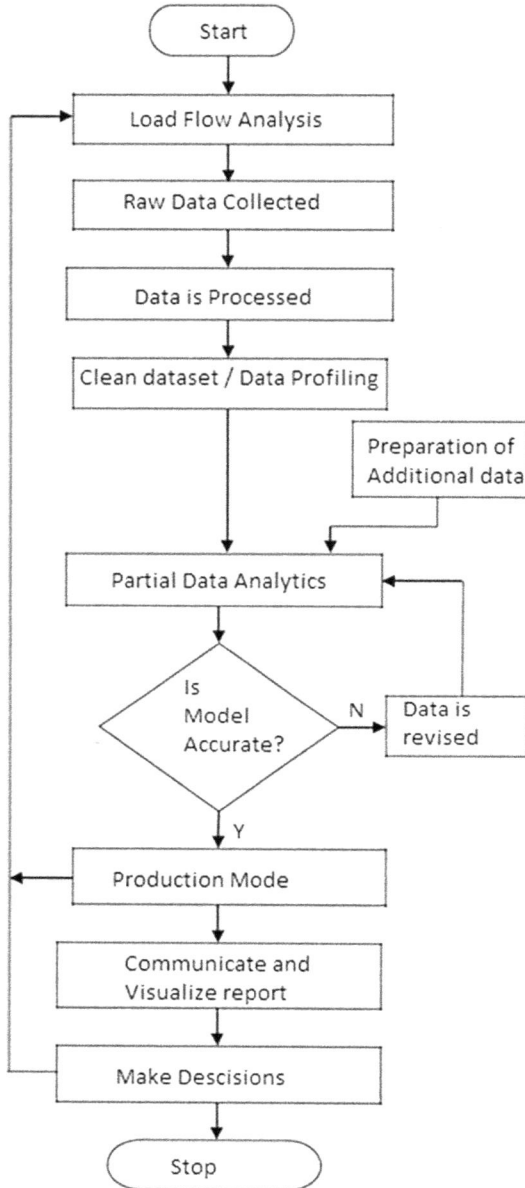

FIGURE 2.1   Flowchart on data analytics-based decision-making.

R-Studio is free to use and is available to its users in two elaborate editions:

- One is R-Studio Desktop, in which the program would be run by local machines as a desktop-based application, and this is for regular use. R-Studio runs on the desktop with all general operating systems in use or on any internet browser connected to R-Studio Server or R-Studio Server Pro.
- R-Studio Server allows multi-platform interoperation by accessing and working with R-Studio using a browser simultaneously while running on a remote Linux server.
- R-Studio codes are written C++ object-based language for programming and utilize a platform called Qt framework for its graphical user interface. Work on R-Studio is reported to have started in December 2010, with the initial beta version (v0.92) being officially declared in February 2011. An advanced Version called R Studio 1.0 was released on November 2016.

2.4.2 Need for R-Studio

The technology advancements to amass data exceeds our abilities to utilize it. People all over the world are turning to R as a trustworthy ally, to make sense of data acquired. Inspired by the breakthroughs of R users in science, education, and industry, R-Studio develops free and open-source tools for teams to scale and share work.

2.5 ANALYTICS FUNCTIONS USING R

2.5.1 Linear Regression in R

Linear regression can also be performed in R using the function 'lm' that is used to build the linear function model. The output has already been saved in the name given to the function; hence, the output is obtained from the summary of the saved name. The summary displays the description of the model, the summary of the residuals, and the description of the coefficient of the models. The coefficient shows the intercept in the first column, the estimate in the second column that estimates the data value of the model, the standard error in the third column, the t value in the fourth column, and the $Pr(<|t|)$ in the fifth column. The bottom of the output displays multiple R^2 and adjusted R^2. The value of multiple R^2 increases on adding independent variables; the value of Adjusted R^2 adjusts the value of multiple R^2 and decreases on adding variables. The value of SSE (Sum of Squared Error) can be found with the function SSE, as in Figure 2.2. In case we have to add variables to the model, it should be saved with a different name, and the variables can be separated using the + symbol. With the addition of variables, the value of the multiple R^2 value increases, representing that the model is probably better than the previous model.

2.5.2 Understanding the Summary of the Model

The coefficient function represents the intercepts (name of the variables in the model) in the first column, the second column provides the estimate that shows the

```
> getwd()
[1] "D:/RSTUDIO/data files/unit 2 data"
> wine = read.csv("wine.csv")
> str(wine)
'data.frame':   25 obs. of  7 variables:
 $ Year       : int  1952 1953 1955 1957 1958 1959 1960 1961 1962 1963 ...
 $ Price      : num  7.5 8.04 7.69 6.98 6.78 ...
 $ WinterRain : int  600 690 502 420 582 485 763 830 697 608 ...
 $ AGST       : num  17.1 16.7 17.1 16.1 16.4 ...
 $ HarvestRain: int  160 80 130 110 187 187 290 38 52 155 ...
 $ Age        : int  31 30 28 26 25 24 23 22 21 20 ...
 $ FrancePop  : num  43184 43495 44218 45152 45654 ...
> model1 = lm(Price ~ AGST, data=wine)
> summary(model1)

Call:
lm(formula = Price ~ AGST, data = wine)

Residuals:
    Min       1Q   Median       3Q      Max
-0.78450 -0.23882 -0.03727  0.38992  0.90318

Coefficients:
            Estimate Std. Error t value Pr(>|t|)
(Intercept)  -3.4178     2.4935  -1.371 0.183710
AGST          0.6351     0.1509   4.208 0.000335 ***
---
Signif. codes:  0 '***' 0.001 '**' 0.01 '*' 0.05 '.' 0.1 ' ' 1

Residual standard error: 0.4993 on 23 degrees of freedom
Multiple R-squared:  0.435,      Adjusted R-squared:  0.4105
F-statistic: 17.71 on 1 and 23 DF,  p-value: 0.000335

> SSE = sum(model1$residuals^2)
> SSE
[1] 5.734875
```

FIGURE 2.2   SSE in R console.

coefficient of the intercepts and each of the variables in the model. The remaining column helps us to determine whether the variable is to be included in the model or its coefficient is significantly different from zero. A coefficient of zero represents that the value of prediction has not changed for the independent variable. If the coefficient is not different from zero by a good margin, then the variable is removed from zero since it does not help to predict the independent variable. The std.error column gives the measure of how much the coefficient varies from the estimated value. The t value column is the ratio of estimate to the std.error. The value is negative if the estimate is negative and positive if the estimate is positive. The larger the t value, the more likely the coefficient will be significant. The last column gives the measure of the possibility of the coefficient to be zero, as shown in Figure 2.3.

   The easiest way to determine the significant variable is to look at the stars at the end of each row. If the variable has three stars, then it has highest level of significance with probability between 0 and 0.001; two stars shows that it is highly significant and has a probability between 0.001 and 0.01; 1 star also indicates significance with a probability between 0.01 and 0.05; a dot indicates that it is almost significant and

```
> model2 = lm(Price ~ AGST + HarvestRain + WinterRain + Age + FrancePop, data=wine)
> summary(model2)

Call:
lm(formula = Price ~ AGST + HarvestRain + WinterRain + Age +
    FrancePop, data = wine)

Residuals:
     Min      1Q    Median      3Q      Max
-0.48179 -0.24662 -0.00726  0.22012  0.51987

Coefficients:
              Estimate Std. Error t value Pr(>|t|)
(Intercept) -4.504e-01  1.019e+01  -0.044 0.965202
AGST         6.012e-01  1.030e-01   5.836 1.27e-05 ***
HarvestRain -3.958e-03  8.751e-04  -4.523 0.000233 ***
WinterRain   1.043e-03  5.310e-04   1.963 0.064416 .
Age          5.847e-04  7.900e-02   0.007 0.994172
FrancePop   -4.953e-05  1.667e-04  -0.297 0.769578
---
Signif. codes:  0 '***' 0.001 '**' 0.01 '*' 0.05 '.' 0.1 ' ' 1

Residual standard error: 0.3019 on 19 degrees of freedom
Multiple R-squared:  0.8294,    Adjusted R-squared:  0.7845
F-statistic: 18.47 on 5 and 19 DF,  p-value: 1.044e-06
```

FIGURE 2.3 Summary of the model.

```
> model3 = lm(Price ~ AGST + HarvestRain + WinterRain + Age , data=wine)
> summary(model3)

Call:
lm(formula = Price ~ AGST + HarvestRain + WinterRain + Age, data = wine)

Residuals:
     Min      1Q    Median      3Q      Max
-0.45470 -0.24273  0.00752  0.19773  0.53637

Coefficients:
              Estimate Std. Error t value Pr(>|t|)
(Intercept) -3.4299802  1.7658975  -1.942 0.066311 .
AGST         0.6072093  0.0987022   6.152 5.2e-06 ***
HarvestRain -0.0039715  0.0008538  -4.652 0.000154 ***
WinterRain   0.0010755  0.0005073   2.120 0.046694 *
Age          0.0239308  0.0080969   2.956 0.007819 **
---
Signif. codes:  0 '***' 0.001 '**' 0.01 '*' 0.05 '.' 0.1 ' ' 1

Residual standard error: 0.295 on 20 degrees of freedom
Multiple R-squared:  0.8286,    Adjusted R-squared:  0.7943
F-statistic: 24.17 on 4 and 20 DF,  p-value: 2.036e-07
```

FIGURE 2.4 Elimination of insignificant variable.

corresponds to a probability between 0.05 and 0.1. No symbol at the end of the row means the variable is not significant in the model. In R, the insignificant variable (variable that has no symbol at the end of the row) is eliminated first and a new model is created in which the output has better significance than the previous model as Figure 2.4. If the model has two variables that are not significant in the output of

```
> cor(wine)
                   Year        Price    WinterRain          AGST HarvestRain
Year         1.00000000  -0.4477679  0.016970024  -0.24691585  0.02800907
Price        -0.44776786  1.0000000  0.136650547   0.65956286 -0.56332190
WinterRain   0.01697002   0.1366505  1.000000000  -0.32109061 -0.27544085
AGST         -0.24691585  0.6595629 -0.321090611   1.00000000 -0.06449593
HarvestRain  0.02800907  -0.5633219 -0.275440854  -0.06449593  1.00000000
Age          -1.00000000  0.4477679 -0.016970024   0.24691585 -0.02800907
FrancePop    0.99448510  -0.4668616 -0.001621627  -0.25916227  0.04126439
                    Age     FrancePop
Year         -1.00000000  0.994485097
Price         0.44776786 -0.466861641
WinterRain   -0.01697002 -0.001621627
AGST          0.24691585 -0.259162274
HarvestRain  -0.02800907  0.041264394
Age           1.00000000 -0.994485097
FrancePop    -0.99448510  1.000000000
```

FIGURE 2.5   Correlation of variables.

the first model, one is removed and a new model is formulated. Now if the variable that is not significant in the first model is significant in the second model, then the variable is said to have multi colinearity and both the variables are highly correlated.

Correlation is the measure of the linear relationship between the variables that have a number between −1 and +1. The correlation rank of +1 corresponds to a linear relationship with a perfect positive relationship, and a correlation rank or score of −1 denotes a negative linear relationship. The middle number 0 indicates no linear relationship. The correlation between the variables can be found using the function cor. The correlation of all the variables can be found with this output matching any one of the columns with any one of the rows, as in Figure 2.5.

The quality of the model can be well determined with the prediction of test data (new data) using the function predict, as shown in Figure 2.6.

2.6   PROPOSED METHODOLOGY

With the advancement in technology, the prediction of voltage and losses has become simple with the use of data analytics that combines the application of the analytical software R-STUDIO. The method of data analytics used in the proposed system is predictive data analytics using R-studio. Prediction becomes easier because of limited lines of code in R-STUDIO. This method is done offline.

The proposed method includes the following steps:

Step 1: Collection of data for various loading conditions of a standard 6-bus IEEE system and IEEE standard 30-bus system by implementing Newton–Raphson method of load flow analysis is done.

Step 2: Sets of sample data are collected for further prediction.

Step 3: The collected data are processed, and the values of errors lesser than significant values, which denote the unnecessary loads, are eliminated using R studio.

Step 4: The obtained data are made to run against a sample data for which the voltage and loss have to be predicted.

```
> winetest = read.csv("wine_test.csv")
> str(winetest)
'data.frame':   2 obs. of  7 variables:
 $ Year      : int  1979 1980
 $ Price     : num  6.95 6.5
 $ WinterRain : int  717 578
 $ AGST      : num  16.2 16
 $ HarvestRain: int  122 74
 $ Age       : int  4 3
 $ FrancePop : num  54836 55110
> wineTest = read.csv("wine_test.csv")
> str(wineTest)
'data.frame':   2 obs. of  7 variables:
 $ Year      : int  1979 1980
 $ Price     : num  6.95 6.5
 $ WinterRain : int  717 578
 $ AGST      : num  16.2 16
 $ HarvestRain: int  122 74
 $ Age       : int  4 3
 $ FrancePop : num  54836 55110
> predictTest = predict(model3, newdata=wineTest)
> predictTest
        1        2
6.768925 6.684910
```

FIGURE 2.6 Prediction with test data.

Step 5: The voltage and loss attributed to the particular load are predicted and obtained as results.

2.6.1 The Standard IEEE Test System with 6 Buses

A sample system, namely, IEEE test system consisting of six buses was taken from University of Washington IEEE Common Data Format, shown in Figure 2.7. The network has three voltage sources and three PQ or load profile buses. Bus-1 is considered as the swing or reference bus, buses 2 and 3 are (PV) or the generator buses, and buses 4–6 are (PQ) commonly known as load buses. The bus data, generator data, and branch data for the IEEE system with these six buses are shown in Tables 2.1–2.3, respectively. The voltage angle (degrees) and the transformer tap ratio of the system are taken as zero. The data given in the following tables are on 100 MVA base.

2.6.2 Formulation of Voltage Prediction for Load Bus 4

The formulation includes the real loads of all the load buses (PL4, PL5, PL6) and the reactive loads of the load buses (QL4, QL5, QL6). In the estimation of voltage in load bus 4, PL5 has no star, which means it is not significant and it is eliminated, as is shown in Figure 2.8.

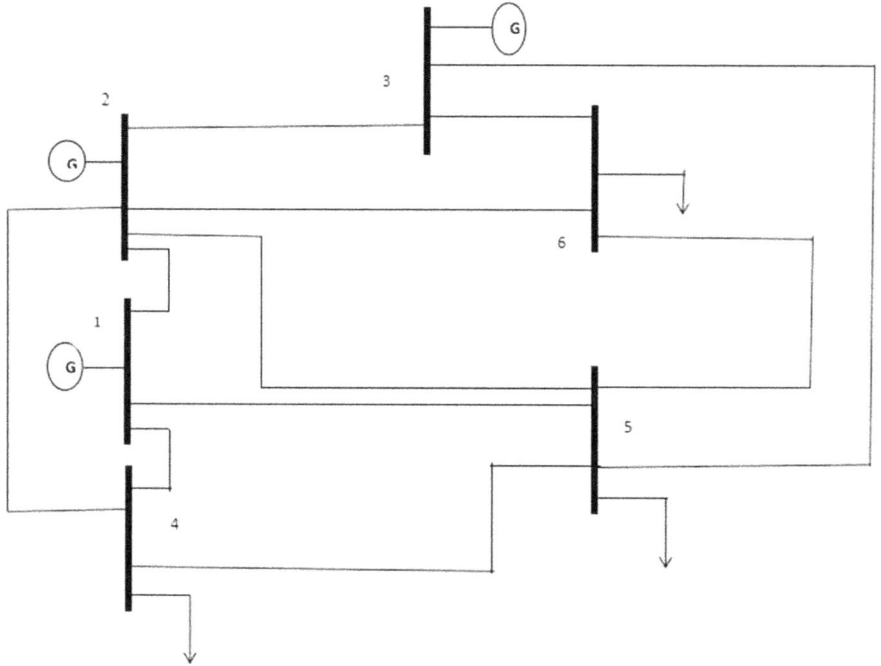

FIGURE 2.7   Connecting line diagram of IEEE system with 6 buses taken for analysis.

TABLE 2.1
Sample Bus Data of IEEE 6 Bus Test System

Bus No.	Bus Type	Voltage Magnitude (p.u.)	Real Power on Load Side (MW)	Reactive Power on Load Side (MVAR)	Maximum Bus Voltage Magnitude (p.u.)	Minimum Bus Voltage Magnitude (p.u.)
1	SB	1.05	0	0	1.1	0.9
2	PV	1.05	0	0	1.1	0.9
3	PV	1.07	0	0	1.1	0.9
4	PQ	1	70	70	1.1	0.9
5	PQ	1	70	70	1.1	0.9
6	PQ	1	70	70	1.1	0.9

With the elimination of PL5, the voltage formulation of the bus changes by the formula

$$V4=lm\left(OV4{\sim}PL4+PL6+QL4+QL5+QL6,data=mainn\right)$$

The formulation includes the real loads of all the load buses (PL4, PL6) and the reactive loads of the load buses (QL4, QL5, QL6).

All the variables of the bus are significant, with two stars, as shown in Figure 2.9.

TABLE 2.2
Data of IEEE 6 Bus System Pertaining to the Generator

Generator – Connected to Bus No.	Real Power Generation (MW)	Max Real Power Generation Limit (MW)	Min Real Power Generation Limit (MW)	Max Reactive Power Generation Limit (MVAR)	Min Reactive Power Generation Limit (MVAR)
1	0	200	50	100	−100
2	50	150	37.5	100	−100
3	60	180	45	100	−100

TABLE 2.3
Branch Data of IEEE 6 Bus System Pertaining to Line Branches

Line No.	Bus No. (Indicating the Lines)		Resistance (R) of the Line	Reactance (X) of the Line	Total Line Charging Susceptance for the Line	Max Apparent Power Flow
	From	To	(p.u.)	(p.u.)	(p.u.)	(MVA)
1	1	2	0.1	0.2	0.04	40
2	1	4	0.05	0.2	0.04	60
3	1	5	0.08	0.3	0.06	40
4	2	3	0.05	0.25	0.06	40
5	2	4	0.05	0.1	0.02	80
6	2	5	0.1	0.3	0.04	30
7	2	6	0.07	0.2	0.05	90
8	3	5	0.12	0.26	0.05	70
9	3	6	0.02	0.1	0.02	90
10	4	5	0.2	0.4	0.08	20
11	5	6	0.1	0.3	0.06	40

2.6.3 FORMULATION OF VOLTAGE PREDICTION FOR LOAD BUS 5

The formulation includes the real loads of all the load buses (PL4, PL5, PL6) and the reactive loads of the load buses (QL4, QL5, QL6).

In the estimation of voltage in load bus 5, PL4, QL4, that has no star and is, thus, not significant, and it is eliminated, as shown in Figure 2.10.

Then the formulation becomes

$$V5=lm\left(OV5{\sim}+PL5+PL6+QL5+QL6,data=mainn\right)$$

The formulation includes the real loads of all the load buses (PL5,PL6) and the reactive loads of the load buses (QL5,QL6).All the variables of the bus are significant, with a maximum of three stars, as shown in Figure 2.11.

```
> V4=lm(OV4 ~ PL4+PL5+PL6+QL4+QL5+QL6,data=mainn)
> summary(V4)

Call:
lm(formula = OV4 ~ PL4 + PL5 + PL6 + QL4 + QL5 + QL6, data = mainn)

Residuals:
      Min         1Q      Median         3Q         Max
-0.0214780 -0.0058997  0.0007713  0.0062436  0.0148987

Coefficients:
              Estimate Std. Error t value Pr(>|t|)
(Intercept)  1.105e+00  2.621e-03 421.476  < 2e-16 ***
PL4         -2.470e-04  7.118e-05  -3.470 0.000649 ***
PL5         -9.289e-05  6.837e-05  -1.359 0.175955
PL6         -1.623e-04  6.284e-05  -2.583 0.010574 *
QL4         -6.396e-04  6.466e-05  -9.892  < 2e-16 ***
QL5         -2.205e-04  6.691e-05  -3.296 0.001176 **
QL6         -2.524e-04  6.168e-05  -4.091 6.43e-05 ***
---
Signif. codes:  0 '***' 0.001 '**' 0.01 '*' 0.05 '.' 0.1 ' ' 1

Residual standard error: 0.007949 on 183 degrees of freedom
Multiple R-squared:  0.9248,    Adjusted R-squared:  0.9224
F-statistic: 375.2 on 6 and 183 DF,  p-value: < 2.2e-16
```

FIGURE 2.8   Elimination of PL5.

```
> V4=lm(OV4 ~ PL4+PL6+QL4+QL5+QL6,data=mainn)
> summary(V4)

Call:
lm(formula = OV4 ~ PL4 + PL6 + QL4 + QL5 + QL6, data = mainn)

Residuals:
      Min         1Q      Median         3Q         Max
-0.0218587 -0.0058469  0.0008139  0.0055070  0.0148673

Coefficients:
              Estimate Std. Error t value Pr(>|t|)
(Intercept)  1.104e+00  2.617e-03 422.012  < 2e-16 ***
PL4         -2.950e-04  6.190e-05  -4.766 3.80e-06 ***
PL6         -2.021e-04  5.574e-05  -3.626 0.000373 ***
QL4         -6.419e-04  6.479e-05  -9.907  < 2e-16 ***
QL5         -2.209e-04  6.706e-05  -3.295 0.001182 **
QL6         -2.509e-04  6.182e-05  -4.059 7.27e-05 ***
---
Signif. codes:  0 '***' 0.001 '**' 0.01 '*' 0.05 '.' 0.1 ' ' 1

Residual standard error: 0.007968 on 184 degrees of freedom
Multiple R-squared:  0.9241,    Adjusted R-squared:  0.922
F-statistic: 447.8 on 5 and 184 DF,  p-value: < 2.2e-16
```

FIGURE 2.9   Prediction of V4.

```
> V5=lm(OV5 ~ PL4+PL5+PL6+QL4+QL5+QL6,data=mainn)
> summary(V5)

Call:
lm(formula = OV5 ~ PL4 + PL5 + PL6 + QL4 + QL5 + QL6, data = mainn)

Residuals:
     Min        1Q    Median        3Q       Max
-0.027859 -0.006967  0.001025  0.007338  0.017039

Coefficients:
              Estimate Std. Error t value Pr(>|t|)
(Intercept)  1.137e+00  3.152e-03 360.778  < 2e-16 ***
PL4         -7.765e-06  8.559e-05  -0.091 0.927816
PL5         -3.890e-04  8.222e-05  -4.731 4.45e-06 ***
PL6         -2.737e-04  7.557e-05  -3.622 0.000379 ***
QL4         -8.991e-05  7.775e-05  -1.156 0.249027
QL5         -8.569e-04  8.045e-05 -10.650  < 2e-16 ***
QL6         -4.451e-04  7.417e-05  -6.001 1.03e-08 ***
---
Signif. codes:  0 '***' 0.001 '**' 0.01 '*' 0.05 '.' 0.1 ' ' 1

Residual standard error: 0.009559 on 183 degrees of freedom
Multiple R-squared:  0.9336,    Adjusted R-squared:  0.9314
F-statistic: 428.8 on 6 and 183 DF,  p-value: < 2.2e-16
```

FIGURE 2.10   Elimination of PL4 and QL4.

```
> V5=lm(OV5 ~ PL5+PL6+QL5+QL6,data=mainn)
> summary(V5)

Call:
lm(formula = OV5 ~ PL5 + PL6 + QL5 + QL6, data = mainn)

Residuals:
      Min         1Q     Median         3Q        Max
-0.0276603 -0.0067478  0.0001973  0.0077045  0.0181953

Coefficients:
              Estimate Std. Error t value Pr(>|t|)
(Intercept)  1.137e+00  3.140e-03 362.080  < 2e-16 ***
PL5         -3.973e-04  7.110e-05  -5.588 8.13e-08 ***
PL6         -2.788e-04  6.783e-05  -4.110 5.94e-05 ***
QL5         -9.003e-04  7.116e-05 -12.652  < 2e-16 ***
QL6         -4.799e-04  6.775e-05  -7.084 2.85e-11 ***
---
Signif. codes:  0 '***' 0.001 '**' 0.01 '*' 0.05 '.' 0.1 ' ' 1

Residual standard error: 0.009542 on 185 degrees of freedom
Multiple R-squared:  0.9331,    Adjusted R-squared:  0.9317
F-statistic: 645.1 on 4 and 185 DF,  p-value: < 2.2e-16
```

FIGURE 2.11   Prediction of V5.

2.6.4 FORMULATION OF VOLTAGE PREDICTION FOR LOAD BUS 6

The voltage magnitude of the load bus 6 for 200 variable loads is estimated using the code

$$V6=lm(OV6{\sim}PL4{+}PL5{+}PL6{+}QL5{+}QL6, data=mainn)$$

The formulation includes the real loads of all the load buses (PL4, PL5, PL6) and the reactive loads of the load buses (QL4, QL5, QL6). In the estimation of voltage in load bus 6, PL4, PL5, QL4 has no star and is not significant, and it is eliminated, as shown in Figure 2.12.

Then, the formulation becomes

$$V6=lm(OV6{\sim}PL6{+}QL5{+}QL6, data=mainn)$$

The formulation includes the real loads of all the load buses (PL6) and the reactive loads of the load buses (QL5, QL6) (Figure 2.13). The magnitude of the load bus voltages for buses 4–6 for 300 variable loads is assessed in the same method as done for 200 variable loads. A comparison of 200 variables and 300 variables is as follows:

1. In the estimation of voltage for V4 with 200 variable loads, only the real load PL5 is eliminated. On estimating with 300 variable loads, the real

```
> V6=lm(OV6 ~ PL4+PL5+PL6+QL4+QL5+QL6,data=mainn)
> summary(V6)

Call:
lm(formula = OV6 ~ PL4 + PL5 + PL6 + QL4 + QL5 + QL6, data = mainn)

Residuals:
      Min        1Q     Median        3Q       Max
-0.026708 -0.007462  0.001290  0.008095  0.020442

Coefficients:
              Estimate Std. Error t value Pr(>|t|)
(Intercept)  1.137e+00  3.560e-03 319.460  < 2e-16 ***
PL4          1.647e-05  9.667e-05   0.170  0.86490
PL5         -1.583e-04  9.285e-05  -1.704  0.08999 .
PL6         -4.522e-04  8.534e-05  -5.299 3.33e-07 ***
QL4         -1.285e-05  8.781e-05  -0.146  0.88386
QL5         -2.877e-04  9.086e-05  -3.166  0.00181 **
QL6         -9.848e-04  8.377e-05 -11.756  < 2e-16 ***
---
Signif. codes:  0 '***' 0.001 '**' 0.01 '*' 0.05 '.' 0.1 ' ' 1

Residual standard error: 0.0108 on 183 degrees of freedom
Multiple R-squared:  0.9051,    Adjusted R-squared:  0.902
F-statistic: 290.9 on 6 and 183 DF,  p-value: < 2.2e-16
```

FIGURE 2.12 Elimination of PL4, PL5, QL4.

```
> V6=lm(OV6 ~ PL6+QL5+QL6,data=mainn)
> summary(V6)

Call:
lm(formula = OV6 ~ PL6 + QL5 + QL6, data = mainn)

Residuals:
      Min        1Q    Median        3Q       Max
-0.027877 -0.007316  0.001330  0.007618  0.022867

Coefficients:
              Estimate Std. Error t value Pr(>|t|)
(Intercept)  1.136e+00  3.528e-03 322.097  < 2e-16 ***
PL6         -5.726e-04  3.677e-05 -15.574  < 2e-16 ***
QL5         -3.010e-04  8.052e-05  -3.738 0.000247 ***
QL6         -9.907e-04  7.677e-05 -12.905  < 2e-16 ***
---
Signif. codes:  0 '***' 0.001 '**' 0.01 '*' 0.05 '.' 0.1 ' ' 1

Residual standard error: 0.01081 on 186 degrees of freedom
Multiple R-squared:  0.9032,     Adjusted R-squared:  0.9017
F-statistic: 578.8 on 3 and 186 DF,  p-value: < 2.2e-16
```

FIGURE 2.13 Estimation of V6.

200 VARIABLE LOADS	300 VARIABLE LOADS
``> V4=lm(OV4 ~ PL4+PL6+QL4+QL5+QL6,data=mainn)
> summary(V4)

Call:
lm(formula = OV4 ~ PL4 + PL6 + QL4 + QL5 + QL6, data = mainn)

Residuals:
      Min   1Q   Median   3Q   Max
-0.0218587 -0.0058469 0.0008139 0.0055070 0.0148673

Coefficients:
              Estimate Std. Error t value Pr(>|t|)
(Intercept) 1.104e+00 2.617e-03 422.012 < 2e-16 ***
PL4   -2.950e-04 6.190e-05 -4.766 3.80e-06 ***
PL6   -2.021e-04 5.574e-05 -3.626 0.000373 ***
QL4   -6.419e-04 6.479e-05 -9.907 < 2e-16 ***
QL5   -2.209e-04 6.706e-05 -3.295 0.001182 **
QL6   -2.509e-04 6.182e-05 -4.059 7.27e-05 ***

Signif. codes: 0 '***' 0.001 '**' 0.01 '*' 0.05 '.' 0.1 ' ' 1

Residual standard error: 0.007968 on 184 degrees of freedom
Multiple R-squared: 0.9241,   Adjusted R-squared: 0.922
F-statistic: 447.8 on 5 and 184 DF, p-value: < 2.2e-16
``` | ```> V4=lm(OV4 ~ PL4+QL4+QL6,data=main)
> summary(V4)

Call:
lm(formula = OV4 ~ PL4 + QL4 + QL6, data = main)

Residuals:
      Min   1Q   Median   3Q   Max
-0.0248800 -0.0041945 0.0006031 0.0055644 0.0150809

Coefficients:
              Estimate Std. Error t value Pr(>|t|)
(Intercept) 1.095e+00 2.043e-03 535.789 < 2e-16 ***
PL4   -4.650e-04 2.502e-05 -18.585 < 2e-16 ***
QL4   -7.953e-04 5.073e-05 -15.688 < 2e-16 ***
QL6   -2.302e-04 4.450e-05 -5.173 4.33e-07 ***

Signif. codes: 0 '***' 0.001 '**' 0.01 '*' 0.05 '.' 0.1 ' ' 1

Residual standard error: 0.007631 on 286 degrees of freedom
Multiple R-squared: 0.9064,   Adjusted R-squared: 0.9054
F-statistic: 923.2 on 3 and 286 DF, p-value: < 2.2e-16
``` |

FIGURE 2.14 Comparison of V5 for 200 and 300 variable loads with PL5 and PL6.

loads PL5, PL6 are eliminated, which are of less significance. The residual std. error of 300 variable loads seems to be lesser than that of 200 variable loads; this increasing accuracy is shown in Figure 2.14.

2. In the estimation of voltage for V5 with 200 variable loads the loads PL4, QL4 are eliminated. On estimating with 300 variable loads, the real load PL4 is eliminated, which is less significant. The residual std. error of 300 variable loads seems to be lesser than that of 200 variable loads; this increasing accuracy is shown in Figure 2.15.

200 VARIABLE LOADS	300 VARIABLE LOADS

```
> V5=lm(OV5 ~ PL5+PL6+QL5+QL6,data=mainn)
> summary(V5)

Call:
lm(formula = OV5 ~ PL5 + PL6 + QL5 + QL6, data = mainn)

Residuals:
      Min         1Q     Median         3Q        Max
-0.0276603 -0.0067478  0.0001973  0.0077045  0.0191953

Coefficients:
              Estimate Std. Error t value Pr(>|t|)
(Intercept)  1.137e+00  3.140e-03 362.080  < 2e-16 ***
PL5         -3.973e-04  7.110e-05  -5.588 8.13e-08 ***
PL6         -2.788e-04  6.783e-05  -4.110 5.94e-05 ***
QL5         -9.003e-04  7.116e-05 -12.652  < 2e-16 ***
QL6         -4.799e-04  6.775e-05  -7.084 2.85e-11 ***
---
Signif. codes:  0 '***' 0.001 '**' 0.01 '*' 0.05 '.' 0.1 ' ' 1

Residual standard error: 0.009542 on 185 degrees of freedom
Multiple R-squared:  0.9331,    Adjusted R-squared:  0.9317
F-statistic: 645.1 on 4 and 185 DF,  p-value: < 2.2e-16
```

```
> V5=lm(OV5 ~ PL5+PL6+QL4+QL5+QL6,data=main)
> summary(V5)

Call:
lm(formula = OV5 ~ PL5 + PL6 + QL4 + QL5 + QL6, data = main)

Residuals:
      Min         1Q     Median         3Q        Max
-0.032515 -0.004950  0.001664  0.007222  0.014237

Coefficients:
              Estimate Std. Error t value Pr(>|t|)
(Intercept)  1.124e+00  2.430e-03 462.693  < 2e-16 ***
PL5         -3.805e-04  6.820e-05  -5.579 5.64e-08 ***
PL6         -2.066e-04  6.397e-05  -3.230  0.00138 **
QL4         -2.169e-04  6.795e-05  -3.190  0.00158 **
QL5         -7.796e-04  7.574e-05 -10.294  < 2e-16 ***
QL6         -3.091e-04  6.839e-05  -4.520 9.07e-06 ***
---
Signif. codes:  0 '***' 0.001 '**' 0.01 '*' 0.05 '.' 0.1 ' ' 1

Residual standard error: 0.009211 on 284 degrees of freedom
Multiple R-squared:  0.9207,    Adjusted R-squared:  0.9193
F-statistic: 659.2 on 5 and 284 DF,  p-value: < 2.2e-16
```

FIGURE 2.15 Comparison of V5 for 200 and 300 variable loads with PL4.

200 VARIABLE LOADS	300 VARIABLE LOADS

```
> V6=lm(OV6 ~ PL6+QL5+QL6,data=mainn)
> summary(V6)

Call:
lm(formula = OV6 ~ PL6 + QL5 + QL6, data = mainn)

Residuals:
      Min         1Q     Median         3Q        Max
-0.027807 -0.007316  0.001330  0.007416  0.022807

Coefficients:
              Estimate Std. Error t value Pr(>|t|)
(Intercept)  1.136e+00  3.528e-03 322.097  < 2e-16 ***
PL6         -5.726e-04  3.677e-05 -15.574  < 2e-16 ***
QL5         -3.010e-04  8.052e-05  -3.738 0.000247 ***
QL6         -9.437e-04  7.677e-05 -12.295  < 2e-16 ***
---
Signif. codes:  0 '***' 0.001 '**' 0.01 '*' 0.05 '.' 0.1 ' ' 1

Residual standard error: 0.01081 on 186 degrees of freedom
Multiple R-squared:  0.9332,    Adjusted R-squared:  0.9317
F-statistic: 578.8 on 3 and 186 DF,  p-value: < 2.2e-16
```

```
> V6=lm(OV6 ~ PL6+QL5+QL6,data=main)
> summary(V6)

Call:
lm(formula = OV6 ~ PL6 + QL5 + QL6, data = main)

Residuals:
      Min         1Q     Median         3Q        Max
-0.032487 -0.005060  0.001954  0.007467  0.020006

Coefficients:
              Estimate Std. Error t value Pr(>|t|)
(Intercept)  1.121e+00  2.702e-03 415.103  < 2e-16 ***
PL6         -5.003e-04  2.963e-05 -16.882  < 2e-16 ***
QL5         -2.802e-04  7.75e-05   -3.615 0.000355 ***
QL6         -8.832e-04  7.277e-05 -12.137  < 2e-16 ***
---
Signif. codes:  0 '***' 0.001 '**' 0.01 '*' 0.05 '.' 0.1 ' ' 1

Residual standard error: 0.01048 on 286 degrees of freedom
Multiple R-squared:  0.8818,    Adjusted R-squared:  0.8806
F-statistic: 711.3 on 3 and 286 DF,  p-value: < 2.2e-16
```

FIGURE 2.16 Comparison of V6 for 200 and 300 variable loads.

In the estimation of voltage for V5 with 200 variable loads, the loads PL4, QL4 are eliminated. On estimating with 300 variable loads, the real load PL4 is eliminated, which is of less significance. The residual std. error of 300 variable loads seems to be lesser than that of 200 variable loads; this increasing accuracy is shown in Figure 2.16.

2.7 ANALYSIS OF RESULTS

The data obtained for the test case IEEE 6 bus system through Newton's method for various loads are assessed using R STUDIO. Error values in the system and the

1	OV4	x4	DIFF4	OV5	x5	DIFF5	OV6	x6	DIFF6
2	1.007	1.009109	-0.00211	1.016	1.018966	-0.00297	1.025	1.02985	-0.00485
3	1.004	1.005438	-0.00144	1.015	1.0179	-0.0029	1.025	1.02985	-0.00485
4	1.012	1.017379	-0.00538	1.016	1.021589	-0.00559	1.025	1.031706	-0.00671
5	1.011	1.015079	-0.00408	1.016	1.020811	-0.00481	1.025	1.030778	-0.00578
6	1.009	1.010478	-0.00148	1.016	1.019254	-0.00325	1.025	1.028921	-0.00392
7	1.007	1.008178	-0.00118	1.016	1.018475	-0.00248	1.025	1.027993	-0.00299
8	1.005	1.007036	-0.00204	1.004	1.00403	-3.00E-05	1.016	1.016615	-0.00062
9	1.004	1.004736	-0.00074	1.004	1.003251	0.000749	1.016	1.015687	0.000313
10	1.003	1.002436	0.000564	1.004	1.002473	0.001527	1.016	1.014759	0.001241
11	1.002	1.000135	0.001865	1.004	1.001694	0.002306	1.015	1.01383	0.00117

FIGURE 2.17 Comparison of voltage magnitude of the proposed method with the conventional method of the IEEE 6 bus system.

unnecessary load buses as seen in the problem are eliminated. The program is then made to run against the sample data generated. The voltage magnitude for a particular load is predicted by the software and is obtained as a result. The predicted voltage is compared with the values of voltage magnitude obtained for the particular load using Newton's method. It can be seen evidently that the voltage values predicted are almost equal to the voltage obtained by the conventional method for a particular load. The values of the predicted voltage have a difference of about two significant digits compared to the conventional method, as shown in Figure 2.17, where OV4, OV5, and OV6 are the voltage magnitudes obtained from the conventional method; x4, x5, and x6 are the voltage magnitudes obtained from the prediction of the proposed model; and DIFF4, DIFF5, and DIFF6 are the differences of the values obtained from the conventional and proposed models.

2.8 CONCLUSION

A simplified approach for the prediction of voltage using R STUDIO is formulated for IEEE 6 bus systems. The result of the prediction when compared with the conventional method seems to be more accurate and also is time-saving. Both results show good convergence. The difficulties of the conventional techniques, such as accuracy, complexity in matrix formation, speed, efficiency, and limited Machine Learning, have been overcome by this methodology. The output of the system seems to be better and more efficient in comparison with that of the conventional method, such as load flow analysis using Newton–Raphson method. The proposed approach on account of its effectiveness, simplified algorithms, and lesser execution time will be suitable for online practical implementations. In the future, it can be extended to studies with other, more complex standard IEEE test systems like IEEE system with 30 bus, 118 bus systems, etc, as the computational complexity is reduced, as is the time taken for solving large formulations or computations.

REFERENCES

1. F. Albuyeh, A. Bose, B. Heath, "Reactive power consideration in automatic contingency selection", *IEEE Trans. Power Appar. Syst. PAS* 101 (1982) 107–112.

 2. S.N. Singh, S.C. Srivastava, "Improved voltage and reactive power distribution factor for outage studies", *IEEE Trans. Power Syst. PWRS* 12(3) (1997) 1085–1093.
 3. P. Araindhababu, G. Balamurugan, "ANN based online voltage estimation", *Appl. Soft Comput.* 12 (2012) 313–319, August 2011.
 4. L.L. Grant, G.K. Venayagamoorthy, "Voltage prediction using a cellular network", *IEEE Conference Publications*, Power and Energy Society General Meeting, 2010
 5. L. Srivastava, S.N. Singh, J. Shama, "Parallel self-organising hierarichal neural network-based fast voltage estimation", IEEE Proceedings –Generation, Transmission, Distribution. *International Journal of Recent Technology and Engineering (IJRTE)*, 8(4), (2020).
 6. C.S. Chang, "Fast power system voltage prediction using knowledge based approach and on-line box data creation", *Proc., IEEE Gener. Trans., Distribution*, 136(2), (March 1989) 87–89.
 7. S.S. Biswass, C.B. Vellaithurai A.K. Srivastava "Development and real time implementation of a synchrophasor based fast voltage stability monitoring algorithm with consideration of load models." In *Industry Applications Society Annual Meeting*, (2013) IEEE. pp. 1–9, 6–11 October
 8. T. Kulworawanichpong, "*Simplified Newton-Raphson Power-Flow Solution Method*", Suranaree University of Technology, Nakhom Ratchasima (2009).
 9. S.M.S. Ghiasi, M. Abedi, S.H. Hosseinian, "A new approach for the estimation of transient voltage profile along transmission line", *Canadian Journal of Electrical and Computer Engineering*, 40(4), (2017) 295–302.
10. X. Zambrano, A. Hern´andez, M. Izzeddine, R.M. de Castro, "Estimation of voltage sags from a limited set of monitors in power systems", *IEEE Transactions on Power Delivery* 32(2) (2016).
11. N.C. Woolley, M. Avenda˜no-Mora, A.P. Woolley, R. Preece, J.V. Milanovic, "Probabilistic estimation of voltage sags using erroneous measurement information," *Electric Power Systems Research* 106, (Jan 2014) 142–150.
12. X. Zambrano, A. Hern´andez, R.M. de Castro, M. Izzeddine, "Comparison of voltage sags prediction methods in power systems," *15th International Conference on Environment and Electrical Engineering (EEEIC)*, Rome, Italy, (June 2015).

3 Intelligent Face Recognition Based on Regularized Robust Coding with Deep Learning Process

Sandhya Swaminathan and Anitha Perla
Malla Reddy College of Engineering

CONTENTS

3.1 INTRODUCTION

Face recognition has attracted researchers for several decades as it is a crucial subject in computer vision and pattern recognition. Consequently, various face recognition techniques, such as Eigenface, Fisherface, and Support Vector Machines, have been established. Yet, robust face recognition remains a challenge because, real-world face images typically consist of variations in lighting, expression, etc. as well as occlusion.

The general method for face recognition is the classification of the features extracted from the image. A widely used classifier is the nearest neighbor (NN) classifier [1], because of its simplicity and efficiency. One major disadvantage of the NN classifier is that only one training image is used to represent the query image. To overcome this limitation, the nearest feature line (NFL) classifier [2] was introduced, which represents the query image with two training images. Subsequently, the nearest feature plane (NSP) classifier [3] was propounded, which represents the query image with three training images. Further, classifiers utilizing more training samples to represent the query face were introduced, for instance, the local subspace classifier (LSC) and the nearest subspace (NS) classifiers, which make use of all the training samples of each class to represent the query face image. It was shown that NFL [2], NSP, LSC, and NS perform better than NN [1]. Yet, all these methods with holistic face features have low recognition rates for facial images with occlusion. These nearest classifiers, like NN [1], NFL [2], NFP, LSC, and NS, attempt to represent the query face image correctly, by classifying it according to which class can represent the query image better than other classes. However, formulating the representation model for classification tasks, such as face recognition, is still a challenge.

Recently, sparse representation (or sparse coding) has been adopted, due to its effectiveness in the image processing field, especially, in face recognition and texture classification. We see face recognition as an issue of classifying among multiple linear regression models and assert that sparse signal representation is the solution to this problem. Based on a sparse representation determined by $l1$-minimization, a general classification algorithm for (image-based) object recognition is presented. Natural images can be coded by structural primitives (e.g. edges and line segments) that are qualitatively analogous in configuration to simple cell receptive fields. Sparse coding represents a signal using a few atoms sparingly chosen from an over-complete dictionary. The sparsity of the coding coefficient may be measured using $l0$-norm, which gives the count of nonzero entries in a vector. The $l0$-norm minimization is an NP-hard problem. Thus, the $l1$-norm minimization, as the nearest convex function to l_0-norm minimization, is used extensively in sparse coding. It has been proven that l_0-norm and $l1$-norm minimizations are comparable if the solution is acceptably sparse. The sparse coding problem can be expressed as:

$$\min_{\alpha} \|\alpha\|_1 \; s.t \; \| y - D\alpha \|_2^2 \le \varepsilon \tag{3.1}$$

where y is the specified signal, D represents the dictionary of coding atoms, α represents the coding vector of y over D, and $\varepsilon > 0$ is a constant. This idea of sparse coding was used in face recognition, and the sparse representation-based classification (SRC [4]) scheme was developed. First, the query image y is coded as a sparse linear combination of all the training samples using Eq. (3.1). Then, sparse representation-based scheme classifies y by determining the class that will result in the minimal reconstruction error of it. One important quality of sparse representation-based classification is its processing of face occlusion and corruption. It presents an identity matrix I as a dictionary. This matrix is used to code the outlier pixels (e.g., corrupted or occluded pixels):

$$\min_\alpha \| [\alpha; \beta] \|_1 \ s.t. y = [D, I]. [\alpha; \beta] \tag{3.2}$$

When Eq. (3.2) is solved, it is observed that SRC [4] is robust for solving face occlusions, such as pixel corruption, block occlusion, and disguise. Equation (3.2) is simply equivalent to: $\min_\alpha \|\alpha\|_1 \ s.t \| y - D\alpha \|_1 < \varepsilon$. This means it employs $l1$-norm to model the coding residual $y - D\alpha$, to enhance robustness to outliers.

The sparse representation-based classification is closely related to the nearest classifiers since SRC [4] also describes the query signal as a linear combination of the training samples. However, it places a sparsity constraint on the representation coefficients and allows across-class representation, which means that significant coding coefficients can be from samples of different classes. This helps overcome the small-sample-size problem in face recognition. Sparse representation-based classification deals with occlusion by assuming a sparse coding residual, as shown in Eq. (3.2). Despite the sparse coding model in Eq. (3.1) being very suitable for image restoration and face recognition, two complications are observed that must be considered for pattern classification tasks such as face recognition. The first issue is whether the $l1$-sparsity constraint $\| \|_1$ is necessary to regularize the solution since the $l1$-minimization requires a large computational cost. The second is whether the term $\| y - D\alpha \|_2^2 \leq \varepsilon$ is effective in characterizing the signal fidelity, especially when y is noisy and/or has many outliers. For the first issue, it has been shown that it is not essential to impose the $l1$-sparsity constraint on the coding vector α and that the $l2$-norm regularization on α performs just as efficiently. Also, SRC [4] owes its effectiveness to its collaborative representation of y over all classes of training samples. For the second issue, it has been shown that the fidelity term has a very high impact on the final coding result. According to *maximum a posterior* (MAP) estimation, describing the fidelity term with $l2$- or $l1$-norm assumes that the coding residual $e = y - D\alpha$ obeys Gaussian or Laplacian distribution. However, for practical uses, especially when occlusions, expression variations, and corruptions exist in the query facial images, this assumption does not hold well. Although fidelity term based on Gaussian kernel is said to be robust to non-Gaussian noise, it may not work well in face recognition with occlusion due to nonuniform occlusion.

To improve the robustness of face recognition to occlusion, corruption of pixels and expression changes, etc., a regularized robust coding (RRC) model is proposed in this chapter. In the regularized robust coding model, it is assumed that the coding residual e and the coding vector α are independent and homogenously distributed and then robustly regress the signal using the MAP principle. Specifically, RRC with $l1$-norm regularization can accomplish high recognition rates as it yields a sparse solution and automatically performs feature selection. Also, it is robust to outliers as it assigns insignificant features with zero weight and useful features with nonzero weight. For practical implementation, the regularized robust coding minimization problem is converted into an iteratively reweighted regularized robust coding (IR^{3}C) problem. Thorough experiments with standard face databases indicate that superior performance is obtained with RRC- $l1$ compared to other sparse representation-based approaches in treating face occlusion, corruption, lighting, expression variations, etc.

3.2   CONCEPTS USED IN EXISTING METHODS

3.2.1   THE MODELING OF REGULARIZED ROBUST CODING (RRC)

The traditional sparse coding model in Eq. (3.1) is represented by the least absolute shrinkage and selection operator (LASSO) problem:

$$\min_\alpha \| y - D\alpha \|_2^2 \ s.t \|\alpha\|_1 \leq \sigma \tag{3.3}$$

where $\sigma > 0$ is a constant, $y = [y_1; y_2; \ldots; y_n] \in R^n$ is the query signal, $D = [d_1, d_2, \ldots, d_m] \in R^{n \times m}$ is the dictionary which has column vector d_j as its jth atom, and $\alpha \in Rm$ is the vector of coding coefficients. For the case of face recognition, the atom d_j is taken as the training facial image (or its dimensionality reduced feature), thus making the dictionary D the whole training data set. If the coding residual $e = y - D\alpha$ follows Gaussian distribution, the solution to Eq. (3.3) will be the maximum likelihood estimation (MLE) solution. If the coding residual follows Laplacian distribution, the $l1$-sparsity constrained MLE solution will be

$$\min_\alpha \| y - D\alpha \|_1 \ s.t \|\alpha\|_1 \leq \sigma \tag{3.4}$$

However, in practical situations, the coding residual does not follow the Gaussian or Laplacian distribution, especially when the query facial image y has occlusion, corruption, etc. Using examples to depict the fitted distributions of residual e by distinct models, we see that Figure. 3.1a shows an uncorrupted clean facial image y_o, and Figure 3.1b and c shows the query images y with occlusion and corruption, respectively. The residual is calculated as $e = y - D\acute{\alpha}$, and to improve the accuracy of the coding vector, we use the clean facial image to compute it using Eq. (3.3): $\acute{\alpha} = \text{argmin}_\alpha \| y - D\alpha \|_2^2 \ s.t \|\alpha\|_1 \leq \sigma$. Modeling the practical and fitted distributions of e using Gaussian, Laplacian, and the distribution model (refer to Eq. (3.15)) concerned with the proposed method are charted in Figure 3.1d.

For a clearer depiction of the tails of the plots, Figure 3.1e translates the distributions into the log domain. It is noted that the empirical distribution of e has a large peak at zero, corresponding to trivial coding errors, and a long tail, associated with the occluded and corrupted pixels. In the case of robust face recognition, the fitting of this tail plays a much more important role than the fitting of the peak. The proposed distribution template, according to Figure 3.1e, can effectively fit the heavy tail of the empirical distribution and thus is a better approximation compared to the Laplacian and Gaussian models. It is also noted that Laplacian performs better in fitting the heavy tail than Gaussian, which is the reason for the better performance of the sparse

(a) (b) (c)

(d) (e)

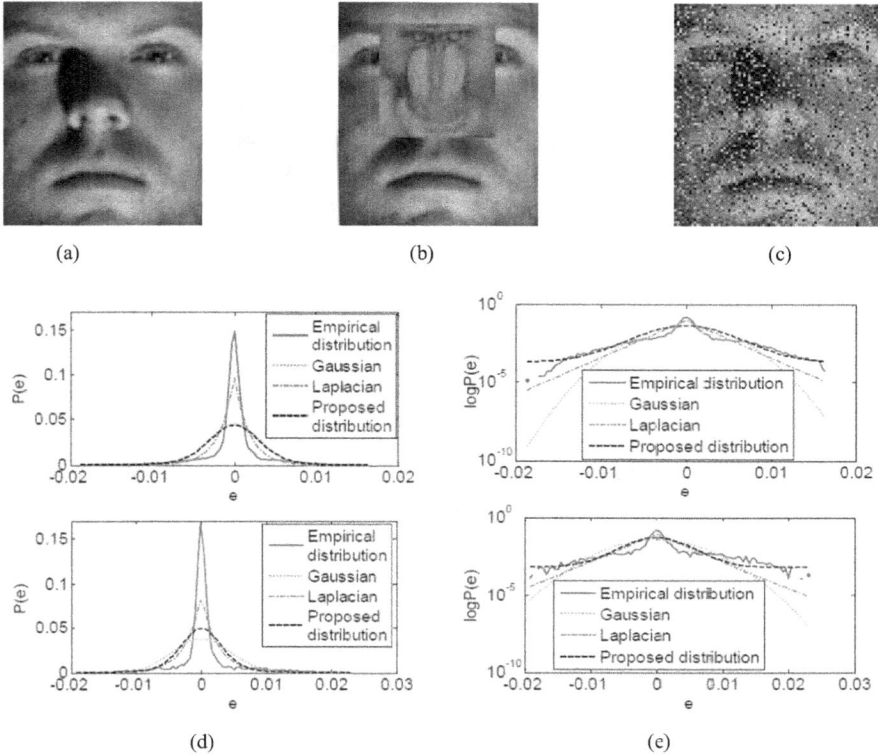

FIGURE 3.1 The empirical distribution of coding residuals e and the fitted distributions by various models. (a) Clean facial image; (b) Occluded test image; (c) Corrupted test image; (d) and (e) representing the distributions (top row: occluded image; bottom row: corrupted image) of coding residuals in linear domain and log domain, respectively.

coding model in Eq. (3.2) (or Eq. (3.4)) than the model in Eq. (3.1) or (3.3) in dealing with face corruption and occlusion.

Rewrite D as $D = [r_1; r_2; \ldots; r_n]$, where r_i represents the ith row of D, while considering e to be $e = y - D\alpha = [e_1; e_2; \ldots; e_n]$, where $e_i = y_i - r_i\alpha$, $i = 1, 2, \ldots, n$. Suppose that $e_1, e_2, \ldots, e_n$ are independent and identically distributed (i.i.d.) and the probability density function (PDF) of e_i is given by $f_\theta(e_i)$, where θ represents the undetermined parameter set that characterizes the distribution, the robust sparse coding (RSC) was expressed as the $l1$-sparsity constrained MLE problem shown below (assume $\rho_\theta(e) = -\ln f_\theta(e)$):

$$\min_\alpha \sum_{(i=1)}^{n} \rho_\theta(y_i - r_i\alpha) \, s.t. \, \| \alpha \|_1 \leq \sigma \tag{3.5}$$

Similar to the aforementioned SRC model, this RSC model assumes that the coding coefficients are sparse and utilizes $l1$-norm to characterize the sparsity. But the $l1$-sparsity constraint increases the complexity of RSC, and it has been proved that the $l1$-sparsity constraint on $\boldsymbol{\alpha}$ is not the reason for the good performance of SRC [4]. In this chapter, a general model, known as regularized robust coding (RRC), is proposed. It is found that the regularized robust coding is far more efficient than RSC.

Observe the face representation problem from a Bayesian estimation point-of-view, specifically, the MAP estimation. When the test image $\boldsymbol{y}$ is coded over a dictionary $\boldsymbol{D}$, the MAP estimation of the coding vector $\boldsymbol{\alpha}$ is given by $\hat{\alpha} = \text{argmax}_\alpha \ln P(\alpha \mid \boldsymbol{y})$.

Making use of the Bayesian formula, we obtain:

$$\hat{\alpha} = \text{argmax}_\alpha \{\ln P(\boldsymbol{y} \mid \alpha) + \ln P(\alpha)\} \tag{3.6}$$

Supposing that the elements e_i of coding residual $\boldsymbol{e} = \boldsymbol{y} - \boldsymbol{D}\boldsymbol{\alpha} = [e_1;\ e_2;\ \dots;\ e_n]$ are independent and identically distributed with PDF $f_\theta(e_i)$, we see that $P(\boldsymbol{y} \mid \boldsymbol{a}) = \prod^n f_\theta (y_i - r_i a)$. Also, supposing that the elements $\alpha_j,\ j = 1, 2, \dots, m$, of the coding vector $\boldsymbol{\alpha} = [\alpha_1;\ \alpha_2;\ \dots;\ \alpha_m]$ are independent and identically distributed with PDF $f_o(\alpha_j)$, there is $P(\alpha) = \prod^m f_\theta(\alpha_j)$ The MAP estimation of the $\boldsymbol{\alpha}$ observed in Eq. (3.6) is given below:

$$\hat{\alpha} = \text{arg max} \left\{ \prod^n f (\boldsymbol{y} - \boldsymbol{r}\alpha) + \sum^m f(a) \right\} \tag{3.7}$$

Assuming $\rho_\theta(e) = -\ln f_\theta(e)$ and $\rho_o(\alpha) = -\ln f_o(\alpha)$, Eq. (3.7) is hence transformed into:

$$\hat{\alpha} = \text{arg min} \left\{ \sum^n \rho(\boldsymbol{y} - \boldsymbol{r}a) + \sum^m \rho(\alpha) \right\} \tag{3.8}$$

The above model is known as regularized robust coding (RRC) due to the fidelity term $\sum^n \rho_\vartheta (y_i - r_i\boldsymbol{\alpha})$ being very robust to outlier pixels, and $\sum^m \rho_o(\alpha_j)$ is the regularization term which depends on the prior probability $P(\alpha)$. It is observed that $\sum^m \rho_o(\alpha_j)$ becomes the $l1$-norm sparse constraint when a Laplacian distribution is assumed for α i.e., $P(\alpha) = \prod_{j=1}^m \exp(-\parallel \alpha_j \parallel_1 / \sigma_\alpha)/2\sigma_\alpha$. Considering the problem of classification, it is preferred that only the representation coefficients related to the dictionary atoms from the identified class have large values. In the beginning, as we do not know the class to which the test image belongs, a rational assumption is that only a negligible percent of representation coefficients have significant values. Thus, we take for granted that the representation coefficient α_j follows generalized Gaussian distribution (GGD). Thus, we have:

$$f_0(\alpha_j) = \beta \exp - \left\{ (\mid \alpha \mid_j /\sigma_\alpha)^\beta \right\} (2\sigma_\alpha \Gamma(1\beta)) \tag{3.9}$$

where Γ represents the gamma function. It is not an easy task to predict the distribution of the representation residual, since there are a variety of image variations to be considered. Therefore, generally, we assume that the unknown PDF $f_\theta(e)$ are symmetric, differentiable, and monotonic with respect to $|e|$. Hence, $\rho_\theta(e)$ has the following features: (1) $\rho_\theta(0)$ is the global minimal of $\rho_\theta(x)$; (2) symmetry: $\rho_\theta(x) = \rho_\theta(-x)$; and (3) monotonicity: $\rho_\theta(x_1) > \rho_\theta(x_2)$ if $|x_1| > |x_2|$. We assume $\rho_\theta(0) = 0$, without loss of generality. Two primary challenges in solving the RRC model are how to decide the distributions ρ_θ (or f_θ), and how to minimize the energy functional. If f_θ is simply taken as either Gaussian or Laplacian, the stated model will reduce to the standard sparse coding problem in Eqs. (3.3) or (3.4). However, as we showed in Figure 3.1, such defined distributions for f_θ are subject to predispositions and hardly robust to outlier pixels. So, the Laplacian setting of f_o renders the minimization inefficient. In order to overcome these drawbacks, the need for f_θ to have a more flexible shape is recognized. In this chapter, f_o is allowed to be adaptive to the input test image y. As a result, the proposed system proves to be more robust to outliers. Therefore, we convert the minimization of Eq. (3.8) into an iteratively reweighted regularized coding problem so that an effective MAP solution of RRC is obtained.

3.2.2 RRC VIA ITERATIVELY REWEIGHTING

Let $F_\vartheta(e) = \sum_{i=1}^{n} \rho_\vartheta(e_i)$. The Taylor expansion of $F_\vartheta(e)$ in the neighborhood of e_0 is given by:

$$\tilde{F}_\theta(e) = F_\theta(e_0) + (e - e_0)^T F'\theta(e_0) + R_1(e) \tag{3.10}$$

where, $R_1(e)$ represents the high-order residual and $F'\theta(e_0)$ denotes the derivative of $F_\theta(e_0)$. Denote by ρ'_θ the derivative of ρ_θ, and there is $F'_\theta(e) = \left[\rho'_\theta(e); \rho'_\theta(e); \ldots; \rho'_\theta(e)/f\right]$ where $e_{0,i}$ is the ith element of e_0.

In order to make $F'_\theta(e)$ absolutely convex to enable easier minimization, the residual term is approximated as $R_1(e) \approx 0.5(e-e_0)^T W(e-e_0)$, where W denotes a diagonal matrix for that the elements in e are independent and there is no cross term of e_i and e_j, $i \neq j$, in $F_\theta(e)$. Since $F_\theta(e)$ attains its minimum value (i.e., 0) at $e = 0$, we also require that its approximation $\tilde{F}_\theta(e)$ reaches the minimum at $e = 0$. Letting $F_\theta(0) = 0$, we obtain the diagonal elements of W as

$$W_{i,i} = \omega_\theta(e_{0,i}) = \rho_\theta^i(e_{0,i})e_{0,i} \tag{3.11}$$

According to the characteristics of ρ_θ, we know that $\rho_\theta^i(e_i)$ will have the same sign as e_i. So $W_{i,i}$ is a nonnegative scalar. Then, $\tilde{F}_\theta(e)$ is

$$\tilde{F}_\theta(e) = 1/2 \left\|W^{1/2}e\right\|_2^2 + b_{e0} \tag{3.12}$$

$b_{e0} = \sum_{i=1}^{n} \left[\rho_\theta(e_{0,i}) - \rho'_\theta(e_{0,i})e_{0,i}/2\right]$ is a scalar constant which depends on $e0$.

Ignoring the constant b_{e_0}, the RRC model in Eq. (3.8) could be approached as shown

$$\hat{\alpha} = \arg\min_\alpha \left\{1/2 \ \|W^{1/2}(y - D\alpha)\|_2^2 + \sum^m \rho_o(a_j)\right\} \tag{3.13}$$

Assuredly, Eq. (3.13) is a local approximation of Eq. (3.8), but it renders the minimization of RRC practicable through iteratively reweighted $l2$-regularized coding, where W is updated using Eq. 3.11).

3.2.3 THE WEIGHTS W

Now, the question of how to calculate the diagonal weight matrix W is addressed. The element Wi, i, i.e., $\omega_\theta(e_i)$, is the weight designated to pixel i of test image y. By intuition, we can conclude that in face recognition, the outlier pixels (e.g., occluded or corrupted pixels) must be assigned small weights, so as to weaken their effect on coding y over D.

The dictionary D is composed of nonoccluded/noncorrupted training face images which could accurately represent the facial parts. Consequently, the outliers will have significant coding residuals. Since the pixel, which has a big residual e_i should have a small weight, outliers are assigned smaller weights than uncorrupted pixels. The inverse proportionality of weights and residuals can be observed from Eq. (3.11), where $\omega_\theta(e_i)$ is inversely proportional to e_i and modulated by () $i\ e\ \rho'_\theta$. Considering Eq. (3.11), it is noted that since ρ_θ is differentiable, symmetric, monotonic and has its minimum at origin. So, we can assume that $\omega_\theta(e_i)$ is continuous, symmetric, and inversely proportional to e_i but bounded (to boost stability). Ensuring no loss of generality, we establish that $\omega_\theta(e_i)\in[0, 1]$. With these considerations, a good choice of $\omega_\theta(e_i)$ would be the following logistic function, which is very commonly used:

$$\omega_\theta(e_i) = \exp\left(-\mu e_i^2 + \mu\delta\right)\Big/\left(1+\exp\left(-\mu e_i^2 + \mu\delta\right)\right) \tag{3.14}$$

where μ and δ are simple positive scalars. Parameter μ manipulates the decreasing rate from 1 to 0, and δ is responsible for the location of demarcation point. Here the value of $\mu\delta$ must be large enough to bring $\omega_\theta(0)$ close to 1 (generally, we let $\mu\delta\geq8$). With Eqs. (3.11), (3.14) and $\rho_\theta(0)=0$, we obtain

$$\rho_\theta(e_i) = -1/2\mu \left(\ln\left(1+\exp\left(-\mu e_i^2 + \mu\delta\right)\right)-\ln\left(1+\exp\mu\delta\right)\right) \tag{3.15}$$

We can see that the above ρ_θ meets all the assumptions and characteristics discussed previously.

The PDF f_θ associated with ρ_θ in Eq. (3.15) is noticeably more flexible than the Gaussian and Laplacian functions in approximating the residual e. It has a longer tail to acknowledge the residuals caused by outlier pixels like corruptions and occlusions, and hence the coding vector α will be robust to the outlier pixels present in y. $\omega_\theta(e_i)$ may also be set as other functions. However, the proposed weight function is the binary classifier formulated using MAP estimation, which can effectively distinguish inliers and outliers. If $\omega_\theta(e_i)$ is set to a constant value like $\omega_\theta(e_i)=2$, it corresponds to the $l2$-norm fidelity in Eq. (3.3); if it is set as $\omega_\theta(e_i)= 1/|e_i|$, it corresponds to the $l1$-norm fidelity in Eq. (3.4); and if it is set as a Gaussian function, it corresponds to Gaussian kernel fidelity. Nevertheless, all these functions are not so robust as Eq. (3.14) to outliers. From Figure 3.2, one can see that the $l2$-norm fidelity regards all pixels equally, regardless of whether it is outlier or not; the $l1$-norm fidelity assigns larger weights to pixels with lower residual values; however, the weight can approach infinity when the residual nears zero, rendering the coding unstable. Both the proposed weight function and the weight function of the Gaussian fidelity lie within the

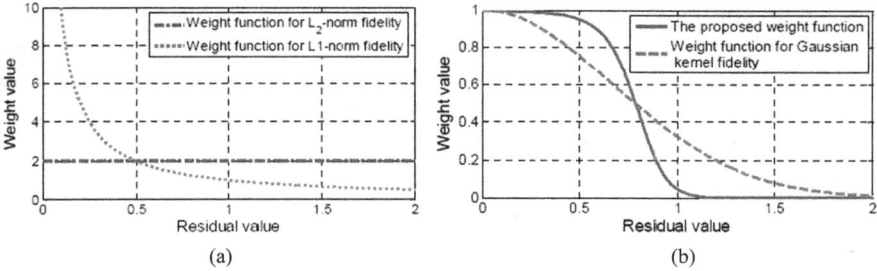

FIGURE 3.2 A subject in Multi-PIE database. (a) Training samples with only illumination variations. (b) Query samples with surprise expression and illumination variations. (c) and (d) show the testing samples with smile expression and illumination variations in Sessions 1 and 3, respectively.

range of [0, 1], and they encounter an intersection point when weight value is 0.5. However, the proposed weight function assigns significant weights to inliers and lower weights to outliers, which implies that it has a higher capability to classify inliers and outliers. The sparse coding models in Eqs. (3.3) and (3.4) are variations of the RRC model in Eq. (3.13) with $\beta = 1$ in Eq. (3.9). The model in Eq. (3.3) can be achieved by letting $\omega_\theta(e_i) = 2$. The model in Eq. (3.4) can be achieved by letting $\omega_\theta(e_i) = 1/|e_i|$. Compared to the models in Eqs. (3.3) and (3.4), the proposed RRC model (Eq. (3.8) or (3.13)) is far more robust to outliers (usually the pixels with big residuals) because it will adaptively designate negligible weights to them. Though the model in Eq. (3.4) also assigns negligible weights to outlier pixels, its weight function $\omega_\theta(e_i) = 1/|e_i|$ is not bounded (i.e., the weights assigned to very small residuals have very big values and drastic changing ratios), rendering it less effective while distinguishing between inliers and outliers.

3.2.4 TWO SIGNIFICANT CASES IN RRC

The minimization of RRC model in Eq. (3.13) can be achieved iteratively, where in each iteration W and α are updated alternatively. By deciding the weight matrix W, the RRC with GGD prior on representation (i.e., Eq. (3.9)) and $\rho_o(\alpha) = -\ln f_o(\alpha)$ can be represented as

$$\hat{\alpha} = \arg\min_\alpha \left\{ 1/2 \parallel W^{1/2}(y - D\alpha)\parallel_2^2 + \sum^m (\lambda^\beta{}_\alpha + b) \right\} \quad (3.16)$$

where $\rho_o(\alpha_j) = \lambda \mid \alpha_j \mid^\beta + b_o, \lambda = (1 / \sigma_\alpha)^\beta$ and $b = \ln(2\sigma_\alpha \Gamma(1/\beta)/\beta)$ is a constant. Identical to the processing of $F_\theta(e) = \sum^n \rho_\theta(e_i), \sum^m \rho_o(\alpha_j)$ can also be approximated by the Taylor expansion. Then, Eq. (3.16) becomes

$$\hat{\alpha} = \arg\min_\alpha \left\{ \left\| W^{1/2}(y - D\alpha) \right\|_2^2 + \sum^M V_{j,j}\alpha_j^2 \right\} \quad (3.17)$$

where W is a diagonal matrix with $V_{j,j} = \rho'(\alpha_j)\alpha_j$ The magnitude of β decides the types of regularization. If $0 < \beta \leq 1$, then sparse regularization is applied; otherwise, nonsparse regularization is used on the representation coefficients. Specifically, the proposed RRC model has two significant cases with two specific values of β.

When $\beta = 2$, GGD degenerates to the Gaussian distribution, the RRC model becomes

$$\hat{\alpha} = \arg\min_\alpha \left\{ \left\| W^{1/2}(y - D\alpha) \right\|_2^2 + \lambda \|\alpha\|^2 \right\} \qquad (3.18)$$

In this case, the RRC model is simply an $l2$-regularized robust coding model. It can be found that when W is given, the solution to Eq. (3.18) is $\hat{\alpha} = (D^T WD + \lambda I)^{-1} D^T Wy$ When $\beta = 1$, GGD reduces to the Laplacian distribution, and the RRC model becomes

$$\hat{\alpha} = \arg\min_\alpha \left\{ \left\| W^{1/2}(y - D\alpha) \right\|_2^2 + \lambda \|\alpha\|_1 \right\} \qquad (3.19)$$

In this case, the RRC model is simply the RSC model, where the sparse coding techniques such as $l1\_ls$ are used to solve Eq. (3.19) when W is given. In this chapter, Eq. (3.19) is solved using Eq. (3.17) by the iterative re-weighting technique. Let $V_{i,j}^{(0)} = v_0^{(0)} = 1$; this implies that in the $(k+1)$th iteration v is set as $V^{(k+1)} = V_0\left(\alpha_j^{(k)}\right) = \lambda \left| \left(\alpha_j^{(k)}\right) + \varepsilon^2 \right|^{-1/2}$, and then $\hat{\alpha}^{(k+1)} = \left(D^T WD + V^{(k+1)} \right)^{-1} D^T Wy$. Here ε is a scalar.

3.3 PROPOSED METHOD

3.3.1 ALGORITHM OF REGULARIZED ROBUST CODING

The minimization of RRC is an iterative process where the weights W and V are updated alternatively. Though we can only have a locally optimal solution to the RRC model, in face recognition we can have an apt initialization to ensure high performance. Now, an IR³C algorithm to minimize the RRC model is discussed.

When a test image y comes, before initializing W, we should initialize the coding residual e of y. e is initialized as $e = y - D\alpha^{(1)}$, where $\alpha^{(1)}$ is the initial coding vector. Because we are not aware of the class to which test image y belongs, a reasonable approximation for $\alpha^{(1)}$ is:

$$\alpha^{(1)} = [1/m; 1/m \ldots 1/m] \qquad (3.20)$$

That is, $D\alpha^{(1)}$ is the mean facial image of all training samples. When IR³C converges, we use the following classification method to classify the face image y:

$$\text{identity}(y) = \arg\min_c \{l_c\} \qquad (3.21)$$

$$l_c = \| W_{\text{final}}^{\frac{1}{2}}(y - D_c\hat{\alpha}_c) \|_2 \qquad (3.22)$$

where D_c is the sub-dictionary corresponding to class c, $\hat{\alpha}_c$ is the final sub-coding vector corresponding to class c, and W_{final} represents the final weight matrix.

3.3.2   The Convergence of IR³C

Equation (3.21) is a local approximation of the RRC in Eq. (3.8), and in each iteration the objective function of Eq. (3.8) reduces by the IR³C algorithm, i.e., in steps 3 and 4, the solved $\alpha^{(t)}$ will lead to $\sum^m \rho_\theta \left(y - r\alpha^{(t)}\right) + \sum^m \rho_0 \left(\alpha^{(t)}\right) < \sum^m \rho_\theta \left(y - r\alpha^{(t-1)}\right) + \sum^m \rho_0 \left(\alpha^{(t-1)}\right)$. Because the cost function of Eq. (3.8) is lower bounded (≥ 0), the iterative minimization process in IR³C will converge. We stop the iteration if the following condition becomes true:

$$\left\| W^{(t+1)} - W^{(t)} \right\|_2 / \left\| W^{(t)} \right\|_2 < \delta_w \qquad (3.23)$$

where δ_w is a small positive scalar.

3.3.3   Complexity Analysis of the Proposed Algorithm

Usually, the complexity of IR³C and SRC is mainly consisted in the coding process, i.e., Eq. (3.18) or (3.19) for IR³C and Eq. (3.1) or (3.2) for SRC. The $l1$-minimization, such as Eq. (3.1) for SRC, has a computational complexity of $O(n2m1.5)$ where n is the dimensionality of face feature, and m is the number of dictionary atoms. It is also found that the commonly used $l1$-minimization solvers like $l1\_magic$ and $l1\_ls$ have a complexity of $O(n2m1.3)$. For IR³C when $\beta = 2$, the coding (i.e., Eq. (3.18)) becomes a $l2$-regularized least square problem. The solution $\hat{\alpha} = \left(D^T W D + \lambda I\right)^{-1} D^T W y$ can be obtained by solving $(D^T W D + \lambda I)\hat{\alpha} = D^T W y$ using conjugate gradient method, which has a time complexity of about $O(k1nm)$ (where $k1$ denotes the iteration number in conjugate gradient method). Consider the case where t iterations are used in IR³C to update W, then the overall complexity of IR³C with $\beta = 2$ is around $O(tk1nm)$. Generally, t is lesser than 15. It is clearly observed that IR³C with $\beta = 2$ has much lower complexity than SRC. For IR³C with $\beta = 1$, the coding in Eq. (3.19) is an $l1$-norm sparse coding problem, which can also be solved using conjugate gradient method. The complexity of IR³C with $\beta = 1$ will be nearly $O(tk1k2nm)$, where $k2$ is the number of iterations required to update V. Generally, $k1$ is lesser than 30 and $k2$ is lesser than 20, and hence $k2k1$ is basically in the similar order to n. Thus, the complexity of IR³C with $\beta = 1$ is approximately $O(tn2m)$. Compared with SRC in case of face recognition without occlusion, although IR³C needs several iterations (usually $t = 2$) to update W, the time it consumes is smaller than or at least comparable to SRC. In the case of face recognition with occlusion or corruption, for IR³C usually $t = 15$; however, SRC's complexity is $O(n2(m + n)1.3)$ because it requires an identity matrix to code the occluded or corrupted pixels, as shown in Eq. (3.2). It can be established that IR³C with $\beta = 1$ has far lower complexity than SRC for face recognition with occlusion. Although many faster $l1$-norm minimization methods have been proposed recently, by adopting them in SRC, the running time required is still larger than or comparable to the proposed IR³C method. In addition, in the iteration of IR³C, we can eliminate any pixel y_i that has negligible weight because this means that y_i is an outlier pixel. Thus, the complexity of IR³C can be brought down further. For example, in face recognition with real disguise on the AR database, about 30% pixels could be deleted.

3.4 EXPERIMENTAL RESULTS

Experiments are performed on standard face databases to quantify and compare the performance of RRC. All the face images are cropped and aligned by using the position of eyes as the reference. The test image (or feature) is normalized, and so is the training image (or feature), to attain unit $l2$-norm energy. For both the AR and the Extended Yale B databases, the locations of the eye are already provided with the databases. For Multi-PIE database, the eyes must be manually located, and the facial region is identified automatically by the face detector. In all experiments, the training data set is used as the dictionary D in coding. The RRC model with $l1$-norm coefficient constraint (i.e., $\beta = 1$ in Eq. (3.19)) is denoted by RRC_L1, and the RRC model with $l2$-norm coefficient constraint (i.e., $\beta = 2$ in Eq. (3.18)) is denoted by RRC_L2. Both RRC_L1 and RRC_L2 are implemented using the IR^{3}C algorithm.

3.4.1 CALCULATION OF PARAMETERS

In the weight function Eq. (3.14), there are two parameters introduced, δ and μ, which need to be computed in the second step of the IR^{3}C algorithm. δ is the parameter used to establish the demarcation point. If the square of residual is larger than δ, then the assigned weight will be lower than 0.5. To ensure that the proposed model is robust to outliers, δ is computed as explained in the following section. Let, $l \leq \tau n f$ where the scalar $\tau \in (0,1)$, and $\leq \tau n f$, outputs the highest integer value which is smaller than τn. We let δ be:

$$\delta = y_1(e)_l \tag{3.24}$$

where for any $e \in \Re^n \psi_1(e)_k$ represents the kth largest element of the set $\{e^2, j = 1, \dots, n\}$.

The scalar μ decides the decreasing rate of weight Wi, i. Here we set $\mu = \varsigma / \delta$, where $\varsigma = 8$ is a constant. While performing the experiments, τ is set as 0.8 for face recognition without occlusion, and 0.6 for face recognition with occlusion. Also, default value of the regularization parameter λ in Eq. (3.18) or (3.19) is taken as 0.001.

In RRC_L1, a parameter ε is used while updating the weight matrix $V : V_{j,j}^{(k+1)} = V_o(\alpha^{(k)}) = b_o |(\alpha^{(k)})^2 + \varepsilon^2|^{-1/2}$. We choose ε as:

$$\varepsilon^{(k+1)} = \min\left(\varepsilon^{(k)}, \psi_2(\alpha^{(k)}) L / m\right) \tag{3.25}$$

where for a vector $\alpha \in \Re^n \psi_2(\alpha)_i$ is the ith largest element of the set $\{|a_j|, j = 1, \dots, m\}$ We set $L = \leq 0.01 m f$. The propounded approximation of ε will make the numerical computing of weight V stable and also ensure the iteratively reweighted least square achieves a sparse solution ($\varepsilon (k+1)$ decreases to zero as k increases).

3.4.2 FACE RECOGNITION WITHOUT OCCLUSION

We first test the performance of RRC in face recognition with variations in lighting, expression, etc., without occlusion. RRC is compared with SRC [4], locality-constrained linear coding (LLC), linear regression for classification (LRC) [5], and the state-of-the-art techniques like NN [1], NFL [2] and linear support vector machine (SVM). In the experiments, Principal Component Analysis (PCA) is employed to decrease the dimensionality of original face images, and the Eigenface features are used for all the methods. Representing the PCA projection matrix by P, the third step of IR^{3}C becomes:

$$\alpha^* = \arg\min_\alpha \left\{ 1/2 \left\| P\left(W^{(t)}\right)^{1/2} (y - D\alpha) \right\|_2^2 + \sum_{j=1} \rho_o\left(\alpha_j\right) \right\} \qquad (3.26)$$

3.4.2.1 Extended Yale B Database

The Extended Yale B database consists of around 2,414 frontal face images of 38 subjects. We made use of the cropped and normalized facial images of size 54×48, which were taken under complex illumination conditions. The database was randomly split into two halves. One half, which contains 32 images for each subject, was marked as the dictionary, and the other half was employed for testing. Table 3.1 displays the plot for recognition rates versus feature dimension by NN, NFL, SVM, SRC, LRC, LLC, and RRC schemes. RRC_L1 yields better results than the other techniques in all dimensions except for the fact that its performance is poorer than that of SVM when the dimension is 30. RRC_L2 works better than SRC, LRC, LLC, SVM, NFL, and NN when the dimension is 150 or greater. The highest recognition rates achieved by SVM, SRC, LRC, LLC, RRC_L2, and RRC_L1 are 97.0%, 98.3%, 96.0%, 97.6%, 98.9%, and 99.8% respectively.

3.4.2.2 AR Database

A subset (with only illumination and expression variations) that consists of 50 male and 50 female subjects was selected from the AR database in this experiment. For every subject, the seven images from Session 1 were utilized for training, with other seven images from Session 2 used for testing. The images were cropped to a size of 60×43. The face recognition rates by the state-of-the-art methods are presented in Table 3.2. Except for the case where the dimension is 30, RRC_L1 yields the highest recognition rates among all methods, while RRC_L2 has the second best

TABLE 3.1
Recognition Rates on the Extended Yale B Database

Dimension	30 (%)	54 (%)	120 (%)	300 (%)
NN[1]	66.3	85.8	90.0	91.6
SVM	**92.4**	94.9	96.4	97.0
LRC[5]	63.6	94.5	95.1	96.0
NFL[2]	89.6	94.1	94.5	94.9
SRC[4]	90.9	95.5	96.8	98.3
LLC	92.1	96.4	97.0	97.6
RRC_L2	71.6	94.4	97.6	98.9
RRC_L1	**91.3**	**98.0**	**98.8**	**99.8**

TABLE 3.2
Recognition Rates on the AR Database

Dimension	30 (%)	54 (%)	120(%)	300 (%)
NN [1]	62.5	68.0	70.1	71.3
SVM	66.1	69.4	74.5	75.4
LRC [5]	66.1	70.1	75.4	76.0
NFL [2]	64.5	69.2	72.7	73.4
SRC [4]	**73.5**	83.3	90.1	93.3
LLC	70.5	80.7	87.4	89.0
RRC_L2	61.5	84.3	94.3	95.3
RRC_L1	**70.8**	**87.6**	**94.7**	**96.3**

performance. The reason that RRC does not work well with very low-dimensional feature is that the coding vector given by Eq. (3.26) is not accurate enough to evaluate W when the feature dimension is too low. However, when the dimension is too low, none of the methods achieves good recognition rate. It is observed that all methods achieve their maximal recognition rates at the dimension of 300, with 93.3% for SRC, 89.0% for LLC, 95.3% for RRC_L2 and 96.3% for RRC_L1.

From Tables 3.1 and 3.2, it can be concluded that when the dimension of feature is not extremely low, RRC_L2 performs similar to RRC_L1, which indicates that the $l1$-sparsity constraint on the coding vector is not so crucial. This is because when the feature dimension is not too low, the dictionary will not be over-complete to a great extent, and hence using Laplacian to model the coding vector is not so different from using Gaussian. As a result, RRC_L2 and RRC_L1 achieve similar recognition rates, but RRC_L2 will have lesser complexity.

3.4.2.3 Multi PIE Database

The CMU Multi-PIE database consists of images of 337 individuals captured in four sessions with simultaneous changes in pose, expression, and illumination. Among these 337 subjects, the face images of all the 249 individual in Session 1 were utilized for training. To make face recognition more difficult, four subsets with both illumination and expression changes in Sessions 1, 2 and 3, were employed for testing. For the training data set, we used the 7 frontal facial images with extreme illuminations {0, 1, 7, 13, 14, 16, and 18} and neutral expression. For the testing data set, 4 frontal images with illuminations {0, 2, 7, 13} and varying expressions were used. Here we make use of the Eigenface with dimensionality 300 as the face feature for sparse coding. Table 3.3 summarizes the recognition rates in four testing sets by the state-of-the-art methods.

From Table 3.3, it is observed that RRC_L1 performance the best in all tests, followed by RRC_L2. Further, all the techniques work best when Smi-S1 is employed for testing since the training set is also from Session 1. From testing set Smi-S1 to Smi-S3, a gradual increase in variations is observed because of the longer data acquisition time interval and the changes in the smiles. The recognition rates of RRC_L1 and RRC_L2 reduce by 21.8% and 25.9%, respectively, and those of NN, NFL, LRC,

TABLE 3.3
Recognition Rates on Multi-PIE Database

Sessions	Smi-S1 (%)	Smi-S3(%)	Sur-S2 (%)	Squ-S2 (%)
NN [1]	88.7	47.3	40.1	49.6
SVM	88.9	46.3	25.6	47.7
LRC [5]	89.6	48.8	39.6	51.2
NFL [2]	90.3	50.0	39.8	52.9
SRC [4]	93.7	60.3	51.4	58.1
LLC	95.6	62.5	52.3	64.0
RRC_L2	96.1	70.2	59.2	58.1
RRC_L1	**97.8**	**76.0**	**68.8**	**65.8**

("Smi-S1": set with smile in Session 3.1; "Smi-S3": set with smile in Session 3.3; "Sur-S2": set with surprise in Session 3.2; "Squ-S2": set with squint in Session 3.2).

SVM, LLC, and SRC decrease by 41.4%, 40.3%, 40.8%, 42.6%, 33.1%, and 33.4%, respectively. This proves that the RRC methods are far more robust to face variations than the competing schemes. It can also be observed that face recognition with surprise and squint expression variations are much more challenging than the case of face recognition with changes in smile expression. In this case, the difference in performance between RRC_L2 and RRC_L1 is large. This is because the dictionary (size: 300×1743) used in this test is very over-complete, and as a result, the $l1$-norm is much more efficient than the $l2$-norm in regularizing the representation of facial images with drastic variations (e.g., expression changes).

3.4.3 FACE RECOGNITION WITH OCCLUSION

One of the most remarkable characteristics of face recognition based on sparse coding is its robustness to face occlusion. Now, we test the robustness of the proposed method to a variety of occlusions, like random pixel corruption, random block occlusion, and real disguise. In the trials for random corrupted images and random block occluded images, we contrast the pitted the RRC methods against SRC, LRC, Gabor-SRC (which can be used only for block occluded images), and correntropy-based sparse representation (CESR), and NN is used as the benchmark method.

In the case of real disguised faces, RRC is compared with SRC, Gabor-SRC (GSRC), CESR, and other standard methods.

3.4.3.1 Face Recognition with Pixel Corruption

Subsets 1 and 2 (717 facial images, normal-to-moderate illumination conditions) of the Extended Yale B database were used for training, and Subset 3 (453 images, extreme illumination conditions) was employed for testing. Extended Yale B database is employed so that the various methods can be compared against the same standards. The dimensions of the images were changed to 96×84 pixels. For every query image, a certain percentage of its pixels was replaced by uniformly distributed random values within the range [0, 255]. The corrupted pixels were arbitrarily chosen for each query image.

The knowledge of corrupted pixel locations is not shared with the algorithm. Figure 3.3 illustrates an example of RRC_L1 and RRC_L2 with 70% random corruption. Figure 3.3a is the original image, and Figure 3.3b shows the query image with random corruption. It can be seen that, even for humans, the corrupted facial images are nearly impossible to recognize. The estimated weight maps of RRC_L1 and RRC_L2 are depicted in the top and bottom rows of Figure 3.3c, respectively, from which we can conclude that corrupted pixels and the pixels in the shadow region have low weights. The top and bottom rows of Figure 3.3d represent the coding coefficients of RRC_L1 and RRC_L2, respectively. Figure 3.3e depicts the reconstructed facial images of RRC_L1 (top row) and RRC_L2 (bottom row). It can be seen that since RRC_L1 tends to produce sparse coefficients; only those dictionary elements

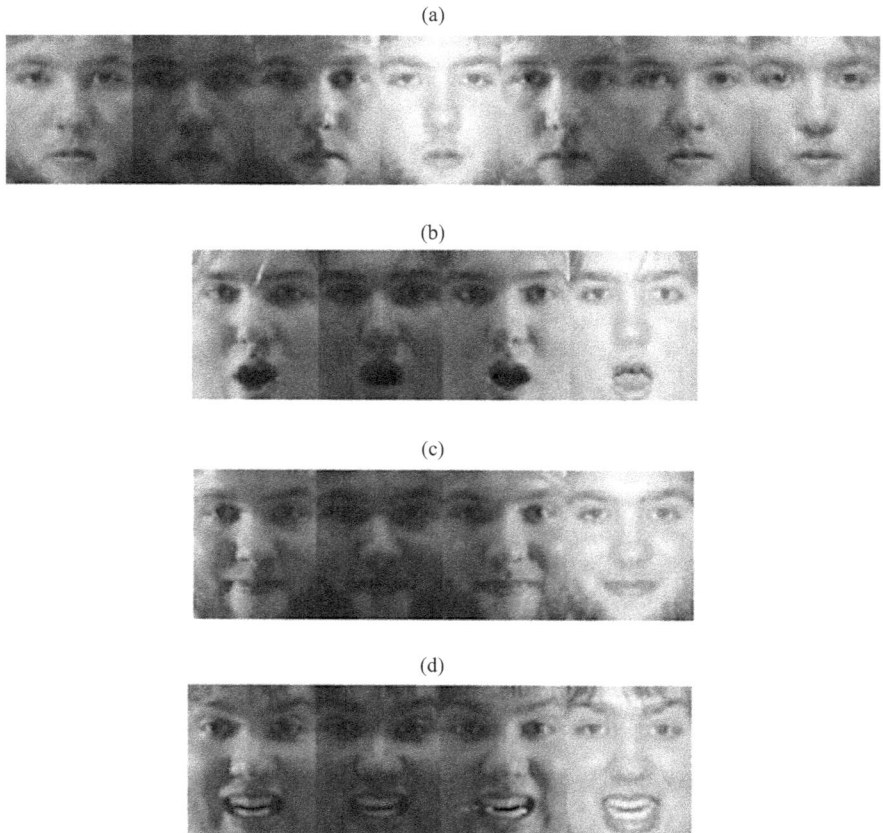

(a)

(b)

(c)

(d)

FIGURE 3.3 Recognition with random corruption. (a) Original image y_0 from Extended Yale B database. (b) Query image y with random corruption. (c) Estimated weight map of RRC_L1 (top row) and RRC_L2 (bottom). (d) Estimated representation coefficients α of RRC_L1 and RRC_L2. (e) Reconstructed images $yrec$ of RRC_L1 and RRC_L2.

TABLE 3.4

The Recognition Rates of RRC, LRC, NN, SRC, and CESR Versus Various Percentages of Corruption

Corruption (%)	0–50 (average) (%)	60 (%)	70 (%)	80 (%)	90 (%)
NN [1]	89.3	46.8	25.4	11.0	4.6
SRC [4]	100	99.3	90.7	37.5	7.1
LRC [5]	95.8	50.3	26.4	9.9	6.2
CESR [7]	97.4	96.2	97.8	93.8	41.5
RRC_L2	100	100	99.8	97.8	43.3
RRC_L1	**100**	**100**	**100**	**99.6**	**67.1**

with an identical label to the query image have large coefficients, and therefore the image reconstructed is identical to the original facial image and is of better visual quality (the shadow that adds difficulty to recognition is eliminated). For RRC_L2, in spite of the coefficients not being sparse, the quality of the reconstructed facial image is high and its performance is comparable to RRC_L1. Table 3.4 shows the results of SRC, CESR, LRC, NN, RRC_L2, and RRC_L1 with images containing various percentages of corrupted pixels. Since all schemes performed well in the cases of 0% to 50% corruptions, the average recognition rates for 0%~50% corruptions are specified. Under these circumstances, RRC_L1, RRC_L2, and SRC were able to classify all the query images accurately. However, CESR proved incapable of recognizing all the query images correctly. Nevertheless, in the case where the percentage of corrupted pixels exceeds 70%, the advantage of RRC_L1, RRC_L2, and CESR over SRC is apparent. Specifically, RRC_L1 performs the best in all cases, with 100%, 99.6%, and 67.1% recognition rates for 70%, 80%, and 90% corruption, respectively. SRC has recognition rates 90.7%, 37.5%, and 7.1% for 70%, 80%, and 90% corruption, respectively. LRC and NN are highly responsive to outlier pixels, with lesser recognition rates than other methods. All RRC methods outperform the CESR method in all cases, which demonstrates that the RRC model could restrain the influence of outlier pixels more effectively. Also, RRC_L2 performs like RRC_L1, which shows that when the feature dimension (whose value is 8064 in this case) is large, $l2$-norm constraint on coding coefficient is as potent as $l1$-norm constraint and has the added benefit of lesser time complexity.

3.4.3.2 Face Recognition with Block Occlusion

Now, the performance of RRC under block occlusion conditions is tested. Subsets 1 and 2 of Extended Yale B are used as the training data set. Subset 3 is used for testing. A random portion of the query image is blocked with an irrelevant image, as shown in Figure 3.4b. The facial image dimensions were modified to 96×84. Figure 3.4 illustrates an instance of occluded facial recognition (with 30% occlusion) using RRC_L1 and RRC_L2. Figure 3.4a and b shows the clean image from Extended Yale B database and the occluded query image, respectively. The top and bottom rows of Figure 3.4c represent the estimated weight maps of RRC_L1 and

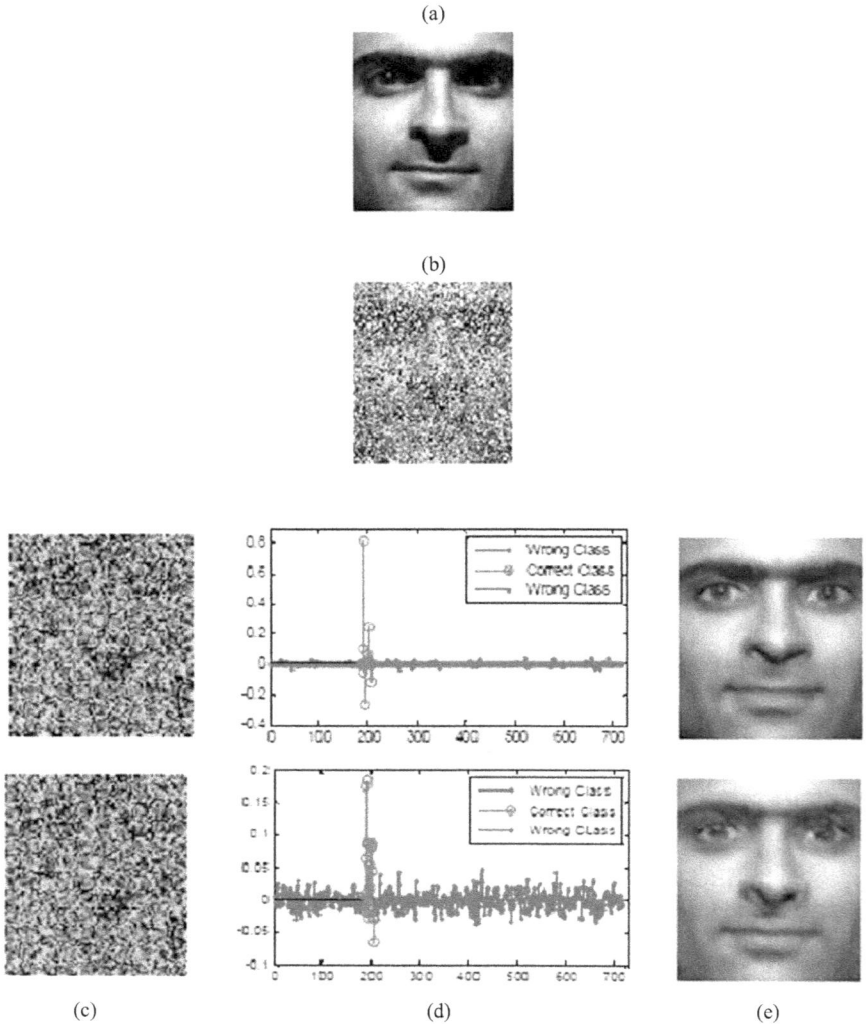

FIGURE 3.4 Face recognition under 30% block occlusion. (a) Original image y_0 from Extended Yale B. (b) Test image y with random corruption. (c) Estimated weight maps of RRC_L1 (top row) and RRC_L2 (bottom row). (d) Estimated representation coefficients α of RRC_L1 and RRC_L2. (e) Reconstructed images y_{rec} of RRC_L1 and RRC_L2.

RRC_L2, respectively, which illustrate how both of them designate large weights (e.g., 1) to the inlier pixels and designate small weight (e.g., 0) to the occluded pixels. The top and bottom rows of Figure 3.4d represent the estimated representation coefficients of RRC_L1 and RRC_L2, respectively. It is observed that RRC_L1 can generate highly sparse coefficients with large values on the elements of the right class; the coefficients by RRC_L2 also have large values on the elements of the correct class. However, by RRC_L2, the obtained coefficients are not sparse. From Figure. 3.4e, it

TABLE 3.5
The Recognition Rates of RRC, LRC, NN, GSRC, SRC, and CESR with Varying Levels of Block Occlusion

Occlusion (%)	0 (%)	10 (%)	20 (%)	30 (%)	40 (%)	50 (%)
NN [1]	94.0	92.9	85.4	73.7	62.9	45.7
SRC [4]	**100**	**100**	99.8	98.5	90.3	65.3
LRC [5]	**100**	**100**	95.8	81.0	63.8	44.8
GSRC [6]	**100**	**100**	**100**	**99.8**	96.5	87.4
CESR [7]	94.7	94.7	92.7	89.8	83.9	75.5
RRC_L2	100	100	100	99.8	97.6	87.8
RRC_L1	**100**	**100**	**100**	**99.8**	96.7	87.4

is observed that both RRC_L1 and RRC_L2 possess excellent image reconstruction quality, adeptly eliminating the block occlusion and the shadow.

Table 3.5 details the precise recognition rates obtained via RRC_L1, RRC_L2, SRC, LRC, NN, GSRC, and CESR when the occlusion percentage varies from 0% to 50%. The values in Table 3.5 indicate that RRC_L2 has the highest accuracy, and RRC techniques achieve higher recognition rates than SRC when the occlusion percentage is greater than 30%. Compared with GSRC, RRC performs well even without utilizing the enhanced Gabor features. CESR performs poorly when compared with SRC in this demonstration. This might be due to the fact that face recognition in the presence of block occlusion is more challenging than in that of pixel corruption, but it proves that CESR cannot accurately identify the corrupted pixels in such block occlusion (i.e., outlier pixels have intensities identical to the inlier pixels). RRC_L2 shows promise as it also has recognition rates that are comparable to RRC_L1 (in fact, RRC_L2 performs even better than RRC_L1 at 40% and 50% occlusion), which supports the fact that the low-complexity $l2$-norm regularization could be as potent as the $l1$-norm regularization for the case of block occlusions.

3.4.3.3   Face Recognition with Real Face Disguise

A subset of 2,599 images from the AR database is used for this demonstration. These 2,599 images correspond to 100 subjects (26 samples per class except for a corrupted image), 50 males and 50 females. We conducted two trials. Images dimensions were modified to 83×60 in the first test and 42×30 in the second test. In the first test, 799 facial images (nearly 8 images per subject) of nonoccluded frontal views with different facial expressions in Sessions 1 and 2 were employed as training data set, and two separate subsets (with sunglasses and scarf) of 200 images (1 image per individual per session, with a neutral expression) were used for testing. Figure 3.5 demonstrates the classification procedure of RRC_L1 with the help of an example. Figure 3.5a depicts a query image with sunglasses; Figure 3.5b and c represents the initial and final weight maps, respectively; Figure 3.5d shows one sample image of the identified individual. Figure 3.5e represents the convergence of the IR^{3}C algorithm to solve the RRC model, and Figure 3.5f depicts the reconstruction error for each class, with the

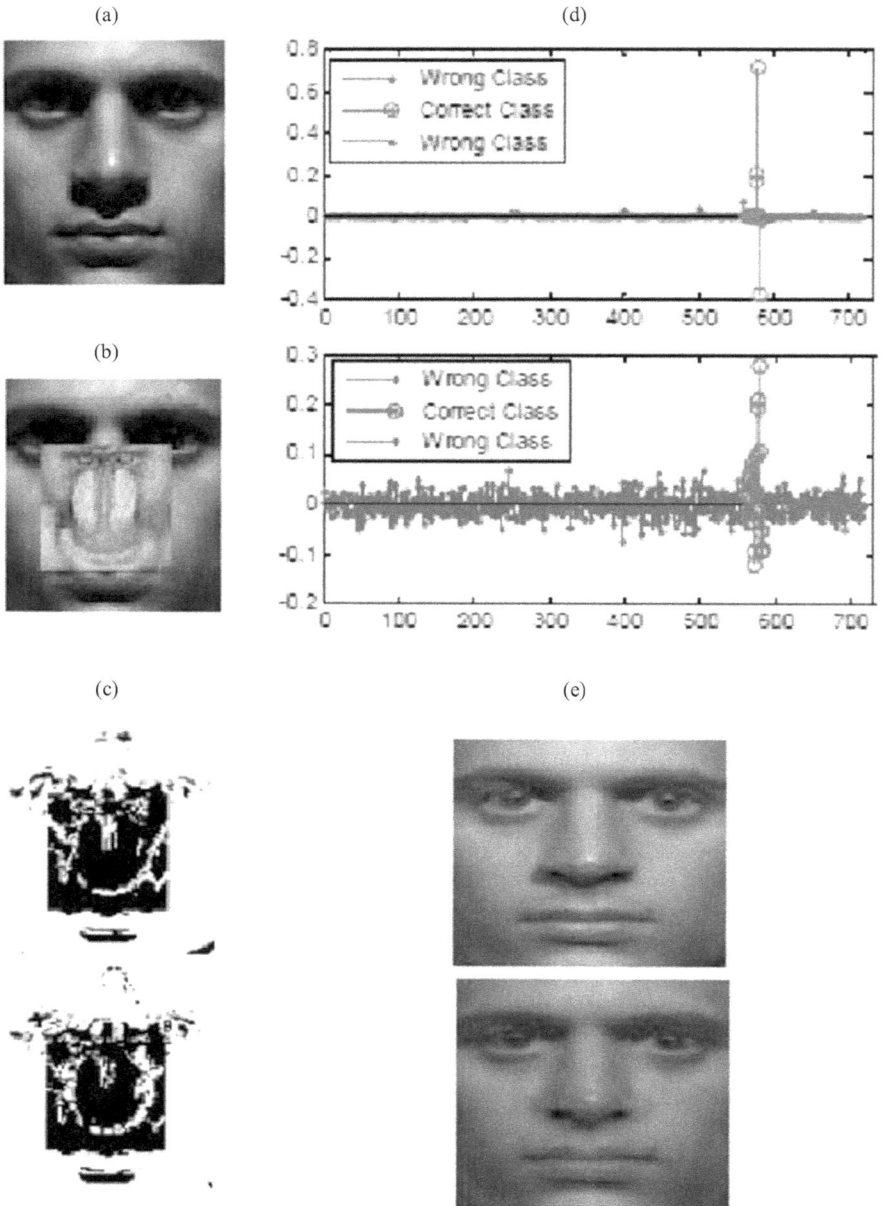

FIGURE 3.5   Face recognition with disguise using RRC_L1. (a) A query image with sunglasses. (b) The initialized weight map. (c) The weight map when IR³C converges. (d) A template image of the identified subject. (e) The convergence curve of IR³C. (f) The residuals of each class by RRC_L1.

right class having the lowest value. The face recognition results achieved by the other methods are provided in Table 3.6. It can be observed that the RRC methods perform much better than SRC, GSRC, and CESR, and RRC_L1 and RRC_L2 achieve comparable results. CESR performs similar to RRC methods in face recognition with sunglass but performs poorly when dealing with a scarf. Just like in the case of face recognition with block occlusion, CESR does not perform well enough for practical robust face recognition. The proposed RRC method remarkably outperforms competing methods. In the second test, we performed face recognition with more complicated disguise (disguise with nonuniform illumination and larger data acquisition interval). Four hundred images (4 neutral images of each individual, with varying illumination and no occlusion) of frontal views in Session 1 were employed for training, whereas the disguised images (3 images, differing in illumination conditions and with sunglasses or scarves per individual per Session) in sessions 1 and 2 were for testing. Table 3.7 lists the results by all the methods. Evidently, the RRC methods perform far better than SRC, GSRC, and CESR. Intriguingly, CESR works well when there is a sunglass disguise but works poorly when there is a scarf disguise, while GSRC shows the opposite. In this demonstration, RRC_L1 performs somewhat better than RRC_L2 on sunglasses, with RRC_L2 marginally better than RRC_L1 on scarf.

TABLE 3.6
Recognition Rates Using Different Methods on the AR Database with Disguise Occlusion

Algorithms	Sunglasses (%)	Scarves (%)
SRC [4]	87.0	59.5
GSRC [6]	93	79
CESR [7]	99	42.0
RRC_L2	99.5	96.5
RRC_L1	**100**	**97.5**

TABLE 3.7
Recognition Rates Using Different Methods on the AR Database with Complex Disguise Occlusion

Algorithms	Session 1 (%)		Session 2 (%)	
	Sunglasses	Scarves	Sunglasses	Scarves
SRC [4]	89.3	32.3	57.3	12.7
GSRC [6]	87.3	85	45	66
CESR [7]	95.3	38	79	20.7
RRC_L2	**99.0**	**94.7**	84.0	**77.3**
RRC_L1	**99.0**	93.3	89.3	76.3

3.4.4 FACE VALIDATION

In practical face recognition system implementations, it is required to reject face images that have no template in the database. These images must be identified as invalid. It is important to note that rejecting invalid images is a bigger challenge than deciding if two facial images are of the same subject. In this section, we test the validation performance of the proposed RRC method. Like all the competing methods, the Sparsity Concentration Index (SCI) is used perform face validation using the coding coefficient. We make use of the large-scale Multi-PIE face database to test face validation performance. All the 249 facial images in Session 1 were collectively labeled as the training set, with the same subjects in Session 2 as query images. The other 88 subjects, i.e. facial images that are different from the training set, were used as the imposter images. For the training data set, the 7 frontal images with extreme illumination condition {0, 1, 7, 13, 14, 16, and 18} and neutral expression were utilized. For the query set, 10 frontal images of illuminations {0, 2, 4, 6, 8, 10, 12, 14, 16, 18} taken with neutral expressions were employed.

Figure 3.6 shows the receiver operating characteristic (ROC) curves of the standard methods: SRC, RRC_L1, RRC_L2, and CESR. It can be seen that CESR performs very poorly, whereas RRC_L2 shows superior performance. For instance, if the

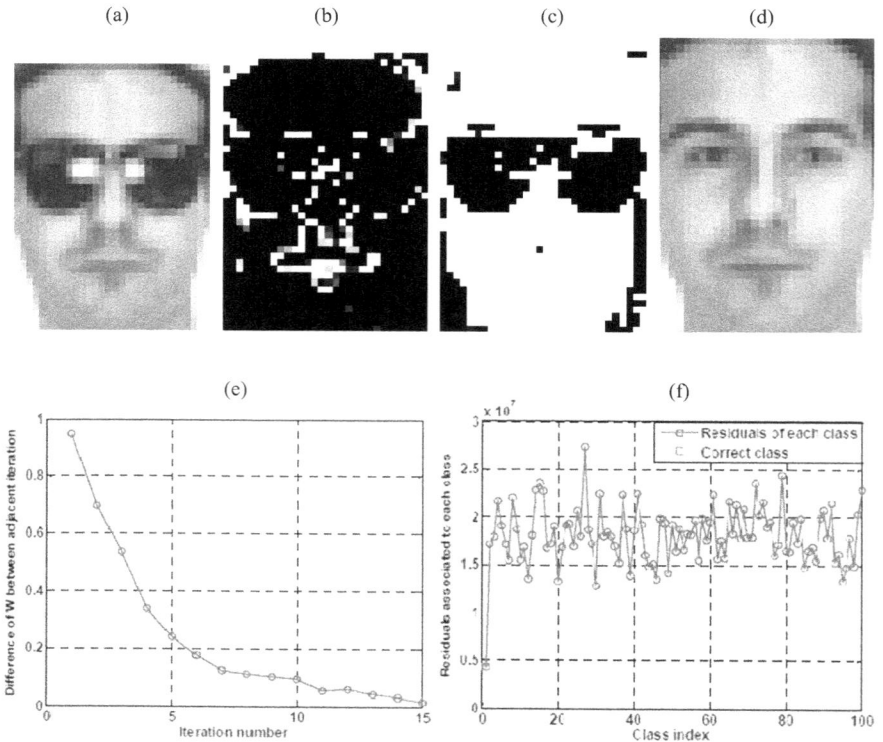

FIGURE 3.6 Subject validation on the large-scale Multi PIE.

false-positive rate is taken as 0.1, then the true-positive rate is 82.6% in the case of CESR, 90.7% for SRC, 93.3% for RRC_L1, and 95.8% for RRC_L2. RRC_L2 with $l2$-norm coefficient constraint validates faces well and outperforms the $l1$-norm coefficient constrained methods, e.g., SRC, RRC_L1, and is also much better than CESR. This might be the case because the $l1$-norm constraint, which strictly insists on the coding coefficients being sparse, will allow precisely one class to represent the input invalid test image and, hence, incorrectly recognize this query image. On the other hand, $l2$-norm constraint does not insist on the coding coefficient being sparse, which permits the representation coefficients of invalid test images to be fairly distributed across different classes. In this manner, incorrect recognition can be avoided.

3.4.5 RUNNING TIME COMPARISON

Along with recognition rate, expense for computation is also a crucial consideration for practical face recognition systems. In this section, the running time of other state-of-the-art techniques, including SRC, GSRC, CESR, RRC_L2, and RRC_L1, is estimated after performing two face recognition experiments (first, without any form of occlusion and, second, with real disguise). The programming environment employed is MATLAB® version 7.0a. The desktop used has a 3.16 GHz CPU and 3.25G RAM. All the methods are implemented using their corresponding codes. For SRC, we make use of $l1\_ls$ and two high-speed $l1$-minimization solvers, ALM and Homotopy, to execute the sparse coding step.

The first demonstration is face recognition without occlusion performed on the AR database with various down-sampled face features. Figure.3.7 compares the running time (Figure 3.7a) and the recognition rates (Figure 3.7b) of the state-of-the-art methods under various feature dimensions. From Figure 3.7a, it can be concluded that RRC_L2, CESR, and SRC (Homotopy) are quicker than other methods. RRC_L1 is

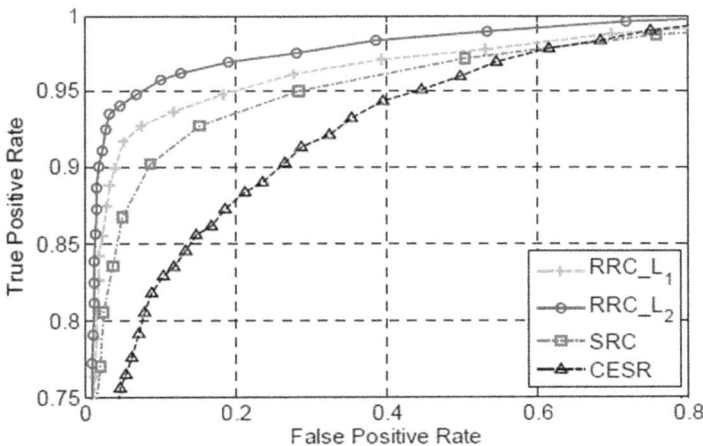

FIGURE 3.7 Running time and recognition rates by the all techniques under different feature dimension in face recognition without occlusion.

TABLE 3.8

The Average Running Time (Seconds) of All Schemes in Face Recognition with Real Face Disguise

Method	Test 1-Sunglass (%)	Test 1-Scarf (%)	Test 2-Sunglass (%)	Test 2- Scarf (%)
CESR [7]	2.50 (99.0)	3.61 (42.0)	0.45 (87.2)	0.47 (29.4)
SRC ($l1\_ls$)	662.15 (87.0)	727.14 (59.5)	38.23 (73.3)	47.73 (22.5)
SRC (ALM)	35.99 (84.5)	36.45 (58.5)	2.34 (72.4)	2.35 (21.7)
SRC (Homotopy)	13.98 (65.0)	13.73 (37.5)	3.56 (60.0)	3.59 (17.3)
GSRC [6]	119.32 (93.0)	118.05 (79.0)	12.95 (66.2)	12.49 (75.5)
RRC_L1	**8.70 (100)**	**8.62 (97.5)**	**2.06 (94.2)**	**2.04 (84.8)**
RRC_L2	2.17 (99.5)	2.04 (96.5)	0.23 (91.5)	0.23 (86.0)

The values in parenthesis represent the average recognition rate.

also relatively more efficient than SRC ($l1\_ls$), which is the slowest method. In the case of the feature of 792 (33×24) dimensions, RRC_L2, CESR, RRC_L1, SRC ($l1\_ls$), SRC (ALM), and SRC (Homotopy) take 0.257, 0.330, 1.450, 8.551, 0.377, and 0.199 seconds, respectively. RRC_L1 achieves the highest recognition rates, closely followed by RRC_L2, as shown in Figure 3.7b. Although CESR is quick, its recognition rates are comparatively lower than other methods. It can be concluded that compared with SRC and CESR, RRC_L2 has high recognition rate with lower computation expense, while RRC_L1 has a much larger recognition rate.

The second demonstration is face recognition performed on faces that have a real disguise. The dictionary has 800 training images, each of size 83×60 in Test 1, and 400 training samples with size 42×30 in Test 2. The recognition rates have been summarized in Table 3.6 (for Test 1) and Table 3.7 (for Test 2). Table 3.8 displays the expected computational expense and recognition rates of different methods on Test1 and Test2. Evidently, RRC_L2 is the fastest method with the least computation time, followed by CESR and RRC_L1. SRC has sizeable computational costs even after making use of fast solvers such as ALM and Homotopy, because of the requirement of an additional identity matrix in order to code occlusion. Considering the recognition rates, SRC performs the worst, and CESR also has a bad recognition rate in recognizing faces with scarves in each test. GSRC solved by $l1\_ls$ has lower time cost than SRC ($l1\_ls$) but is still not fast enough to qualify as one of the better-performing techniques. When both the recognition rate and running time are considered, RRC_L1 and RRC_L2 perform the best. RRC_L1 achieves the highest recognition rates in most cases, at a faster speed than SRC and GSRC. RRC_L2 is the quickest in all cases, and it has the second best recognition rate.

3.4.6 PARAMETER DISCUSSION

Now, consider the effect of parameter δ in RRC on the final face recognition rate. As described below Eq. (3.14), the parameter δ is an important parameter for

(a)   (b)

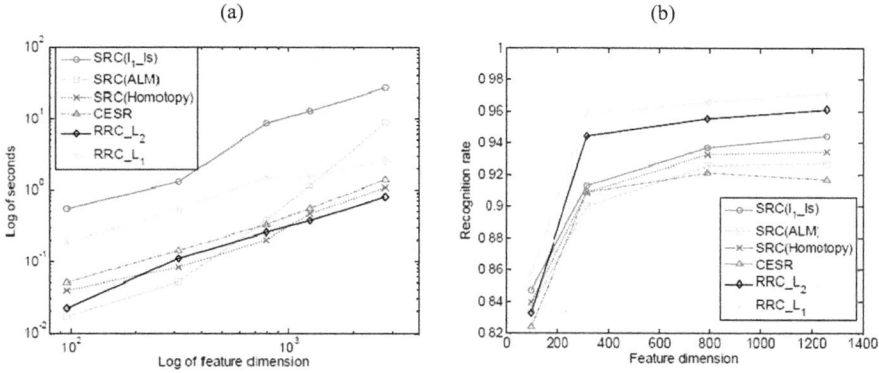

FIGURE 3.8   Recognition rate versus τ in estimating δ of the weight function of RRC.

distinguishing between inliers or outliers (if the residual's square of a pixel is larger than δ, its weight will be lesser than 0.5; otherwise, its weight is greater than 0.5). The parameter τ is used to estimate δ, as described in Eq. (3.24). Hence, it is essential to discuss the selection of τ. We perform the experiment with assorted levels of random pixel corruption to interpret the selection process of τ for RRC. Figure 3.8 shows the graph for the recognition rates of RRC_L1 versus varying values of τ for 0%, 30%, 60%, and 90% pixel corruption. It can be seen that for low and moderate corruption levels (i.e., 0%~60%), RRC_L1 could lead to excellent performance (i.e., greater than 95% recognition rates) in a broad range of τ. For all percentages of pixel corruption, we find that superior performance is achieved when $\tau = 0.5$. Usually the domain of τ might be set within the range [0.5, 0.8]. This is practical because at least 50% samples should be trusted when there is a high percent of outliers.

3.5   CONCLUSION

This chapter proposed an RRC model and an IR³C algorithm for robust face recognition. One significant advantage of RRC is its robustness to a variety of outlier pixels (e.g., occlusion, corruption, expression, etc.) by seeking an approximate MAP solution to the coding problem. By assigning the weights to the pixels, adaptively and iteratively, according to their coding residuals, the IR³C algorithm is able to robustly identify the outliers and thereby diminish their effects on the coding process. Also, it was shown that the *l*2-norm regularization entails a much lower computational cost when compared with *l*1-norm regularization, without compromising on the performance in RRC. The proposed RRC methods were assessed on face recognition under a variety of conditions, like nonuniform illumination, expression variation, occlusion, and corruption. The experimental results suggest that RRC performs remarkably better than various state-of-the-art techniques. More specifically, RRC with *l*2-norm regularization could realize very large recognition rates while offering the benefit of low computational costs, thus proving it to be a good candidate model for practical robust face recognition systems.

REFERENCES

1. James, A. P., and Dimitrijev, S. (2012). Nearest neighbor classifier based on nearest feature decisions. *The Computer Journal.*
2. Jiang, J., Hu, R., Han, Z., and Lu, T. (2013). Nearest feature line embedding for face hallucination. *Electronics Letters.*
3. Lu, J., and Tan, Y.-P. (2011). Nearest feature space analysis for classification. *IEEE Signal Processing Letters.*
4. Melek, M., Khattab, A., and Abu-Elyazeed, M. F. (2018). Fast matching pursuit for sparse representation-based face recognition. *IET Image Processing.*
5. Naseem, I., Togneri, R., and Bennamoun, M. (2010). Linear regression for face recognition. *IEEE Transactions on Pattern Analysis and Machine Intelligence.*
6. Yang, M., and Zhang, L. (2010). Gabor feature based sparse representation for face recognition with Gabor occlusion dictionary. *Proceedings of the 12th European conference on Computer Vision.*
7. R. He, W.S. Zheng, and B.G. Hu (2011) Maximum correntropy criterion for robust face recognition," *IEEE Transactions on Pattern Analysis and Machine Intelligence*, 33(8).

4 Big Data Analysis, Interpretation, and Management for Secured Smart Health Care

V. Sucharita and P. Venkateswara Rao
Narayana Engineering College

Pellakuri Vidyullatha
KoneruLakshmaiah Education Foundation

CONTENTS

4.1   INTRODUCTION

Many smart applications have been developed in light of the smart city development. For example, smart health care, smart building, smart energy, smart education, smart living, and smart mobility. In this chapter, Big Data analytics in secure smart health is proposed. Numerous data analytics methods are applied to analyze the data to realize smart healthcare applications.

Currently, it can be said that data is increasing exponentially and many people are accessing it. This has led to investigating and visualizing the information successfully in the health care industry by using the various techniques of Big Data analytics. These huge records that are analyzed play an important role and offer many advantages in dealing with Big Data. Various techniques are described in this process that improve the health care by considering the diverse information.

This chapter presents the introduction to Big Data and its role in different applications of health care. The usage of the Big Data techniques and architecture assists in maintaining speedy growth of data in the health care sectors. Mobile health care apps have been developed that allow patients to send a query to providers. These apps are equipped with instructions of first aid, and patient is given further treatment as an emergency section or may be directed to other specializations [1]. From various users the health care system will collect and analyze biomedical signals like Blood pressure etc. Usually, smart health care application is installed on the smartphone, and all data with respect to health is synchronized into the cloud service for storage, retrieval, and analysis. Big Data can be captured with the information technology. Valuable information can be achieved from huge data sets via data mining. To handle Big Data in health care, Machine Learning and security mechanism will be used. In the healthcare industry, a considerable amount of work has been done using the Big Data. In this chapter, security and privacy of the data are discussed as are various techniques like encryption, decryption activity monitoring, and validation. It also discusses about the strategies to be followed by the users' for accessing the data when Big Data is applied to the health care sector.

The patients who are undergoing medical treatments for certain diseases that are specific is matching will reduce the side effects caused unnecessarily, further which the quality of the treatment can be improved by avoiding the unnecessary treatment.

Nowadays, data is growing rapidly at an uncontrollable manner due to the development of cloud and internet [2].

In databases, 2.5 petabytes of data are imported in an hour every day. Nine hundred million objects and 250 million photos are handled by Facebook [3]. Big Data is defined as the voluminous data collected from various sources like web applications, digital repositories, and mobile devices, which cannot be managed very easily by the existing tools [4]. Big Data has a large amount of data and is the process of storing

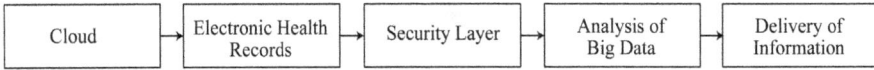

FIGURE 4.1 Health information system.

and managing data with volume, velocity, veracity, and variety. Voluminous data is analyzed in terabytes [5,6]. Big Data helps to predict the spread of the epidemics; current health status can be predicted through various moves which can be tracked through location data of the phone. It can predict the spread of the virus. Big Data in healthcare systems is very important for decision-making, which has immediate impact on patient care. The Big Data analytics in health care systems include trial reporting, drug safety analytics, patient analytics, and drug pricing and promotion analytics. Ahemed et al. [7] proposed a method for secure smart health care in Big Data analytics, as shown in Figure 4.1. To give benefit to the patients, this proposed method manages and analyzes huge amounts of data. In this method, five components are described. They are the cloud environment, HER, security layer, Big Data analysis, and information delivery [55]. The cloud environment provides different services to authorized users and sometimes allows sharing of the data. Patients' data from various locations are integrated by using HER. Various security issues like confidentiality and authentication will be taken care of by the security layer, which uses encryption algorithms for data protection and also provides data security on the networks also. The Big Data analytics layer is used to analyze and get insights from the raw data. Finally, the health care information is distributed by providers in various locations.

4.2 5 V'S OF BIG DATA IN HEALTH CARE SYSTEMS

Different dimensions of Big Data are basically used for problem solving. In what way the five dimensions are applied to health care is being discussed in this section [23]. The five dimensions are shown in Figure 4.2. All the features of Vs in terms of healthcare are shown in Figure 4.3.

4.2.1 VOLUME

If the five Vs of Big Data are understood, they can be applied to health care. The data grows severely in health Enterprises, i.e. volume. In health care systems, data will be in petabytes and terabytes. The data include patients' personal information, images of scanning and x-ray, sensor readings, etc. It is needed to analyze and manage this complex data, which are in various forms. Various health information systems like CDSS, EHRs PACS, etc. generates patient data in huge amounts. The digital data of health care is approximately 500 petabytes; in 2021, it is expected to reach more than 25 exabytes.

4.2.2 VARIETY

Various forms of structured, unstructured, and semi-structured data exist [43]. The structured data, like clinical data from different health care centers, are easy to store,

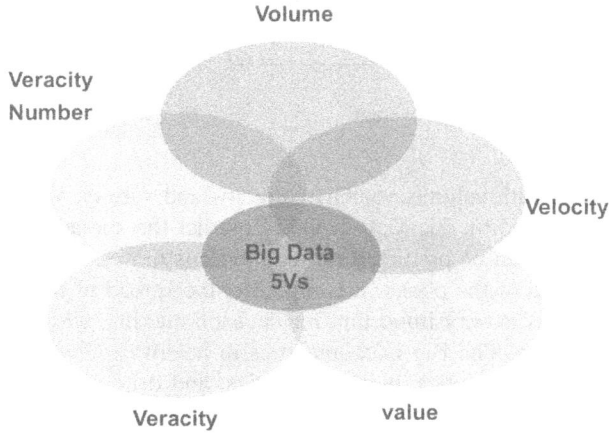

FIGURE 4.2   Five V's of Big Data.

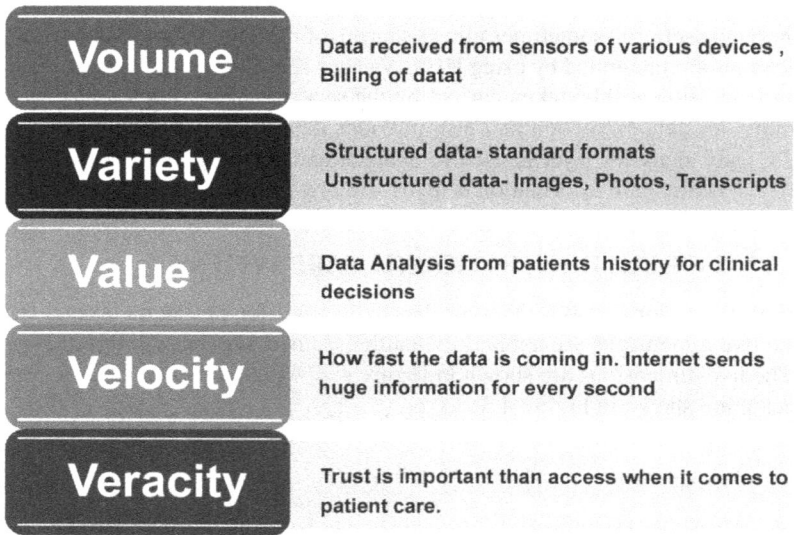

FIGURE 4.3   Features of V's.

retrieve, modify, and analyze by system. Doctors' prescriptions, case sheets, officer medical records, images, and scanning reports may be semi-structured or unstructured.

4.2.3 VELOCITY

Data available in health enterprises systems is correct. If the data is modified continuously on a normal basis, then the data available in health care systems may not be accurate. The huge amount of the data stored must be retrieved very quickly to compare and to take accurate decisions on the existing real-time data. To detect and prevent the diseases in early stages, Big Data analytics.

4.2.4 VERACITY

To achieve the results effectively with the Big Data analytics, data with quality is required. As there are different forms of the data like structured, unstructured, and semi-structured [48], the quality of the data changes from one form to another form. The data quality comes into picture when the patient's life is related to having the right information. It will be very difficult if the data in the three forms mentioned is not accurate.

4.2.5 VALUE

The last V that represents value is something different as it represents the preferred outputs of processing the Big Data. People are always concerned about extracting and collecting data that give accurate value. People require quality data. Storing the health care data for diseases on untrustworthy storage may affect the data in future but can reduce expenditure. In health care sectors, value means promoting patients health by reducing cost.

4.3 SUCCESSFUL RELATED WORKS

To present the outlines in healthcare, a conceptual architecture was introduced by the study of Raghupathi and his co-authors [8]. The architecture has four levels to get an accurate and deep understanding by applying Big Data tools like map reduce, Hadoop, etc. [49,52].

Kim et al. [9] have used a technology called complex event processing to manage ERP systems for Big Data in real-time systems for transactions in health care systems.

The different opportunities of Big Data in health sector like prediction [45] of diseases in advance to avoid risks, taking preventive measures, reducing cost in medical field. Accurate information is provided by analyzing huge records in the studies [10].

Data sets were analyzed for the quality of health care services in India from 1950 to 2015 [11].

A patient-centric health care framework was proposed by the authors that helps the physicians to manage patient using the filtering techniques to find similarities [12].

Kurian et al have proposed a method for detecting chronic diseases like cancer, diabetes, and morbid obesity. To reduce the cost of diagnosis, Big Data analysis was used [13].

To predict the type of the patient suffering from diabetes, Saravana et al. used Map reduce and Hadoop. The results of this study will help the medical experts for giving the better treatment [14].

An application regarding E-Health service to diagnose heart disease was proposed by Abhinaya. The data mining methods along with Hadoop were used for developing the architecture of the system [15].

To predict type 2 diabetics in older adults and women, authors proposed to diagnose diabetes using MHealth application from data through cloud [16].

A model for predicting type 2 diabetes and the risk factors associated with it based on Big Data and Machine Learning has also been proposed [17].

Real-time monitoring using Hadoop was proposed by one study for analyzing how the health care system can be improved by delivering information on time to the right person [18].

The study by Sukumar et al. has considered the quality of data in healthcare sector to manage data collection from various sources and to manage erroneous data [19].

Luo et al.'s study discussed the opportunities and threats associated with the applications of Big Data in health care sectors [20,47].

Archenaa at al. assessed huge amounts of data in health care industries and the government and made a real-time analysis [21].

To develop architecture for observing and checking diseases in real time and to take decisions efficiently [22,50], after reviewing several research papers, it is observed that nowadays in the health care industries, it is not the size of the data that is the main issue but the privacy and security of the data that is very important. Many of the authors in different research studies concentrated on the framework, architecture, interoperability, and using Big Data for analyzing various patterns, but not many concentrated on the security. Maintaining security and privacy of the records in the healthcare sectors is the most important and challenging issue.

4.4 IMPACT OF BIG DATA IN HEALTH CARE

In health care industries, Big Data analytics have a very good effective impact. Industry in view of the patient through five different modes as shown in Figure 4.4. They are right care, right provider, right value, right innovation, and right living [23].

4.4.1 RIGHT LIVING

It makes the patient live healthier. The craving for good health ends in right living. Patients can, thus, take care of their health by analyzing the results of Big Data for taking good decisions for improving their status of health.

4.4.2 RIGHT CARE

Proper treatment will be given to ensure safety to the person suffering from the disease at the right time. All the caregivers should also have the same data. This is very important for the right care.

4.4.3 RIGHT PROVIDER

The professionals must ensure the treatment proposed to the patient. The care providers gives better and more effective treatments by using the Big Data as huge amount of data is available; thus, good analysis is possible, aiding decision-making.

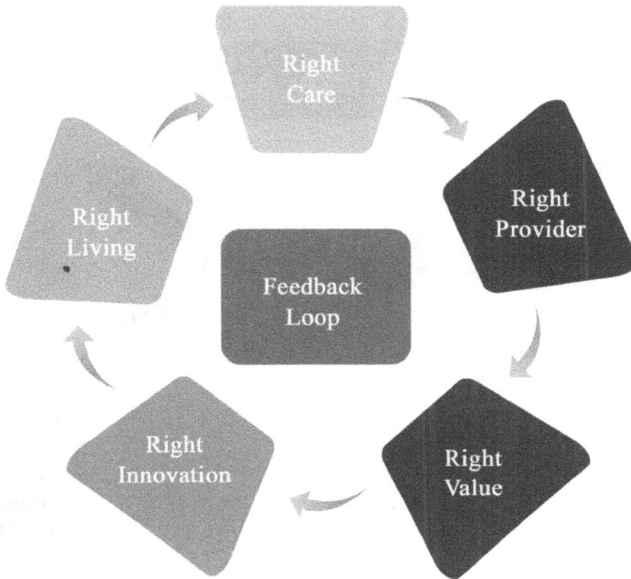

FIGURE 4.4   Five roadways in view of the patient.

4.4.4   RIGHT INNOVATION

This approach aims at predicting new problems and providing better treatments. It tries to improve the new safety methods for the health care development.

4.4.5   RIGHT VALUE

The value and quality of the healthcare must be preserved by using the latest technology with optimized resources, guaranteeing the cost-effective care, as shown in Figure 4.5.

The goal of the Big Data in health care is to provide good and personalized care to the right patient through the right intervention at the right time.

4.5   APPLICATIONS OF BIG DATA

Business have changed the way they operate because of the new technologies and developments in this century. Big Data helps us to understand and analyze what the customers want from the company in each and every aspect. That is, Big Data has been used increasingly in various professional sectors, social sectors, and industrial sectors. Some of the sectors where Big Data is used is shown in Table 4.1.

4.6   BIG DATA SOURCES FOR HEALTH CARE

A. *The Internet of Things (IoT)*: One important source for data is IoT. It has already had an influence in the energy and home monitoring and transport

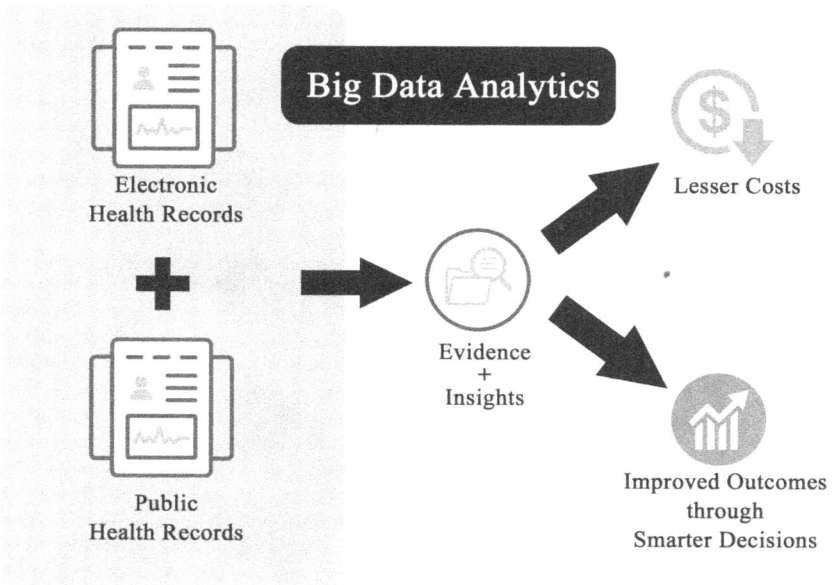

FIGURE 4.5   Improved outcomes using Big Data.

TABLE 4.1
Big Data Applications in Various Sectors

Category	Applications
Healthcare sectors	Reduction of cost in treatments, Identification of various diseases, Precautionary measures.
Public sectors	Reduction of taxes, Protection of environment, Safety of public
Education sector	Interested learning mode of students, Tracking the performance of the students, Assessment of the students in digital platform.
Transportation	Controlling traffic, Intelligent transport systems, Congestion control.
Insurance sectors	Predicting the behavior of the customer, Evaluate the risk of the Customer, Managing policy premiums.
Entertainment	Manage data, Measure the performance of content
Banking	Analyzing the business, customer patterns, and competitors
Fraud detection	Detecting credit cards misuse

sectors. In health care, IoT is connected with the number of things. All the platforms related to IoT are considerably cheap and cost-effective. The advantage of using IoT devices is that a doctor can easily monitor and measure different parameters of his client despite being in different locations, be it the workplace or at home. Through early treatment, a patient need not be hospitalized, thus leading to reduced expense and the right treatment at the right time. Peoples' heart rate, weight, blood pressure, and stress levels

TABLE 4.2
Various Diseases and Diagnostic Methods

Type of Disease	Method of Diagnosis	Measures through IoT
Heart-related diseases	Pattern matching	Patterns of ECG
Hypertension	Frequency based	Blood pressure
Obesity	Scale based	Body weight, blood pressure
Stress index	Frequency based	Sensors for measuring stress
Infectious diseases	Scale based	Sensors for measuring temperature

can be tracked through wearable devices nowadays. Many applications are available for smart phone users to track exercise, sleep quality, etc. Some diagnostic techniques must, however, be used before analyzing the disease severity, as shown in Table 4.2.

B. *Electronic Health Records*: Transforming the offline patients data in to an online storage form is a great boon for medical practice. The records of the patients, which were on paper earlier, are moving to Electronic health records, or EHRs. Implementation of EHRs continues to expand. Large amounts of data can be stored easily, which represent the massive number of patients encountered and data generated. The transformation of clinical data into knowledge for improving the patient care is the goal of health care information systems

C. *Insurance Providers*: Including private plan claims, health plan related to government sectors, and pharmacy.

D. *Clinical Data*: Data taken from various physicians regarding order entry, prescriptions, reports, lab reports, insurance claims, medical shop receipts, and other administrative data.

E. *Social Media*: Posts on social media, such as Tweets, blogs, status updates on Facebook, and various other platforms and information on web pages provide evidence of a person's health.

F. *Web Knowledge*: Patient-related general information can be obtained from websites. Less data will be obtained about emergency situations; however, some specific information can be obtained from articles in medical journals.

4.7 MINING USING BIG DATA

The use of health care systems is increasing worldwide and new technologies are using huge amounts of data for predictive analysis. The behavior of a person can be studied through the sensors of mobile phone using Big Data. Mobiles are provided with various sensors, like location sensors, motion sensors, and touch sensors. Some information is collected whenever a person uses a sensor. Thus, surveys based on mobile usage gives data on a person's behavior. The biometric sensors can track heartbeat, blood pressure, sleep activity, etc. [24]. Wireless mobile body area networks are used very frequently. About 8.2 million health and wellness wireless sensor networks are being used, per the new report by research firm ON World, which

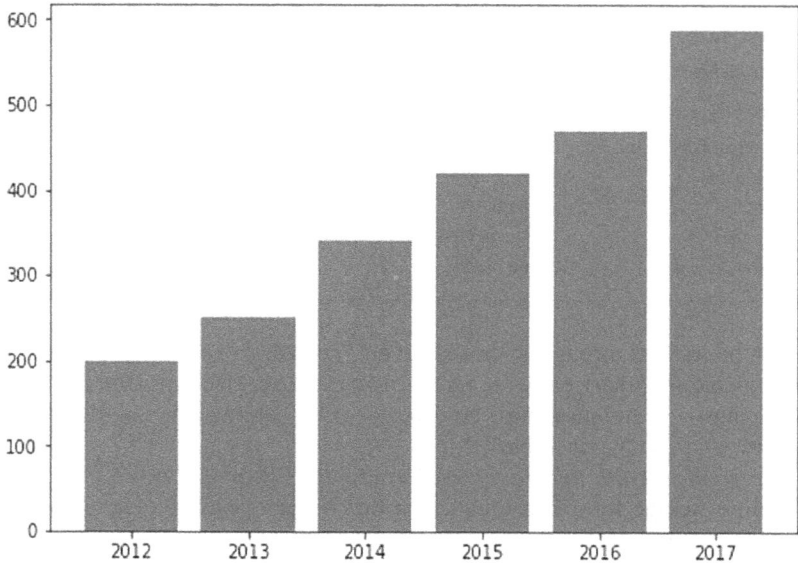

FIGURE 4.6   Health and fitness.

is been shipped throughout the world in the year 2017, producing $16.3 billion as annual income. As shown in Figure 4.6, 60% of 200 evaluated are there in marker in the year 2012 [25,26]. The number of fitness usage and health sensors have risen to 552 during the period between 2012 and 2017, as shown in Figure 4.7.

4.8   BIG DATA ANALYTICS DRAWBACKS IN HEALTH CARE SECTORS

Although the health care sectors have been improved, there are still many limitations.

Privacy and security of the data are not guaranteed. Efficient security algorithms are lacking. The data coming from various sources like medical care center, pharmacies, and hospitals are all in different formats. Unification requires significant expenditure.

Another limitation is the data quality because data is collected in various forms (structured, unstructured, etc.). All the information in the different forms should be brought together into one meaningful format.

People with expertise and knowledge are required for Big Data analytics. To maintain staff and resources for this is a big investment for companies. Also, health care industries must be convinced about the advantages of using Big Data analytics. Staff with technical expertise must then be hired. Thus, only big organizations can maintain such Big Data.

4.9   PRIVACY AND SECURITY IN BIG DATA

Privacy and security are very important for secure smart health care using Big Data. Privacy is the ability to safeguard the sensitive information in the health sector. It

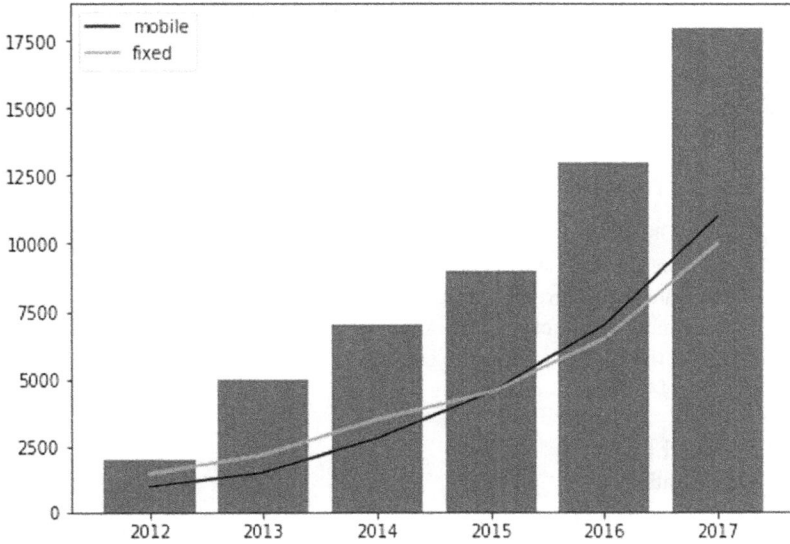

FIGURE 4.7   Global health WSN systems shipment.

TABLE 4.3
Differences between Security and Privacy

Security	Privacy
Security is the integrity, confidentiality, and availability of the data	Privacy is to what extent the information can be convened to others.
To prevent the data from network vulnerabilities, various methods like encryption and firewalls are used	Without the permission of the patient, the organization cannot sell its patient data to a third party.
Provides confidentiality	It is mainly concerned with the right to safeguard patient information.

concentrates on implementing authorization requirements to provide security to the personal data of the patients. It must be shared and used in the right way. Data must be protected from malicious attacks, and there should not any chance for data theft with the intent to make profit [27,56]. Both security and privacy are very important. The differences between security and privacy are given in Table 4.3.

4.9.1   SECURITY IN HEALTH CARE

To deliver proper care, the huge amount of data that is stored and maintained by the health care sector needs to be secure. Security of the huge data is the need for the future. Health care sectors must be more liable to publicly disclosed data breaches.

The attackers use techniques to find out data that are important and sensitive and put it in the public domain, thus causing a data breach. It is very important for the organizations to execute data security by protecting assets and satisfying the healthcare compliance codes available.

4.9.2 LIFE CYCLE OF BIG DATA SECURITY

Security in Big Data is explained in three ways: data security, information security, and control of access [28]. Health care systems must use security measures to protect Big Data, associated software and hardware, and also personal and clinical information. To ensure that good decisions are made about cost-effectiveness, retention, and reuse of the data, the life cycle must be established [29]. The main phases of life cycle shown in Figure 4.8 are as suggested by Vazanl [31], and these include the following:

1. Collection of data
2. Transformation
3. Modeling
4. Knowledge creation

4.9.2.1 Collection of Data

The basic step is gathering data. We consolidate data from various sources. We should identify different data formats in which we can collect the data. Since we are dealing with health information of the people, it is important to ensure to collect data from trustworthy parties. As we are dealing with sensitive information of users, we should consider data privacy requirements in this phase. We should identify appropriate

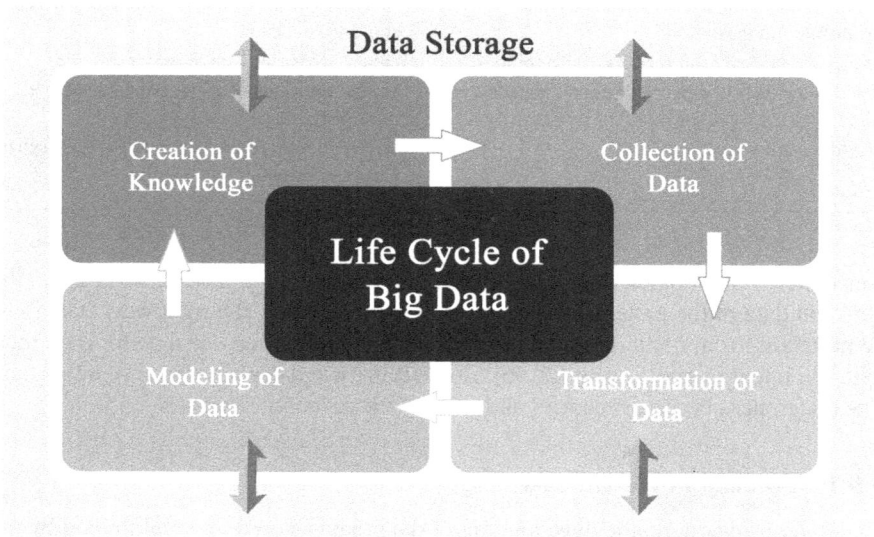

FIGURE 4.8   Life cycle of Big Data in health care.

security measures to protect the data and information systems from unauthorized access, misuse, destruction, tampering, and disclosure to other parties.

4.9.2.2 Transformation

Now that we have gathered data, classification of the data is the next step based on the patterns. We need to undertake the necessary transformation in such way that we can perform meaningful analysis [44]. To improve the quality of data, we need to filter, enrich, and transform data before the data modeling phase; if needed, we can remove noise, duplicates, outliers, values that are missing, etc. We need to keep in mind that data collected may have sensitive information; we have to take necessary measure during data transformation and storage. To ensure data security, we should maintain different access levels and access controls to restrict unauthorized access to data. We can define security measures such as partitioning of data, anonymization, and permutation.

4.9.2.3 Data Modeling

In the previous phases, we have collected the data, transformed it, and stored it securely. As part of this phase, we perform data processing analysis and generate knowledge. Different supervised data mining techniques can be used, such as classification techniques and clustering, to derive the features and predictive models [46,51,53]. We use a combination of other learning techniques to make the final model more accurate and robust [54]. As in other phases, data security is vital is this phase too; we should have secure processing environments as we are extracting sensitive data. Hence, the data mining processes, servers, and network components should be setup with the necessary security configuration in order to allow only authorized personnel access to data to protect it from potential security attacks and security breaches. This helps in assessing vulnerabilities and fixing and reducing the risks.

4.9.2.4 Knowledge Creation

The data modeling phase provides us the information and valuable insights that can aid us visualize the data and make the right decisions. This knowledge created is very sensitive as we are dealing with health-related information of patients. It is important to protect this data in this competitive world. Health care organizations are well aware of the legal regulations to protect user's personal data, as per the data privacy rules and regulations. The primary objective in this phase is to regularly verify access controls and ensure security compliance.

4.9.3 Technologies to Provide Security and Privacy

In Big Data different technologies are used for ensuring security and privacy of healthcare data.

4.9.3.1 Authentication

It is the act of administering confirmation that the claims made are true. Authentication plays an important role in the organization for accessing corporate networks accurately, protecting the user's identity, etc. To prevent attacks by middlemen, some authentication is required. The cryptographic protocols, like secure sockets layer and transport layer security, provide security on the internet. Secure Socket Layer can be used for authenticating the server by using the mutually trusted certification authority. In healthcare systems, security must be provided based on the customers' identity. It should also be assessed at each and every entry point and access point. Kerberos and Hashing techniques like SHA-256 [30] can be used to implement authentication.

a. *Data Encryption*: It is the process of converting the plain text to cipher text. To prevent the unauthorized access of confidential and sensitive data, encryption is used. Data breaches like packet sniffing can be avoided using encryption. Encryption of the data protects and helps maintains the ownership of data. The encryption scheme used by the health organizations must be certified. Also, the staff of health care organization and the patients must find it easy to use. Although several encryption algorithms are available, the best and most efficient encryption algorithm must be selected to provide data security. Also, the keys used at both the ends should be minimized. There are various encryption algorithms like DES, IDEA. AES, and RC4 [30,32,33]

b. *Masking*: Masking of the data is a process of creating a similar structure by hiding original data with modified content. The main reason for undertaking data masking is to protect the sensitive or personal data. The actual data is protected by developing a substitute for situations where the real data is not required. It replaces the important data with a value that cannot be identified. It is not like encryption. Therefore, the original value will not be capable of being derived from the masked value.

c. *Access Control*: Information system can be accessed only by authorized users, and this monitored by access control policy. Privileges will be granted for users such as staff or the patients themselves. There are different mechanisms for access control, like role-based access control and attributed-based access control. In electronic health records, these two models are commonly used.

d. *Accounting*: This is the identifying of the events happening in the network to enable catching the intruders. All the health care activities done by the users are recorded by maintaining the logs of user access to data. There are some security measure to provide safety to the healthcare providers, i.e. detection and prevention of the intrusion [34]. An architecture for the monitoring of the security is developed via analyzing DNA traffic [35]. In this method storing, retrieving and modifying data in the distributed environment is done using data correlation scheme to identify whether the packet flow is harmful or not.

4.9.4   PRIVACY OF BIG DATA IN THE HEALTH CARE INDUSTRY

Patient privacy is major issue in Big Data analytics because of threats that are continuing for a prolonged period and attacks on information systems. As a result, it is a challenging issue for organization to address these critical issues. In a Forbes report, it was mentioned that target corporation had sent a baby care coupons to a girl without the knowledge of her parents [36]. In such a case, privacy of data is very important and needs to be taken into consideration. As privacy is very important, some laws of data protection are followed in some countries [42]. A few of the conventional methods are now briefly mentioned. To ensure the privacy of the patient, these traditional techniques used but their disadvantages led to the arrival of new methods.

1. *De-identification*: It is conventional process to prevent the revelation of hidden data by simply throwing out particulars of the patient, either by the first technique or by the alternate statistical procedure in which unnecessary identifiers are removed. However, there is every possibility that one can get additional data even from de-identified data in the Big Data. Hence, this method is not enough for shielding Big Data confidentiality. It is the need of the hour to build up and efficient privacy-preserving algorithm to assist alleviating the threat of re-identification. K-anonymity, T-closeness, and l-diversity concepts are some that help to augment this conventional process [37–41].
2. *k-anonymity*: In this method, if k increases, the probability of re-identification decreases. But there is a possibility of distortions and loss of information due to k-anonymization. Besides, in k-anonymization, the sensitive attribute (like Disease) is exposed if the quasi-identifiers are used to connect with other openly accessible information to identify individuals. Different methods are planned to enumerate data loss that do not reveal the exact values of data.
3. *L-diversity*: This is one of the cluster types of anonymizations that ensures privacy in data sets, and it is an extension of k-anonymity that uses the approaches of suppression and generalization to decrease the data representation granularity to k-variant data records. An L-diversity approach can hold an imitation of k-anonymity method. The main issue in this method is focused on the range of attribute sensitivity. To generate the data as L-diverse over sensitivity of attribute and misleading information to be infused and may enhance the security then L-diversity approach is become sensitive to imbalance and can't avoid discloser of attribute property.
4. *T-closeness*: It facilitates advance changes to L-diversity group-placed anonymization. The T-closeness approach either equal distance by treating the attribute values. The major improvement of this method is in providing the disclosure of attribute, but if there is an increase of variety of data, then the re-identification rises.
5. *HybrEx*: This approach is for privacy and provides assurance in cloud computing. The structure of HybrEx is shown in Figure 4.9. It uses public types of clouds for the organization's nonsensitive information with that computation classified as public; furthermore, if the applicant wishes to access both private as well as public data, then the application itself gets partitioned and executed on both represented clouds. It examines the sensitivity of data earlier to an execution of task which gives safety. The limitation of HybrEx is that it does not consider the keys in the map phase, which is generated by public clouds and private clouds.

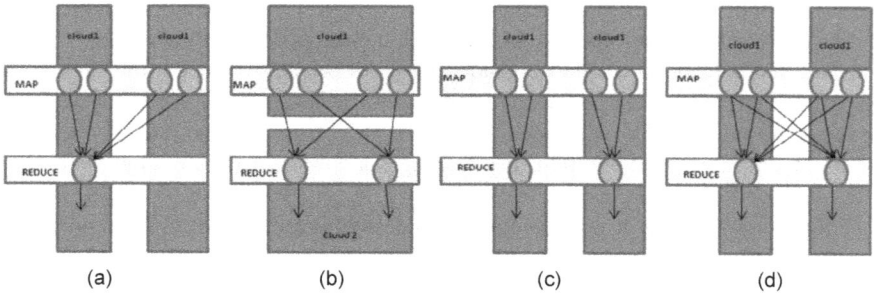

FIGURE 4.9   HybrEx map reduce.

6. *Identity-Based Anonymization*: The above-mentioned techniques have issues of skillful integration of anonymization and protection of privacy; these techniques of Big Data are used to analyze data usage to give assurance to authorized users. To significantly aid storage on the Cloud, Intel proposed an open architecture for anonymization that allows different tools for re-identifying the log records of web. For achievement of architecture, the enterprise data have different features to the standard examples in anonymization study. In spite of concealing the various personal IDs Intel found that anonymized data is defenseless in the face of correlation attacks. By observing these vulnerabilities, they decided the information of user Agents that are strongly correlated to individual users.

4.10   CONCLUSIONS

This chapter provides an Introduction to Big Data as well as information on analysis, interpretation, and management for secure and smart health care. The techniques of traditional data processing cannot handle Big Data generated in the medical field. All the limitations of the traditional data processing can be overcome by Big Data analytics. Data analytics has the capacity to develop epidemic control plans, scan clinical data, perform disease surveillance, etc. Efficiency, scalability, and reliability are the main advantages of using Big Data in health care systems; but these have challenges too. The challenges include storing, capturing, searching, and analyzing data; quality; privacy and security; and integration of heterogeneous databases. The main intention is to propose a secure and smart healthcare frame work using Big Data. It has been identified from literature that existing Big Data applications can revolutionize health care industries and provide good information at less cost. In this chapter, security and privacy measures for Big Data have been proposed by using data security and privacy layers. It provides additional security features like monitoring the activity, masking of data, homomorphic encryption, etc. The proposed framework provides uniqueness in maintaining the security of the patient's data. With the advancement in technologies, it is not only easy to collect data but also to convert it into appropriate forms to enable critical insights that can be used to provide better health care.

REFERENCES

1. Panda M., Ali S.M, and Panda S.K. (2017). Big data in health care: A mobile based solution. In *Big Data Analytics and Computers.*
2. Ozgur C., Kleckner M., and Li Y. (2015). "Selection of statistical software for solving big data problems: A guide for businesses, students, and universities." *Sage Open*: 1–12.
3. Sagiroglu S. and Sinanc D. (2013). "Big data: A review." *Presented in International Conference: Collaboration Technologies and Systems (CTS).* IEEE Xplore.
4. Hansen M. M., Miron-Shatz T., Lau A.Y.S., and Paton C. (2014). "Big data in science and healthcare: A review of recent literature and perspectives." *Yearbook of Medical Informatics*: 1–6.
5. McAfee A. and Brynjolfsson E. (2012). "Big data: The management revolution." *Harvard Business Review.*
6. https://www.cognizant.com/industries-resources/healthcare/Big-Data-is-the-Future-of-Healthcare.pdf.
7. Youssef A.E. (2014) "A framework for secure healthcare systems based on big data analytics in mobile cloud computing environments." *International Journal of Ambient Systems and Applications* 2(2):1–11.
8. Raghupathi W. and Raghupathi V. (2014) "Big data analytics in healthcare: promise and Potential." *Health Information Science and Systems* 2(3): 5–10.
9. Kim M.J. and Yu Y. S. (2015) "Development of real-time big data analysis system and a case study on the application of information in a medical institution." *International Journal of Software Engineering and Its Applications* 9(7): 93–102.
10. Patel S. and Patel A. (2016) "A big data revolution in health care sector: Opportunities, challenges and technological advancements." *International Journal of Information Sciences and Techniques* 6:155–62.
11. Dayal M. and Singh N. (2016) "Indian health care analysis using big data programming tool." *Procedia Computer Science* 89: 521–27.
12. Chawla N.V. and Davis D.A. (2013) "Bringing big data to personalized healthcare: A patient-centered framework." *Journal of General Internal Medicine* 28(3): 660–5.
13. Kuriyan J. and Cobb N. (2013) "Forecasts of cancer and chronic patients: Big data metrics of population health." *Cornell University Library*: 1–26.
14. Saravana N.M.K., Eswari T., Sampath P., and Lavanya S. (2015) "Predictive methodology for diabetic data analysis in big data." *Procedia Computer Science* 50: 203–8.
15. Abinaya K. (2015) "Data mining with big data e-health service using map reduce." *International Journal of Advanced Research in Computer and Communication Engineering*, 4(2): 123–127.
16. Wang L. and Alexander C.A. (2016) "Big data analytics as applied to diabetes management." *European Journal of Clinical and Biomedical Sciences* 2(5): 29–38.
17. Razavian N., Blecker S., Schmidt A.M., McLallen A.S., Nigam S., and Sontag D. (2015) "Population-level prediction of type 2 diabetes from claims data and analysis of risk factors." *Big Data* 3(4): 277–282.
18. Shinde K.V. (2016) "A real time monitoring system in healthcare with hadoop." *'Research Journey' International Multidisciplinary E-Research Journal*, Special Issue-I: 15–19.
19. Sukumar R., Ramachandran N., and Ferrell R. K. (2015) "Big Data' in health care: How good is it?" *International Journal of Health Care Quality Assurance*: 2–9.
20. Luo J., Wu M., Gopukumar D., and Zhao Y. (2016) "Big data application in biomedical research and health care: A literature review." *Biomedical Informatics Insights* 8: 1–10.
21. Archenaa J. and Anita E.M. (2015) "A survey of big data analytics in healthcare and government." *Procedia Computer Science.* 50: 408–13.

22. Boukenze B., Mousannif H., and Haqiq A.(2016) "A conception of a predictive analytics platform in healthcare sector by using data mining techniques and hadoop." *International Journal of Advanced Research in Computer Science and Software Engineering* 6(8):65–70.
23. Asri H., Mousannif H., Al Moatassime H., and Noel T. "Big data in healthcare: challenges and opportunities" ©2015 IEEE.
24. "Sleeptracker® by MotionX | Sleeptracker.com." [Online]. Available: http://www.sleeptracker.com/. [Accessed: 10 April 2015].
25. "Report: 18.2M health sensors will ship in 2017 | mobihealthnews." [Online]. Available: http://mobihealthnews.com/21101/report-18-2mhealth-sensors-will-ship-in-2017/. [Accessed: 08 April 2015].
26. "Prediction: Wearables to lead the 515 million sensors to ship in 2017 mobihealthnews." [Online]. Available: http://mobihealthnews.com/22752/prediction-wearables-to-lead-the-515-million-sensors-to-ship-in-2017/#disqus_thread. [Accessed: 08 April 2015].
27. Wang L.and Ann Alexander C. (2019). "Big data analytics in Healthcare Systems" *International Journal of Mathematical, Engineering and Management Sciences* 4(1): 17–26.
28. Kim S.-H., Kim N.-U., Chung T.-M. "Attribute relationship evaluation methodology for big data security." In: *2013 International Conference on IT Convergence and Security (ICITCS)*, IEEE: 1–4. Doi: 10.1109/icitcs.2013.6717808.
29. "Data-driven healthcare organizations use big data analytics for big gains." IBM White Paper February.2013.
30. Shafer J., Rixner S., and Cox A.L. (2010, March). "The hadoop distributed filesystem: balancing portability and performance." In: *Proceedings of 2010 IEEE International Symposium on Performance Analysis of Systems &Software (ISPASS)*, WhitePlain, NY, 122–33.
31. Yazan A., Yong W., and Raj Kumar N. (2015). "Big data life cycle: threats and security model." In: *21st Americans Conference on Information Systems*.
32. "Federal Information Processing Standards Publication 197." Specification for the advanced encryption standards (AES). 2001.
33. Somu N., Gangaa A., and Sriram VS.(2014). "Authentication service in hadoop using one time pad." *Indian Journal of Science and Technology* 7:56–62.
34. Linden H., Kalra D., Hasman A., and Talmon J. (2009) Inter-organization future proof HER systems – a review of the security and privacy related issues. *International Journal of Medical Informatics* 78:141–60.
35. Marchal S., Xiuyan J., State R., and Engel T. (2014). "A big data architecture for large scale security monitoring." In: *Big Data (BigDataCongress)*, Anchorage, AK, 56–63.
36. Hill K. (2010). *How Target Figured Out a Teen Girl was Pregnant Before Her Father Did*. Forbes, Inc.
37. Li N., et al.(2007). "t-Closeness: privacy beyond k-anonymity and L-diversity." In: *Data Engineering (ICDE) IEEE 23rd International Conference*. Abouelmehdi et al. (2018). *Journal of Big Data* 5:1, 18 of 18.
38. Ton A. and Saravanan M. "Ericsson research." http://www.ericsson.com/research-blog/-data-knowledge/ big-data-privacy-preservation/2015.
39. Samarati P. (2001). "Protecting respondent's privacy in microdata release." *IEEE Transactions on Knowledge and Data Engineering* 13(6):1010–27.
40. Machanavajjhala A., Gehrke J., Kifer D., and Venkitasubramaniam M. (2006). "L-diversity: Privacy beyond k-anonymity." In: *Proceedings 22nd International Conference Data Engineering (ICDE)*, 24.
41. Samarati P. and Sweeney L.(1998). *Protecting Privacy When Disclosing Information: k-Anonymity and Its Enforcement through Generalization and Suppression*. Technical Report SRI-CSL-98-04, SRI Computer Science Laboratory.

42. Data Protection Laws of the World. 2017 LA Piper. http://www.dlapiperdataprotection. com.
43. Prakash K.B. and DoraiRangaswamy M.A. (2019). "Content extraction studies for multilingual unstructured web documents."Doi:10.1007/978-3-319-74808-5_58. Retrieved from www.scopus.com.
44. Kolla B.P. and Raman A.R. (2019). "Data engineered content extraction studies for Indian web pages."Doi:10.1007/978–981-10–8055-5_45. Retrieved from www.scopus. com.
45. Sakhare N.N. and SagarImambi, S. (2019). "Performance analysis of regression based machine learning techniques for prediction of stock market movement." *International Journal of Recent Technology and Engineering* 7(6): 655–62.
46. Banchhor C. and Srinivasu N. (2019). "Holoentropy based correlative Naive Bayes classifier and MapReduce model for classifying the big data." *Evolutionary Intelligence.* doi: 10.1007/s12065-019-00276-9
47. Vidyullatha P., VenkateswaraRao P., and Sucharita, V. (2018). "Study on potential of big visual data analytics in construction arena." *International Journal of Engineering and Technology(UAE)*7: 652–56.
48. Varish, N., Pal A.K., Hassan R., Hasan M.K., Khan A., Parveen N., Banerjee D., Pellakuri V., Haqis A.U., and Memon I. (2020). "Image retrieval scheme using quantized bins of color image components and adaptive tetrolet transform." *IEEE Access*8.
49. Alange N. and Mathur A. (2019). "Small sized file storage problems in hadoop distributed file system." *Proceedings of the 2nd International Conference on Smart Systems and Inventive Technology*, ICSSIT 2019.
50. Ismail M., Vardhan V.H., Mounika, V.A., and Padmini K.S. (2019, June). "An effective heart disease prediction method using artificial neural network." *International Journal of Innovative Technology and Exploring Engineering* 8 (8): 1529–1532.
51. Prasad K.R., Mohammed M., and Noorullah R.M. (2019). "Visual topic models for healthcare data clustering." *Evolutionary Intelligence*: 1–18.
52. Bisoyi S.S., Mishra P., and Mishra S. (2018). "Extracting global exceptional frequent pattern from distributed data sources: A map reduce approach." *Journal of Advanced Research in Dynamical and Control Systems.*
53. Mehrotra S., Kohli S., and Sharan A. (2019). "An intelligent clustering approach for improving search result of a website." *International Journal of Advanced Intelligence Paradigms*, 12(3–4): 295–304.
54. Parveen N. (2019, December). "Disciplinary control system based on naïve bayes classification technique using Jsp servlets." *International Journal of Engineering and Advanced Technology (IJEAT)*, ISSN: 2249–8958, 9(2): 2251–56, Doi: 10.35940/ijeat. B3410.129219.
55. Kaura P., Sharmab M., and Mittal M. (2018). "Big data and machine learning based secure healthcare framework." *International Conference on Computational Intelligence and Data Science (ICCIDS 2018) Procedia Computer Science* 132: 1049–59.
56. Abouelmehdi K., Beni-Hessane A., and Khalouf H. (2018). "Big healthcare data: Preserving security and privacy". *Journal of Big Data*5:11.

5 Big Data Handling for Smart Healthcare System

A Brief Review and Future Directions

Arnaja Banerjee, Yashonidhi Srivastava, and Souvik Ganguli
Thapar Institute of Engineering and Technology

CONTENTS

5.1 INTRODUCTION

In today's world, with a huge population of different kinds of people, different healthcare issues arise. Therefore, the smart healthcare system is a very important factor in building a smart city. If the system can support and manage all the data and records, only then it can be called a smart healthcare system. In recent years, there has been a very large increase in the modernization of healthcare techniques and equipment. Today, devices can provide direct access to the patient's past health record, allowing medical staff to monitor patients and consult with them remotely. All these actions have resulted in the generation of a large amount of data, and it is very important to keep track of these data. These data can be used to detect the diseases of a patient based on symptoms and by comparing the trends of health in recent years and previous medical history. Hence, storing, sharing, and processing these large and complex data efficiently have become extremely important, but it can be very challenging and inefficient at the same time. This is the reason Big Data analytics and intelligent techniques are used to manage these data in a very effective way [1].

Proper management of data helps in research and discovery of further different ways of advanced treatments, expanding the diagnostic sector of smart healthcare. The technologies of Big Data analytics, like business intelligence, database, and cloud computing, have made storage and management work easier. The record of a patient stored by a hospital regarding health history and personal information is also accessed by the insurance companies and can be accessed by other hospitals. This ability to share data can be achieved through cloud computing. In the cloud system, many devices are connected to a single storage platform and can share data. This has automatically increased the accessibility of data. One of the important features of the cloud is that data sharing can be practiced on different levels. Data can be shared among many organizations or it can remain limited to a single organization. This level of sharing of information has increased the security of data. Database and business intelligence have helped in storing data in an ordered and logical way. Business intelligence includes many useful features like online analytical processing, reporting analytics, data mining, and complex event processing. All these features have made it possible for us to manage the zettabytes of data [2]. Internet of things (IoT) is yet another field that has contributed to smart healthcare. IoT devices are used to check blood pressure and control oxygen tanks and other devices (like the controlling of monitoring devices and automatic nebulizers) and have improved the quality of healthcare [3].

Although the system is improving day by day, there are still many challenges that are yet to be addressed like an update of data because replacing old data with new correct information is equally important and essential. These problems can be resolved in the future when the healthcare sector gets completely automated by applying Big Data principles into the concepts of IoT, machine language, and deep computing [1,2].

The rest of the chapter is structured in the following manner. Section 5.2 discusses the different artificial intelligence (AI) techniques in the healthcare system. Also, some selective diseases are taken up in this section to deliberate the role of Machine Learning (ML) techniques to support the healthcare sector. In Section 5.3, emphasis is placed on the IoT-enabled healthcare units and the importance of various sensors being employed as IoT devices, while Section 5.4 deals with the storage and security issues of the private medical data of the patients. Section 5.5 narrates the usage of cloud computing in Big Data storage in the context of smart healthcare systems. Section 5.6 concludes the chapter showing some future directions of research in the years to come.

5.2 ARTIFICIAL INTELLIGENCE IN THE HEALTHCARE SECTOR

In today's world of digitalization, healthcare systems have rapidly developed and have seen a growth in the interest of vast amounts of data that are being generated by computers. To manage and handle this data in a smart and efficient manner, the use of machine language and AI has increased over the years. AI can treat, cure, and diagnose by interacting with the data set of patients. AI has also helped in managing severe diseases, aided in drug research, and helped in disease prediction.

5.2.1 Deep Learning in Smart Healthcare Applications

ML is part of AI. ML is a series of algorithms that is useful to develop programs. A broader concept of ML is deep learning, and it helps to put abstraction in the input data so that the data becomes easier to grasp. In the recent past, there has been extensive research and a great deal of work done in the field of healthcare and medicine and, thus, a huge amount of data was generated. Deep learning provides an abstraction to this generated data and makes it simpler and more efficient to understand the information [4].

5.2.2 Machine Learning in Smart Healthcare

These days, computer software has become smarter and easier to handle, which has helped in speeding up the rate of research. ML is one such software that has been utilized in the research domain. ML together with AI has helped in addressing some very important problems in the healthcare system related to human genetics and drug development [5]. ML can process huge amounts of data and filter out important information from it. Regression analysis, clustering analysis, decision tree, and Bayesian network are some of the common ML methods [6].

5.2.3 APPLICATION OF MACHINE LEARNING IN THE FIELDS OF HEALTHCARE

The healthcare system is becoming more efficient and refined with time; at this point, it has spread to the fields of healthcare that were once considered to just be the human expert territory [7]. It has been very important to store data of patients for the smooth working of the healthcare system. Presently, data of patients are being kept as her, which stands for Electronic Health Record. EHR is a digital platform where data of patients, like electrocardiograms, videos, and images, are stored in an organized manner. In earlier days, the data were stored in handwritten documents. With the help of ML software, like the OCR, these documents were successfully converted to a digital form. These abilities have helped us to increase the rate of development of the smart healthcare system [8]. The application of ML through the digital platform EHR is shown in Figure 5.1.

Applications of ML are now applicable to every stage of healthcare. ML is employed in patient data collection, treating them, examining and analysing the data, and suggesting a possible cure for them. In modern medicine, the use of AI has

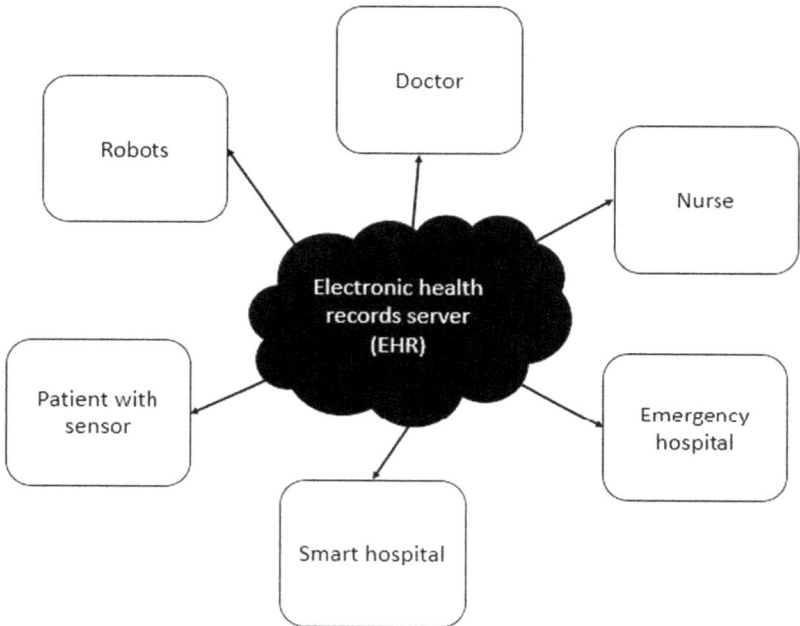

FIGURE 5.1 Application of ML using the EHR server.

increased a lot as medical information is digitalized and a huge amount of volume of tools and robotics have gotten associated with it [5]. There are many applications of ML. AI in the field of cardiology, oncology, and hematology has progressed a lot in recent years, as discussed in the next subsection.

5.2.3.1 Cardiology

Cardiology is the field of medicine that deals with the heart. ML and AI became a popular choice as these algorithms can be improvised for the real work. These algorithms are made without assuming data; thus, it gives more accurate solutions. Therefore, cardiovascular medicine can benefit from using AI and ML [9]. Since there is a huge amount of detail and vast records of the patients in the domain of cardiology, the cardiologists use ML to compare the data and draw an outcome. Figure 5.2 depicts the functions of AI in cardiovascular medicine.

5.2.3.2 Oncology

Oncology is a sector that deals with cancer treatment, diagnosis, research, and pre-vention. The data collected after a lot of research is now fed into the EHR of a patient and deep learning helps to understand the trend and diagnose the underlying diseases of a patient. The symptoms are compared with the predictive model [10].

5.2.3.3 Hematology

Hematology is the study of blood and blood diseases. Whenever our health is affected, the main diagnostic method in pathology will be based on blood and, therefore, we can say that it is one of the most important body fluids. In general, white blood cell is used for the treatment of hematological diseases. In hematology, ML has helped in the diagnosis and treatment of diseases like leukemia, lymphoma, etc. [11].

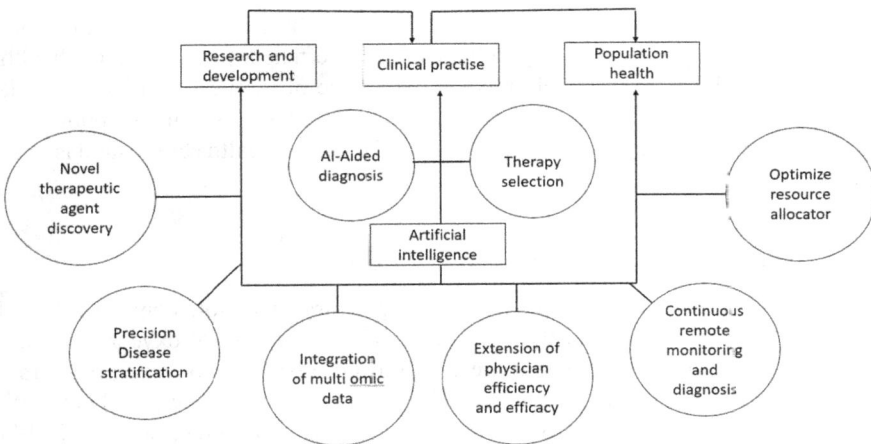

FIGURE 5.2 Role of AI in cardiovascular medicine.

5.2.3.4 Machine Learning and Patients in the intensive care unit

One of the most common reasons for the death of patients is due to sepsis in the intensive care unit. AI can help to overcome this factor. Artificial Intelligence Sepsis expert algorithms can be used for its determination at an early stage so that precautions can be taken accordingly. According to one of the studies, using the knowledge of IoT and ML algorithms, many diseases (one of them being sepsis) can be predicted at a very early stage by comparing the symptoms; thus, they can be treated accordingly [12].

5.3 INTERNET OF THINGS AND THE HEALTHCARE SYSTEM

IoT is a group of computers and machines that are inter-related and possess the ability to share data without any human interaction. IoT uses technologies such as radiofrequency identification (RFID) and wireless sensor network (WSN) to deliver results. Hardware, middleware, and presentation are the three constituents of a system functioning on IoT, and these are detailed below:

- *Hardware*: These are made up of devices like a communication device and other devices.
- *Middleware*: These are the processing and storage parts of these devices.
- *Presentation*: It is the tool used for interpretation of the data processed by the middleware [13].

There are many applications of IoT like location sensing and sharing of location information, mobile asset tracking, fleet management, environment sensing, appliance control, disaster recovery, secure communication, industry use, and smart healthcare.

5.3.1 IoT in the Field of Smart Healthcare

Modernization is occurring at a very fast rate in the realm of smart healthcare. Novel devices and equipment are helping in increasing the efficiency of the present healthcare system. IoT has been largely recognized as one of the best solutions to make smart healthcare systems more effective and efficient. There are many applications of IoT in the field of smart healthcare [14]. An IoT-enabled health care system is, thus, provided in Figure 5.3.

5.3.2 Wearable Sensors and Central Nodes

Physiological conditions can be measured using wearable sensors controlled by IoT devices. These sensors can easily measure blood pressure, blood oxygen level, and the other important conditions that are essential for a healthy body, i.e., pulse rate, body temperature, and respiratory rate. These wearables have integrated computer sensors that can track the health records and sync them with other devices like the central nodes. The central nodes receive information and forward the information to some external device after processing it.

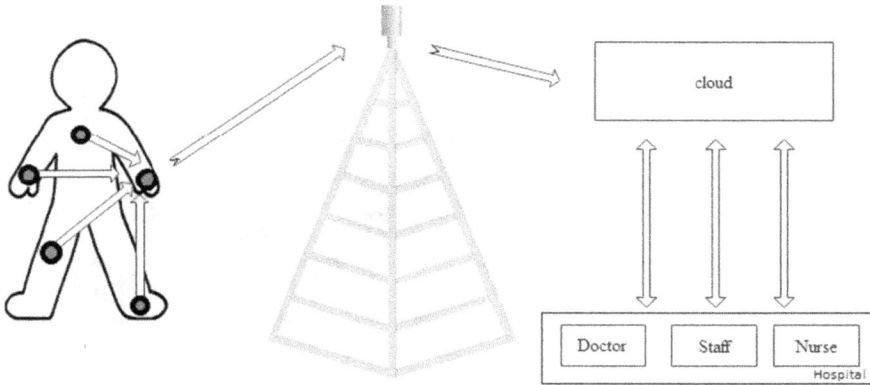

FIGURE 5.3   An IoT-based healthcare system.

5.3.2.1   Pulse Sensors

The pulse sensor is one of the most common wearables used around us. Pulse rate can be measured from different body parts like chest, wrist, earlobes, etc. It can be used to prevent some very serious conditions like cardiac arrest. An extensive amount of research has been conducted to enhance the sensing efficiency of these pulse sensors. These are available in different forms like wristwatches and fingertip sensors. The most accurate type of sensor is the PPG or photoplethysmography sensor. It uses a light-emitting diode (LED) to detect the pulse rate.

Efficient methods for sensing pulses have been researched in the recent past. As a result, several varieties of sensors were acquired, applied, and analyzed lately in works that discuss pressure, PPG, ultrasonic, and RF sensors. Light is transmitted to the artery with a photodiode and is reflected after blood has absorbed the required amount of light. Finally, the pulse rate can be recorded after measuring the amount of change in light. Figure 5.4 shows the working of the PPG. As PPG has turned out to be an efficient way to measure pulse rate, it has found utility in the field of healthcare [14].

5.3.2.2   Respiratory Rate Sensor

The rate of respiration can be defined as the number of breaths taken by a patient in a minute. It can be used to detect many conditions like tuberculosis, hyperventilation, lung cancer, asthma, etc. Different types of sensors like a nasal sensor, ECG, EDR, fiber optic, and stretch sensors have been developed throughout the years to improve the process to measure respiratory rate. For example, to detect asthma, fiber optic sensors can be used as one of the most common symptoms of asthma is wheezing and optical sensors are sensitive enough to detect the vibrations while wheezing. There are many different types of sensors available, but the main factor to choose the suitable one is the wearability, and the best sensor regarding this point is the stretch sensor. These are the sensors in which the property changes when the tensile force changes; for example, during contraction and expansion of the diaphragm during inhalation.

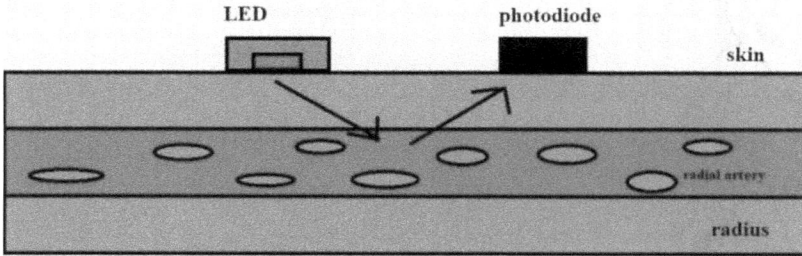

FIGURE 5.4   The working of PPG.

5.3.2.3   Body Temperature Sensor

Body temperature sensors are the diagnostic tools that could be utilized to measure the temperature of a person and can be used for the determination of severe conditions like heat stroke and fevers. Many types of sensors have been developed recently like a negative temperature coefficient (NTC)-type temperature sensor and positive-temperature-coefficient (PTC) sensors. These types of sensors work best when they are put closer to the human body.

5.3.2.4   Blood Pressure Sensor

The pressure exerted on the walls of arteries as a result of the blood that flows through is called blood pressure. It is important to maintain a certain range of blood pressure as high values can result in hypertension that can cause a heart attack. The integration of blood pressure with the IoT devices has saved a lot of lives. Pulse transit time can be measured by using sensors like electrograms on a wrist, chest, or ear. The pulse transit time depends on several factors like the stiffness of arteries and the density of blood. Therefore, it is not exactly a suitable method to measure blood pressure. As no system has been accurately developed to measure blood pressure, there happens to be tremendous scope for further work and research.

5.3.2.5   Pulse Oximeter

The function of a pulse oximeter is to measure the amount of oxygen level in the blood. Oxygen level in blood is an indicator of the functioning of the respiratory system. Diseases like hypoxia can be detected using oximeters. These wearables are generally worn on fingers and are connected to the hospital computers. A pulse oximeter uses PG signals to detect the oxygen level. One red and one infrared lights are transmitted through the skin and reflected after hemoglobin has absorbed the required amount of light. The photodiodes are used to measure the amount of not used light, and the difference between the emitted and received light gives us the

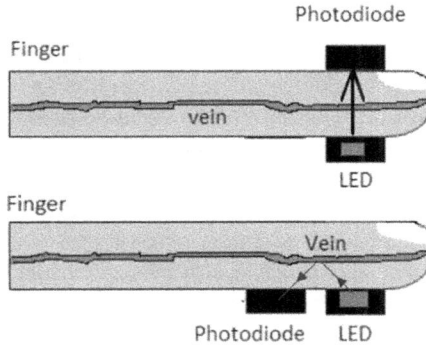

FIGURE 5.5 The working of an IoT-enabled pulse oximeter.

amount of oxygen content in the blood. The working of an IoT-based pulse oximeter is depicted with the help of a diagram in Figure 5.5.

5.3.3 Communication Standards in IoT-Based Healthcare System

Communication technology is the technology to interact between the devices. There are two types of communication standards prevailing in the literature, namely,

- Short-range communication
- Long-range communication [14].

5.3.3.1 Short-Range Communication

Short-range communication is the communication that is used to interact between the IoT devices. This type of communication is used to communicate among the sensor nodes. The most commonly used short-range communications in the field of IoT are the Zigbee and Bluetooth low energy (BLE).

5.3.3.1.1 Bluetooth Low Energy (BLE)

BLE is an energy-efficient method used to connect small devices of IoT wearables with smart processing devices like computers. Star topography is an example of a networking system that is used in BLE to communicate between sensors and the central node.

BLE has a range of up to 150 m and a data rate of 1 Mbps in an ideal case, which is more than sufficient for use in healthcare devices. The power consumption of BLE devices is also very low; so, it is an ideal process to use in the healthcare sector. Security in BLE has been increased in multiple ways. A numeric comparison method termed as LE secure connections as well as an algorithm called Elliptical Curve Hellman–Diffie provides multiple levels of security layers. Encryption is another feature of BLE that is used to protect data from hackers. Earlier diagnostic tools for Alzheimer's and blood pressure monitoring systems used classic Bluetooth devices. Therefore, BLE is one of the most suitable options to use in the smart healthcare system.

5.3.3.1.2   Zigbee

Zigbee is a type of short-range communication method that can be used to interact with one central and many sensing nodes. It generally uses mesh networking to connect the nodes. Zigbee has a range of up to 30m and a data rate of 20 Mbps. The power consumption of Zigbee is only 1 mW. In the healthcare system, for the wearables, the XBee 1 mW happens to be an efficient choice as its requirements are small in the range and this facilitates on-body communications. Zigbee has already been used in medical fields like biomedical and medical equipment designing for Alzheimer detection. There are many security features available in ZigBee. It provides facilities like encryption and other kinds of security keys. Although there are security measures available, still the security measures are easy to exploit comparatively. Therefore, the application of Zigbee in smart healthcare should be done with extra security mechanisms. Therefore, we can say that due to the security bridges, BLE is preferred for wearables over Zigbee.

5.3.3.2   Long-Range Communication

Long-range communication is communication that is used to communicate with computers several kilometers away. These are longer than Bluetooth or Wi-Fi connections. Sigfox, LoRa WAN, and NB IoT are the common types of long-range communication. These types of communication are suitable for a huge amount of hospital-related functions like receiving an emergency call, getting health updates, and maintaining health records. Hence, the transmission of data to the cloud network could be carried out in the most optimized way by this method.

Lo-Ra provides a long-range and low-power communication interface. It covers a bandwidth of 125 kHz. LoRaWAN uses a star topology and communicates only when needed, such as after an event. LoRaWAN ensures the passing of several messages to a particular network in the same duration as it has a very big capacity to store node networks. Sigfox has more capacity than LoRaWAN, but it can still be used for long connections. The security provided by LoRaWAN is of standard quality and encoding helps to eliminate the risk of middleman attacks, but if illicitly a key was found, then it could be used to decode all the message. Therefore, LoRaWAN can be applied to smart healthcare with extra security measures. NB IoT provides long-range communications with low power and operates in LTE. Interference risk in NB IoT reduces drastically if operating on licensed bands, although the huge cost is one potential disadvantage. Another concern is the reusability factor of the LTE. It is observed that the energy consumption allows minimum interaction by the system wearer and is comparatively low. Therefore, it can be concluded that it is suitable and applicable for the smart healthcare system as it is secure, energetic, and compatible with several devices [14].

5.3.4   CHALLENGES FACED BY IoT

There are certain challenges faced during the implementation of IoT in the field of health care, which are described below.

5.3.4.1   Challenge in Architecture

As IoT involves many numbers of smart interconnected smart devices, the interaction between these devices should be kept open and flexible. Rigid end-to-end solutions should be avoided. All the devices carry different volumes (the architecture cannot be the same) of data, which results in different networking patterns. Therefore, the architecture should try to be similar and bring uniformity in reference, much needed in IoT devices.

5.3.4.2 Hardware and Technical Challenge

The devices needed for IoT need to function properly but, at the same time, it should have a low power consumption. The high cost of these parts is another challenge that can be faced by the hardware developers. In IoT, there are different devices for which there are different networking architecture that makes things more complex. These devices have different applications, and each has different security problems and their respective solutions. These complexities may increase problems that can create barriers or problems [15].

5.3.4.3 Security Challenges

Information on users and the protection of their privacy is a very important issue in IoT. As the IoT devices are connected wirelessly, there can be severe security bridges and attackers may pretend to be the legitimate user and misuse data, and this can lead to system damage. This is the reason just a simple username and authentication password cannot give the required amount of safety needed. Therefore, a session key is required for a safe connection. After authentication, the most important security consideration is the middleman attack. To tackle the problem of spoofing, Global Positioning System (GPS) can be embedded in these IoT devices to get the location of a user. Security can be increased by creating different algorithms within the sensors and devices [16].

5.4 STORAGE AND PRIVACY OF MEDICAL DATA

5.4.1 IoT-Based Storage Systems

Intending to protect the patient's healthcare data, a smart, IoT-based Big Data storage system was introduced. It ended up saving storage space in the Big Data storage system and also provided relevant information for regular and emergency cases. The IoT network generated the files that get relocated to the storage. The data could also be shared among healthcare workers, making it easier to access the entire record of the patient when needed. The proposed system has provided the solution to the lack of one's medical history or data in the case of an emergency. The healthcare worker assigned to deal with first-aid remains unaware of the patient's medical history as they were not allowed to access their past medical records. A two-fold system with access control was, hence, discussed that was self-adaptive. For regular usage, the data could be accessed using keys. The system was made password-protected to be accessed in emergencies. Aimed for the removal of duplicate files, deduplication was also discussed. Its feature namely accessibility to every authorized individual had also been discoursed. The proposed system proved secure, efficient, and trustworthy [17].

MetaFog-Redirection (MF-R) and Grouping and Choosing (GC) architectures were introduced based on IoT to store and successfully process medical records stored as Big Data. The architecture proposed provided better results than cloud-based computing as it provided a more efficient computing platform. The system processes Big Data in a distributed order to store and assess a large amount of sensor data. Logistic regressions were implemented using Apache Mahout. It was trained to utilize the past medical records of the patients along with the data accumulated from sensors along with wearable medical devices. Current sensor data including BP monitoring, heart rate, and blood sugar level also helped in deciphering the patient's heart status [18].

5.4.2 Blockchain Technology for Data Security

Wearable technologies encompassing sensors and other devices have gained a lot of appreciation and acceptance in the recent past. It has improved the healthcare monitoring process exponentially but has also increased the probable privacy risk to the patient. The security concerns come into consideration during data transfer and transactions. Another concern is the delayed medical help that may result in the loss of life. A novel system utilizing blockchain models was discussed due to their distributed nature as well as other features concerning data security. The system introduced talked of blockchain structures that included the concept of a shared data stack from wearable, embedded, stationary, as well as health monitoring devices, which kept track of all past transactions. The process of data collection began from the patient being monitored to the use of associated healthcare device. The information hence obtained would get processed utilizing blockchain technology and then further be made available for the healthcare worker. The features in regard to the security of data have their roots in cryptographic primitives. It has made the transaction and storage of data safe and anonymous [19]. The process of blockchain technology is well explained in Figure 5.6.

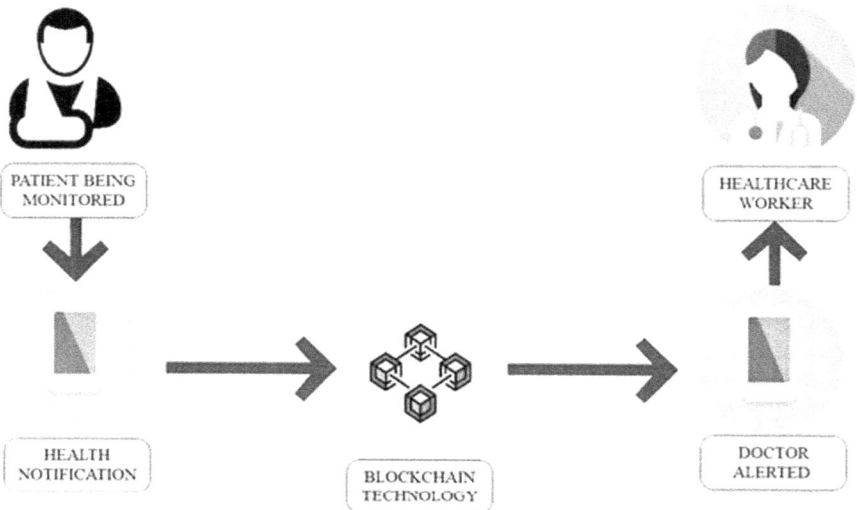

FIGURE 5.6   Blockchain process for data security.

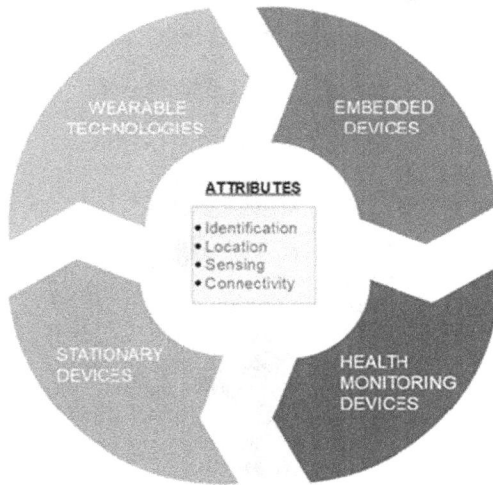

FIGURE 5.7 Different application areas of the blockchain framework.

Some of the attributes of blockchain technology include the identification, location, sensing, and connectivity. The other fields of applications like wearable technologies, embedded devices, stationery, and health monitoring devices are also shown in Figure 5.7.

The limited capacity of IoT devices, as well as limitations of the computing power, had brought about the need for a new system. Traditionally, data was stored in a centralized manner, probably in the hospital's database, which could result in a privacy breach. A privacy-preserving proposal based on blockchain technology was hence introduced where data was kept encrypted. Medical conflicts also decreased remarkably as the diagnosis conducted could no longer be altered. The system, henceforth, found its applicability in smart healthcare systems [20].

The sharing of medical data in a safe scenario has been necessary to improve healthcare services. The patient at times does not have the authority to protect their data. This has resulted in privacy breaches and has also compromised valuable data over time. An app termed as the Healthcare Data Gateway (HGD) was proposed to provide the patient with the platform to own, control, and share their records. The app was based on blockchain and suggested efficient methods to prevent violation of privacy while keeping the patient's data secure. The data retrieved from the data usage layer involving physicians, companies, government, and researchers would be passed onto the management layer. After being processed through the management layer, the records would reach the storage layer where it would be assessed utilizing blockchain cloud technology. The introduction of Indicator Centric Schema (ICS) made it practical to efficiently arrange the data and also suggested a solution to enable third parties to access the data without the threat of a possible breach [21]. The basic operation of the Healthcare Data Gateway (HGD) system is thus provided in Figure 5.8 to make the applications more understandable.

The storage of healthcare data has observed a shift to cloud storage in the recent past. Cloud storage provides the comfort of accessing the complete medical history of

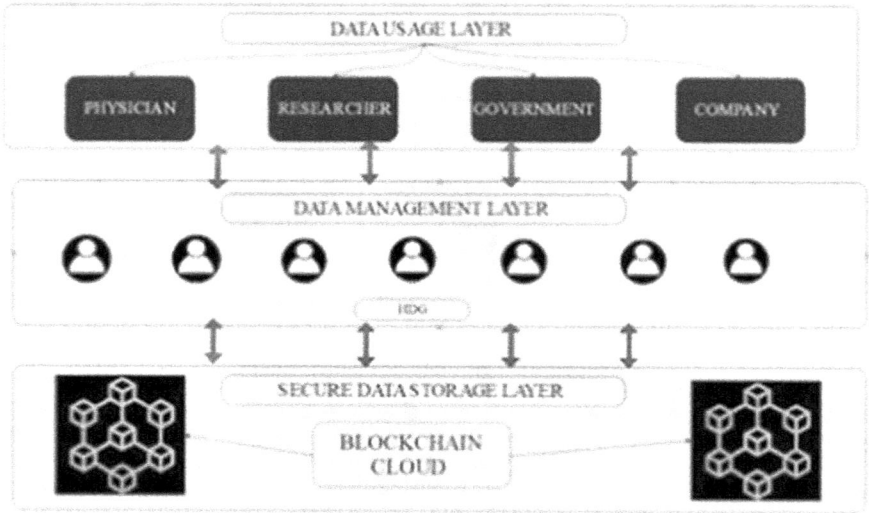

FIGURE 5.8 Structure of the HDG system.

the person efficiently. It was also economic, practical, and time-saving. Nonetheless, the security concerns that arose from storing the records in cloud-based storage could not be ignored as it put a lot at risk. To protect the healthcare data stored in cloud systems, a blockchain technique was proposed. The potential challenges regarding its applicability and implementation have also been discussed with emphasis on the easy accessibility of data without putting the patient's privacy at risk [22].

5.4.3 OTHER ASSOCIATED TECHNOLOGIES

Big Data comprises a complex structure that is difficult to store, analyze, and apply. The complications result in security concerns that were discussed, emphasizing the differences between privacy and security as well as its requirement in the domain of Big Data. The aspect regarding privacy was dealt with taking into consideration the existing technologies of HybfEx, k-anonymity, L-diversity, and their applications. The systems brought into use for the protection of data at different levels, including data generation along with data storage, and data processing have also been discussed. We also further reviewed the privacy systems and highlighted the challenges in the process. Techniques to improve privacy were also presented, including the concepts of hiding a needle in a haystack, plots of anonymization, differential privacy, as well as the process of making data streams anonymous. A comparative study between the existing technologies to preserve and protect data was also expounded on [23].

5.4.4 WIRELESS BODY AREA NETWORKS (WBANS)

Wireless body area networks (WBANs) have gained appreciation in their recent past after collaboration with cloud computing (CC). The amalgamation of the two (S-CI) has enabled real-time monitoring along with assessment of probable threats to the patient's security. The possible techniques to make the (S-CI) system more protected have been discussed, including multibiometrics, key generation, hash functions, and chaotic maps. A framework had also been discussed for physiological parameters as well as the security and privacy of the individual. WBANs have been extensively utilized for the monitoring of patients suffering from diseases like Alzheimer's, epilepsy, and Parkinson's. The monitoring facilitates the generation of Big Data hence putting the privacy of the patient at risk. Therefore, the need to make the system secure along with its importance has been mentioned [24].

5.4.5 MODELS FOR ASSESSMENT AND SECURITY THREATS

Invasion into healthcare data had occurred over multiple instances where the password had been guessed, resulting in the security failure of IoT devices. Vulnerabilities of passwords and methods to strengthen them have been introduced, keeping in mind the patient's personal information along with their medical history [25].

Medical data remain at high risk of exposure as it goes through multiple processes of collection, transmission, and storage, followed by the provision of data to healthcare workers. The Big Data turned out to have 4 primary and 35 secondary risk indicators. The proposal of a fuzzy evaluation system was discussed to assess the risk in Big Data security disclosures in urban computing. It was found that the probable security threat during the storage and sharing of data was significantly higher than the risk while collecting or transmitting the Big Data. The effectiveness of the evaluation system had also been discussed [26]. The methods adopted for data collection, their transmission, and, finally, their secured storage is thus narrated with the help of Figure 5.9.

5.4.6 ENABLING SECURITY USING FOG COMPUTING

Telemedicine has been a growing field that has resulted in the accumulation of Big Data over time. To diagnose and monitor the patient, the data obtained from techniques of the EMR, including X rays, ultrasounds, and CT scans, were added to the database. Thus, its need to be stored in the healthcare cloud has been discussed. The theft attacks to access healthcare data in the cloud and its dangers also have been discussed. An authenticated key protocol has been proposed to keep the healthcare data secure, further facilitated utilizing a fog computing technique. The implementation of a decoy technique has also been deliberated, enabling secured access to the Big Data accumulated [27].

5.5 SMART HEALTHCARE: BIG DATA AND CLOUD STORAGE

5.5.1 SAFE DATA STORAGE

The professionals and the policymakers in the field of healthcare have been highly cautious in regard to patient safety in the recent past. The expansion of the industry had resulted in an abundance of information, termed as Big Data. Therefore, handling

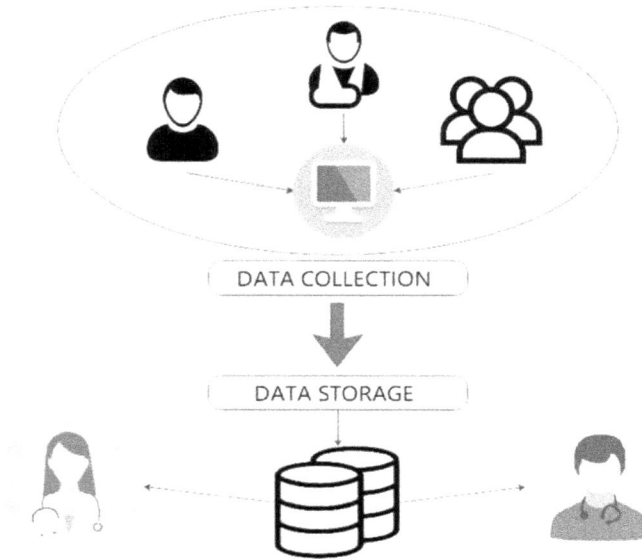

FIGURE 5.9   Process of Big Data collection, transmission, and storage.

and managing Big Data became a matter of concern. The need was quenched by cloud computing. Several systems have been developed utilizing cloud computing for the management of Big Data. This chapter discusses those mechanisms along with their benefits and drawbacks. The mechanism utilized has also been thoroughly analyzed, and the challenges in the system have been highlighted. Better techniques for the processing of Big Data incorporating the technology of cloud computing are also explained [28].

The challenges of Big Data processing systems and their inability to keep up with great advancements in the amount of digital healthcare data have been highlighted. Analytical solutions to bring about efficient decision-making in regard to patient care had been proposed. Its impacts on the improvement of the quality of life as well as the optimization of services have also been discussed. The chapter also deliberates upon the dispute of having to ensure the confidentiality of Big Data while making it available and extractable on such a large scale. The ability to incorporate cloud computing to facilitate the sharing of Big Data with medical workers while keeping the data safe and manageable has also been discussed along with the need for more such trusty platforms for the optimal storage of Big Data [29].

5.5.2   MOBILE DEVICES AND CLOUD COMPUTING

With the increasing usage of mobile devices, a collaboration between mobile and cloud computing was proposed. The integration expands the abilities of a mobile device in terms of memory, CPU power, and the life of the battery. The limitations of the device could also be eradicated by the proposed amalgamation. The objective

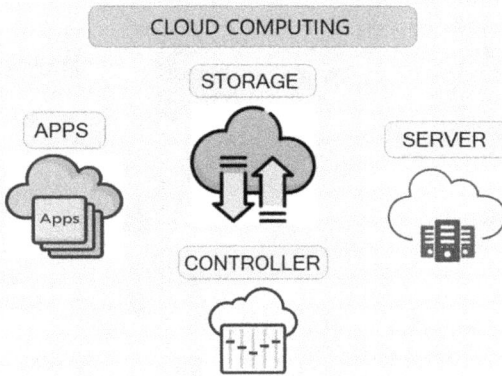

FIGURE 5.10 Mobile with cloud computing features for Big Data applications.

FIGURE 5.11 System of networked healthcare.

of mobile cloud computing for the extraction of Big Data possessing mass, veracity, velocity, and types has been discussed. A cloudlet-based mobile computing system has been introduced to significantly improve the processing of Big Data. The applications and techniques applied for the proposed framework have been explained with emphasis on the outlook of the design as well as the system of networked healthcare [30]. The usage of cloud computing in the realm of Big Data handling with the help of smartphone apps is shown in Figure 5.10.

The networked health care model having the capability of cloud facility to store Big Data is provided with the help of a diagram representation in Figure 5.11.

Mobile devices, over time, have developed into providers of tools that can store, collect, manage, and analyze a vast amount of data. This had a significant impact on the expansion of the realm of Big Data, including semi-structured, structured, and unstructured data produced by the healthcare industry. Other characteristics of data, including its continuous or discontinuous nature, time, quality, and inconsistencies, were also taken under consideration. The analytics segment concerning Big Data has gradually been gaining importance to improve the care delivery process and, in the process, exploration of diseases. The chapter has discussed the inclusion of large storage systems facilitated by the incorporation of Big Data with cloud computing. A framework for Big Data healthcare have also been proposed, and its applications

FIGURE 5.12 Different characteristics of the Big Data.

have been discussed [31]. The various features of the Big Data are represented with the aid of a network diagram (Figure 5.12).

5.5.3 5G SMART SYSTEMS IN HEALTHCARE UNITS

Advancements in Big Data technologies including wireless networks like 5G networks, analytics of Big Data, IoT, and AI have brought about the development of a smart diabetes monitoring system. To reduce the suffering of diabetes patients, the system focused on the diagnosis as well as the treatment of diabetes. The proposed system has classified the applications into Diabetes 1.0 and Diabetes 2.0. The paper also discussed sustainability as well as the affordability of the proposed system. Smart 5G systems were introduced in collaboration with technologies like ML and Wearable 2.0 in regard to the Big Data generated from the patient's records. The system of smart diabetes testbeds had also been discussed with the use of smart clothing and smartphones associated with Big Data clouds. The experimental results determining the ability of the system for the diagnosis and treatment of the patients had also been included [32].

5.5.4 HEALTHCARE CYBER-PHYSICAL SYSTEMS

The introduction of new technologies had made the healthcare process more advanced but also posed the issue of the management of information from various fields, i.e., Big Data. The challenge of data being created quickly and having different characteristics and then being stored in different formats was discussed, and the issue was termed as the Big Data problem. To bring about convenient services for the patients, a cyber-physical system was proposed. The system was termed as a healthcare cyber-physical system and was targeted for the improvement of patient-centric healthcare services. The proposed system has a layer functioning to accumulate data followed by data management and a service layer. The emergence of better managerial approaches and processes were brought along by the segregation of data into three layers. Remarks were also made on the utilization of Big Data and cloud computing technologies and its objectives to make more efficient healthcare systems [33].

5.5.5 MULTI-STRUCTURED PATIENT DATA

Clinical reports, wearable devices, body sensors, and doctor's references have led to large amounts of multi-structured patient reports and information. The data keeps getting added up for the future even if the health report was still at initial stages. A cloud-based, Big Data collection system was proposed to bring about the analysis and management of structured as well as the unstructured data generated in healthcare systems. A model to predict the patient's future health based on their current reports had also been introduced. Simulations in cloud computing had provided trustworthy and efficient results [34].

The introduction of new devices, as well as the advancements in the healthcare sector, had resulted in huge volumes of Big Data. While necessary, organizations had been having trouble analyzing the Big Data. To bring about a system to process and analyze the data to bring about faster and more efficient decision-making, the technology of cloud computing was adopted. To enable scalability and reduce management issues, the Hadoop framework was adopted to trigger the facilitation of large-scale data sets. The collection of data would be facilitated by the healthcare apps on the cloud followed by its storage and analysis. Lately, applications have been floated over the internet in contrast to the traditional methodologies. This demanded the requirement for real-time data to further improve healthcare. The impacts of the management and analysis of data on large-scale cloud platforms were also discussed [35]. The cloud framework for healthcare applications involving Big Data is visualized in Figure 5.13.

The healthcare apps store, visualize, and also help in analyzing a huge amount of medical data through smartphones, laptops, and desktops, as seen in Figure 5.14.

5.5.6 WEARABLE HEALTHCARE SYSTEMS

With the advancement in technologies adopted for the improvement of the healthcare sector, technologies including IoT, cloud computing, and ML have gained an appreciation in the direction of the accumulated Big Data. In cloud-integrated systems, Quality of Experience (QoE) and Quality of Service (QoS) were being applied heavily. A wearable healthcare system was thus proposed, incorporating methodologies to improve QoE and QoS for the future of the healthcare industry. The proposed system consisted of smart clothing, including sensors and electrodes. The present

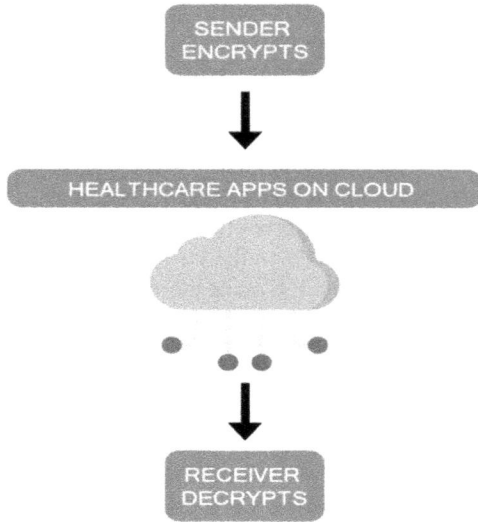

FIGURE 5.13   Cloud framework of Big Data for healthcare applications.

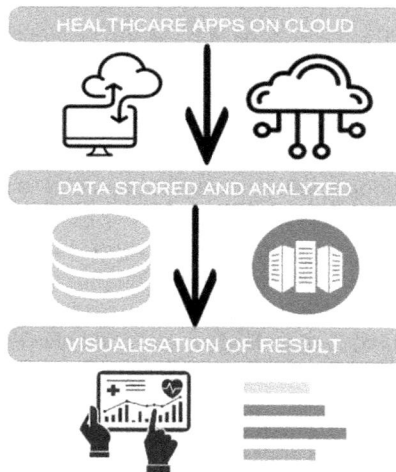

FIGURE 5.14   Apps concerning healthcare systems on the cloud.

condition of the patient is then monitored by utilizing a cloud-based machine intelligence system [36].

5.5.7   DATA INTEGRATION AND SHARING

A framework was introduced for the healthcare information systems (HISs) that provided integration as well as easy availability and sharing of data among healthcare providers and patients. The proposed framework includes concepts of data analytics and mobile cloud computing. The working of the framework involved the collection of EMRs from various organizations. The data would then be stored over a cloud platform, providing fast access to EMRs from several platforms. The framework proposed also utilized measures to improve the security and confidentiality of data [37].

5.6 CONCLUSIONS

The chapter provides a comprehensive overview of the role of Big Data in smart healthcare systems. The chapter addresses the various methods of artificial healthcare intelligence. In this portion, some specific diseases are also used to address the role of ML techniques in supporting the healthcare sector. The chapter also stresses the importance of the healthcare systems enabled by IoT and of the use of different sensors as IoT equipment, while the security and safety issues of the patients' private data are also considered. The utilization of cloud computing in the context of smart healthcare systems in Big Data storage has also been explored in this chapter. With the present pandemic situation due to COVID-19 engulfing the entire world, the smart healthcare system of handling Big Data is the need of the hour. Researchers have already started exploring ways to come up with a solution for its early diagnosis, monitoring, and treatment applying AI- or IoT-based systems. Our consolidation of facts regarding Big Data analysis in smart healthcare systems will further instigate more researchers to work in this area.

REFERENCES

1. Hashem, I. A. T., Chang, V., Anuar, N. B., Adewole, K., Yaqoob, I., Gani, A., ... Chiroma, H. (2016). The role of big data in smart city. *International Journal of Information Management, 36*(5), 748–758.
2. Allam, Z., & Dhunny, Z. A. (2019). On big data, artificial intelligence and smart cities. *Cities, 89*, 80–91.
3. Babar, M., & Arif, F. (2017). Smart urban planning using Big Data analytics to contend with the interoperability in Internet of Things. *Future Generation Computer Systems, 77*, 65–76.
4. Coeckelbergh, M. (2010). Health care, capabilities, and AI assistive technologies. *Ethical theory and moral practice, 13*(2), 181–190.
5. Datta, S., Barua, R., & Das, J. (2019). Application of Artificial Intelligence in modern healthcare system. In *Alginates-Recent Uses of This Natural Polymer*. IntechOpen, London.
6. Chen, M., Li, W., Hao, Y., Qian, Y., & Humar, I. (2018). Edge cognitive computing based smart healthcare system. *Future Generation Computer Systems, 86*, 403–411.
7. Yu, K. H., Beam, A. L., & Kohane, I. S. (2018). Artificial intelligence in healthcare. *Nature Biomedical Engineering, 2*(10), 719–731.
8. Hsiao, C. J., & Hing, E. (2012). *Use and Characteristics of Electronic Health Record Systems among Office-Based Physician Practices, United States, 2001–2012 (No. 111)*. US Department of Health and Human Services, Centers for Disease Control and Prevention, National Center for Health Statistics, Hyattsville, MD.
9. Johnson, K. W., Soto, J. T., Glicksberg, B. S., Shameer, K., Miotto, R., Ali, M., ... Dudley, J. T. (2018). Artificial intelligence in cardiology. *Journal of the American College of Cardiology, 71*(23), 2668–2679.

10. Bibault, J. E., Giraud, P., & Burgun, A. (2016). Big data and machine learning in radiation oncology: state of the art and future prospects. *Cancer letters*, *382*(1), 110–117.
11. Zini, G. (2005). Artificial intelligence in hematology. *Hematology*, *10*(5), 393–400.
12. Nemati, S., Holder, A., Razmi, F., Stanley, M. D., Clifford, G. D., & Buchman, T. G. (2018). An interpretable machine learning model for accurate prediction of sepsis in the ICU. *Critical Care Medicine*, *46*(4), 547.
13. Gubbi, J., Buyya, R., Marusic, S., & Palaniswami, M. (2013). Internet of Things (IoT): A vision, architectural elements, and future directions. *Future Generation Computer Systems*, *29*(7), 1645–1660.
14. Baker, S. B., Xiang, W., & Atkinson, I. (2017). Internet of Things for smart healthcare: Technologies, challenges, and opportunities. *IEEE Access*, *5*, 26521–26544.
15. Chen, S., Xu, H., Liu, D., Hu, B., & Wang, H. (2014). A vision of IoT: Applications, challenges, and opportunities with china perspective. *IEEE Internet of Things Journal*, *1*(4), 349–359.
16. Yeh, K. H. (2016). A secure IoT-based healthcare system with body sensor networks. *IEEE Access*, *4*, 10288–10299.
17. Yang, Y., Zheng, X., Guo, W., Liu, X., & Chang, V. (2019). Privacy-preserving smart IoT-based healthcare big data storage and self-adaptive access control system. *Information Sciences*, *479*, 567–592.
18. Manogaran, G., Lopez, D., Thota, C., Abbas, K. M., Pyne, S., & Sundarasekar, R. (2017). Big data analytics in healthcare Internet of Things. In *Innovative Healthcare Systems for the 21st Century* (pp. 263–284). Springer, Cham.
19. Dwivedi, A. D., Srivastava, G., Dhar, S., & Singh, R. (2019). A decentralized privacy-preserving healthcare blockchain for IoT. *Sensors*, *19*(2), 326.
20. Xu, J., Xue, K., Li, S., Tian, H., Hong, J., Hong, P., & Yu, N. (2019). Healthchain: A blockchain-based privacy preserving scheme for large-scale health data. *IEEE Internet of Things Journal*, *6*(5), 8770–8781.
21. Yue, X., Wang, H., Jin, D., Li, M., & Jiang, W. (2016). Healthcare data gateways: Found healthcare intelligence on blockchain with novel privacy risk control. *Journal of Medical Systems*, *40*(10), 218.
22. Esposito, C., De Santis, A., Tortora, G., Chang, H., & Choo, K. K. R. (2018). Blockchain: A panacea for healthcare cloud-based data security and privacy?.*IEEE Cloud Computing*, *5*(1), 31–37.
23. Jain, P., Gyanchandani, M., & Khare, N. (2016). Big data privacy: A technological perspective and review. *Journal of Big Data*, *3*(1), 25.
24. Masood, I., Wang, Y., Daud, A., Aljohani, N. R., & Dawood, H. (2018). Towards smart healthcare: Patient data privacy and security in sensor-cloud infrastructure. *Wireless Communications and Mobile Computing*, Article ID 2143897, 23 pages.
25. He, D., Ye, R., Chan, S., Guizani, M., & Xu, Y. (2018). Privacy in the Internet of Things for smart healthcare. *IEEE Communications Magazine*, *56*(4), 38–44.
26. Jiang, R., Shi, M., & Zhou, W. (2019). A privacy security risk analysis method for medical big data in urban computing. *IEEE Access*, *7*, 143841–143854.
27. Al Hamid, H. A., Rahman, S. M. M., Hossain, M. S., Almogren, A., & Alamri, A. (2017). A security model for preserving the privacy of medical big data in a healthcare cloud using a fog computing facility with pairing-based cryptography. *IEEE Access*, *5*, 22313–22328.
28. Rajabion, L., Shaltooki, A. A., Taghikhah, M., Ghasemi, A., & Badfar, A. (2019). Healthcare big data processing mechanisms: the role of cloud computing. *International Journal of Information Management*, *49*, 271–289.
29. Nepal, S., Ranjan, R., & Choo, K. K. R. (2015). Trustworthy processing of healthcare big data in hybrid clouds. *IEEE Cloud Computing*, *2*(2), 78–84.

30. Lo'ai, A. T., Mehmood, R., Benkhlifa, E., & Song, H. (2016). Mobile cloud computing model and big data analysis for healthcare applications. *IEEE Access*, *4*, 6171–6180.
31. Alexandru, A., Alexandru, C., Coardos, D., & Tudora, E. (2016). Healthcare, big data and cloud computing. *Management*, *1*, 2.
32. Chen, M., Yang, J., Zhou, J., Hao, Y., Zhang, J., & Youn, C. H. (2018). 5G-smart diabetes: Toward personalized diabetes diagnosis with healthcare big data clouds. *IEEE Communications Magazine*, *56*(4), 16–23.
33. Zhang, Y., Qiu, M., Tsai, C. W., Hassan, M. M., & Alamri, A. (2015). Health-CPS: Healthcare cyber-physical system assisted by cloud and big data. *IEEE Systems Journal*, *11*(1), 88–95.
34. Sahoo, P. K., Mohapatra, S. K., & Wu, S. L. (2016). Analyzing healthcare big data with prediction for future health condition. *IEEE Access*, *4*, 9786–9799.
35. Rallapalli, S., Gondkar, R., & Ketavarapu, U. P. K. (2016). Impact of processing and analyzing healthcare big data on cloud computing environment by implementing hadoop cluster. *Procedia Computer Science*, *85*, 16–22.
36. Chen, M., Ma, Y., Li, Y., Wu, D., Zhang, Y., & Youn, C. H. (2017). Wearable 2.0: Enabling human-cloud integration in next generation healthcare systems. *IEEE Communications Magazine*, *55*(1), 54–61.
37. Youssef, A. E. (2014). A framework for secure healthcare systems based on big data analytics in mobile cloud computing environments. *International Journal of Ambient Systems and Applications*, *2*(2), 1–11.

6 Big Data Analysis for Smart Energy System

An Overview and Future Directions

Tanya Srivastava and Abhimanyu Kumar
Thapar Institute of Engineering and Technology

Swadhin Chakrabarty
Regent Education and Research Foundation

Souvik Ganguli
Thapar Institute of Engineering and Technology

CONTENTS

6.1 INTRODUCTION

An intelligent energy grid promotes both power and data exchange between producers and costumers, which facilitates the management of the delivery of energy in terms of economic quality, security, as well as protection. The nonstop deployment of sensors, wireless networking, and cloud computing technology handles vast volumes of data that are rapidly collected. This system allows energy users and suppliers to become more competitive in the electricity and dynamic energy management (DEM) markets. The main problem with a smart grid is how to make the best of its participation of consumers to reduce electricity costs. Power management, demand-side electricity management, and the incorporation of clean energy plants, inovloves an immense number of data sets. The collection and interpretation of these data provide more in-depth perspectives, which can allow experts to understand/improve power grid service to deliver improved efficiency. The word "Big Data" implies the use of predictive analytics. It is also related to consumer experience or other specialized forms that come from data-derived interest, although sometimes it refers to a particular size of data collection [1,2].

Smart grids are electric networks that allow the use of energy storage to suit generation requirements in a high-performance, bidirectional energy supply. The energy sources distributed, such as renewable sources of energy, storage methods, and other latest-generation technologies, provide inexpensive and efficient electrical services for the benefit of residential, commercial, industrial, and transport infrastructure. They include distributed energy sources (DES). This calls for strategic monitoring and control of various sections in the generation and transmission lines of power, heat, electricity, transport, and other critical services, such as natural gas. By having minimum impacts on the environment, it takes care of energy quality, safety, reliability, and security of the system. Networks that manage local load are important parts of micro-energy grids. Governments across the globe are investing extensively these days in intelligent energy networks to ensure efficient utilization and availability of electricity [1,3].

Moreover, it also allows for better forecasting of power outages and recovery. Finally, it further promotes several heterogeneous technologies, such as projects in the renewable energy sector, networking of electric vehicles, and intelligent housing system in the smart grids. It offers greater protection and privacy for separate components within intelligent energy networks and enables secure integration of intelligent metering, web, and mobile applications. Several types of sensors are used in the smart grid, which converts the physical parameter into its equivalent electrical signal, which is converted to data in the digital communication format. The synchronization of DES has been achieved by using advanced digital information and communication technologies to supply electrical demand efficiently, sustainably, and economically. For performing this, a large amount of data has to be handled to achieve the desired performance. Data might become a major tool and also a key driver of the grid in the future. A major challenge is to maintain the value of 4 Vs of data (volume, variety, velocity, and veracity) by using limited resources (time, hardware, human, etc.) to achieve the desired performance. To operate the grid, supervisory control and data acquisition systems (SCADA) [1,3] are necessary.

The introduction of Big Data changed the conventional energy industry by introducing various monitoring methods to increase stability, reliability, and efficiency. It also takes care of environmental issues by utilizing nonconventional energy sources like solar photovoltaic (PV) cells; waves and tidal energy; fuel cell; biogas; etc. BDA is a data-driven tool that aims to determine the statistical characteristics (particularly correlations) stipulated by the statistical parameters. The modeling of Big Data has also been applied in other diverse fields, such as quantum systems, financial systems, biological systems, and cellular communication networks. In the power system, supervised learning is useful in data analysis. The main components are the deduced functions and the analytical models; these functions/models, created by artificial training, give a specified parameter as the predictor [4,5].

The smart energy grid can be represented as the internet of various smart devices and sensors across the whole grid. In the intelligent grid, the key point is the sophisticated measuring system, namely, the advanced metering infrastructure

(AMI). AMI uses a wide variety of intelligent meters and other measuring devices for end-users. A significant amount of data can thus be collected regarding the power output, transmission, distribution, and consumption using the metering and feedback elements. Environmental data, such as sun angle, wind speed, and temperature, play a major role in fostering intelligent energy management. The data for the Geographical Information System or GIS is also associated with spatial and attributable data resources, and thus constitutes an inherent part of Big Data. The collection of data from different sources needs different data acquisition systems, which will be well coordinated to achieve effective database integration for BDA. The level of data gathering and analysis in intelligent energy systems is quite quick, ranging from 5- to 15-minute intervals to sub-second intervals. From an energy consumption database, a proper marketing strategy can be developed for better consumer handling [6].

The rest of the chapter is structured in the following way. Section 6.2 discusses several Big Data analysis (BDA) techniques involved in load and price forecasting. Section 6.3 focusses on some other issues of smart grid, like developing cloud-based software for smart grid, associating smart gid with the Internet of Things (IoT), automatic demand response, and real-time pricing of smart grid environment applying Big Data methods, etc. Section 6.4 deals with the involvement of BDA in smart cities and intelligent metering systems. Section 6.5 emphasizes the role of Big Data in many industrial applications. In Section 6.5, some miscellaneous applications of Big Data are considered such as in internet of energy, maintenance systems, and also improvement of social and environmental sustainability.

6.2 BIG DATA APPLICATIONS IN ENERGY AND PRICE FORECASTING

In [7], a novel load forecasting tool for the short term was presented. In the first step, a clustering method was conducted to differentiate the pattern of loads daily with the aid of the data from the smart meter for each of the individual loads. An association analysis was utilized to obtain the factors with critical influence; then, in the next turn, a decision tree was used for classification. This was followed by appropriate forecasting models selected for various load patterns. Last but not least, the complete forecasted system load was determined with the help of the accumulation of the forecasting outcomes of the individual loads.

Similarly, a load forecasting method was presented in [8], which was based on a parallel random forest (RF) algorithm. Data such as wind speeds, temperature, and load data were analyzed by the parallelization RF algorithm. This model was designed and implemented on Hadoop. There were tests performed to check its efficiency with data sets of varying sizes, and the model proved that it had a higher prediction accuracy than that of a decision tree.

With smart grid integration, several renewable energy resources have been utilized such as wind and solar. This has resulted in a very complex and varied power system that has especially affected the short-term forecasting. To combat this, a short-term forecasting technique was stated in [9]. Initially, cluster analysis was performed to classify daily load patterns and then, a decision tree was applied for the establishment of classification rules. Later on, appropriate forecasting models were selected for varying load patterns. The forecasting technique involved the total system forecasting. Case studies performed using actual load data demonstrated that the proposed new architecture will maintain consistency within the necessary limits of short-term load forecasting.

The challenge of making systems more flexible and adaptive demands more effort. A hybrid evolutionary model based on fuzzy logic with optimization was presented in [10]. The authors stated that the combinatorial task of selecting fuzzy rules and weights was better handled by the bio-inspired optimizer. They employed GRASP Evolutionary Strategies (GES), which is inspired by a mixture of two heuristic approaches: evolutionary strategies and GRASP procedure. This was tested using real data. The proposed approach showcased its potential to be used over micro- and large grids. The time taken by it was also considerably low.

Another application of forecasting, as detailed in [11], was to be used in event-organizing venues. An application like this is highly needed since these events have an extremely complicated scenario resulting from large variations in consumption. The significance of data sets was also explored, especially in extremely short time intervals. Neural networks (NNs) outperformed support vector regression (SVR) when it was worked out on the daily collected data, but their performance was comparable in other smaller time intervals. In continuation of the above discussion, an electricity generation forecasting system was proposed in [12]. A Machine Learning (ML) model was employed to train the system to make it have the ability to predict power consumption.

As discussed before, the dependency of load and the weather conditions provide an excellent avenue to fine-tune predictions. Thus, an IoT-enabled deep learning system was introduced in [13], which can automatically extract features for the given data to forecast the expected load demand. The prime merit of their model is the fact that it is a two-step forecasting scheme, which increases its prediction accuracy. It also analyzed the effect of many factors such as area, climate, and social conventions. The simulations that were performed showcased that their method outperformed many existing approaches.

The chief advantage of the smart grid is its capability to forecast electricity pricing, and it makes the whole technology cost-effective. The major challenge in this application is the problem of feature redundancy. Therefore, a novel electricity pricing forecasting model was presented [14]. This model was a culmination of three modules, viz. hybrid of RF and Relief-F algorithm as well as feature selector based on grey correlation analysis (GCA), integration of the PCA, differential evolution (DE)-based SVM classifier, to forecast the price classification.

Likewise, in [15] a Big Data-based short-term forecasting technique was suggested, where the energy management problem was observed in the Stackelberg game. Some wind power forecasting methods were also introduced that were beneficial to the microgrid management. Simulations highlighted the success of their model.

As the forecasting technology grew, so did the vast amount of data collected, making it difficult for the algorithms to accommodate such enormous data sets. To combat the same, a novel model combining CNNs and K-means clustering was introduced in [16]. This improved the scalability, and the tests showcased the model's effectiveness too.

An improved SVM method was suggested in [17], which was based on Hadoop. It employed fuzzy C clustering of quantum-behaved particle swarm optimization. This was done to analyze the previous data collected. A cloud platform was also introduced to store and match the load data and the influencing factors. This not only decreased the response time of the method but also improved the accuracy of the forecasting. They concluded that the improved SVM outperformed the normal SVM in terms of time and accuracy.

CNN based on the bagging model was employed for load forecasting in [18] for a short duration. It was utilized for training, and later on employed on the the real time industry data to fine tune their model. The bagging forecasting model then works on these assembled weak forecasting models. Spark is utilized for their learning and assembly procedures. Experimental findings showed the method's efficacy.

Fu et al. [19] suggested employing a multi-stage load forecasting procedure based on a power grid planning platform. A dynamic topology model of "Point-Line-Plane-Solid" four-level coordination was proposed, which worked on the provincial power network. Further, a database system was constructed and optimized by analysis of the data collected. This provided a concise and effective load prediction.

Similarly, a multiple linear regression (MLR) method was presented in [20] to aid in load forecasting for Sulawesi Island in Indonesia utilizing hourly data for the same. The authors also focused on the relative ease of understanding and development of their method. One key drawback worth noting was that their model was extremely sensitive to fluctuation in temperature. Singh et al. [21] applied data clustering, pattern mining, and Bayesian network to predict the usage. Li et al. [22] proposed a method of data-based linear clustering to resolve the issue of long-term load prediction in some developed towns induced by load fluctuations. The linear clustering approach was used to prepare the modeling of a load data set with an interval. To forecast their future load, an optimum autoregressive integrated motion average (ARIMA) model for each cluster was built. Finally, all ARIMA predictions were summarized to obtain the device load forecasting result.

The intelligent grid, part of the smart city, offers new chances for improved energy consumption efficiency, which can lead to substantial cost savings and have a major environmental contribution. The nature of the study of power engineering was profoundly modified by grid advances and the liberalization of the electricity markets. Large amounts of data produced by the intelligent grid can be processed online and provide timely and precise power demand predictions. Thus, a short-term power load forecast was rendered by an online SVR method [23].

A method was proposed to process and store vast quantities of data using graphical technologies. To satisfy loads, weather settings, and outages in a defined time, the query language Cyphers and the graph data database route definition were then used. The method was used to model predictions of load for a certain time from the details as data. The performance of the transmission system was obtained using the 24-hour load forecast [24].

In [25], an intelligent adaptive switching module was presented for boosting the parity between the performance of the PV system with the regular load curve. A scaling approach had been developed to provide a precise PV output forecast based on collected data from sensors. For data processing and analysis, the approach suggested was the sensor implementation in the application of renewable energy systems and Big Data technologies. In this regard, the data logger was used to analyze and predict PV performance using SVM and regression, finding that error was significantly improved for prediction when additional parameters were being used.

For the use of a two-stage prediction analysis to efficiently run their power system, a model was proposed for educational buildings in [26]. The electrical charging data was obtained from a university campus for five years. By using the motion average method, the load pattern was considered. The regular electric charge was predicted using RF and, finally, its efficiency was evaluated with the cross-validation time series.

In [27], the effect of economic and meteorological influences on characteristics of the load was taken into consideration. The total load decomposed into the basic charge of economic operation and weather-responsive load, influenced by meteorological considerations. Besides, two models were used for the linear regression and RF regression in large data technology. Finally, it was used to intelligently correct the prediction results by the wavelet neural network (WNN) algorithm. A method of data analysis was suggested in [28] that analyzed the power Big Data smartly and carried out the power grid load prediction. The procedure involved the required data from every platform database, acknowledging the customer's load forecast request, and performing the large data load forecasting by modifying the gray model of chaotic genetic algorithm.

A hybrid method called WLANFIS (constituted of WLS, NN, and adaptive neuro-fuzzy inference system) was proposed in [29] and used to predict the short-term load. A refined data set was taken by NN and WLS, which helped to find the best membership function number and forms. It also helped to get the powerful fuzzy set ranges for a single membership function used by the fuzzy scheme. The WLS provided estimates of the actual scenario, while NN modeled the demand profile for nonlinearity and was tested both for IEEE 14 and 30 bus systems and on data sets obtained from real-world sources.

A hybrid electric load forecasting model was showcased in [30]. An optimization module was used, based on the genetic wind-driven optimization (GWDO) algorithm to adjust the model parameters. The work narrated in [31] deliberated upon an effective procedure of prediction. This new method had two key contributions: first, was to enhance the original predictive accuracy algorithm; and the second, to turn it into the Big Data sense, with important scalability performance. The algorithm also used the distributed computing architecture of Apache Spark and was ready to use with just a few setting changes. The experiments involving the application of an algorithm to real-world electrical demand data were carried out using physical and cloud clusters.

6.3 BIG DATA IN SMART GRIDS

Significant energy assets are aimed at generating electricity; thus, the electricity sector is growing worldwide. A smart grid is an improved form of the classical electrical grid that includes the use of IT, electronics, and automation. There is an escalating need to move towards a sensor-embedded smart grid. For this very reason, [32] described building a cloud-based platform for BDA to help envision the future smart grid. Dynamic demand response, abbreviated as D2R, model was employed, which performed demand-side management and relieved peak load. These decisions are guided by data-rich smart grid data analytics and mining, which must be scaled up for many buildings and customers. Similarly in [33], a technique was presented for a grid having cloud storage features for analyzing the Big Data set containing customer's power usage patterns, location's weather information, etc.

The real-time processing ability of the smart grid can make it more intelligent. For demand response and real-time pricing, an Apache spark was developed as a platform that seamlessly combined data processing requirements for batch, real-time, as well as iterative processes [34]. The three major obstacles in power systems from the perspective of BDA have been addressed in [35]. The authors not only suggested Big Data applications in distribution systems but also showcased their tremendous importance in the planning and service processes of the systems. Finally, data privacy and protection issues with the use of BDA in distribution systems had been tackled by anonymizing usage data sets by removing identifiable information.

Taking the discussion about the challenges further, [36] surveyed new developments in big energy data analytics. Likewise, [37] highlighted the vulnerabilities in the smart grid by a detailed overview and survey to illustrate research problems associated with the issues in view of IoT. Another review of the subject matter, later, was done in [38], which pointed out the future prospects in this field.

Price forecasting is another necessary aspect as it makes the smart grid cost-effective. Realizing the limited capability of existing price forecasting methods to handle enormous data, an assessment of the notable aspects of BDA in power systems was provided in [39]. The barriers, according to them, that should be prioritized, are addressing the discarded data and siloed data, support for real-time analytics, and balancing of integrated and disintegrated systems. Potential solutions to the problems of large volume/variety of data, acquisition, processing of cost, complexity, and storage requirements are proposed in [40]. A case study of load profiling using energy consumption data is considered in [41]. Thus, applications like fault detection, power quality monitoring, load forecasting, load disaggregation, and load profiling are also apparent.

6.4   ROLE OF BDA IN SMART CITIES AND SMART METERS

The rapid population increase, especially in urban areas, has led to an increase in the requirement of services and infrastructure for fast communication and connectedness. The current era of IoT and Big Data analytics presents a considerable potential in smart future cities for such demanding urban environments. Smart cities employ a cloud environment capable of information intelligence as laid out in [42]. In view of BDA, two elements necessary for creating a generic cloud service are (i) cloud service design and (ii) reusability of current techniques.

The design principles on which the system architecture is generally based are scalability, reducing latency/delay, and open system principles being adopted so that new technologies can be easily extended with the link and, at the same time, new databases can be integrated. Data acquisition and management and processing analysis provide the other two major pillars that should be dealt. Cloud computing is, thus, a great step for effective city data analysis. Big Data could be utilized for congestion management for monitoring traffic performance, accident reduction, prediction of traffic speed and volume, and road maintenance [43]. With the evolution of Big Data, certain new challenges arise related to their performance, sources, security, and cost-effectiveness [44].

Researchers have also noted the use of IoT in BDA; thus, [45] proposed an IBDA framework for the real-time analysis of the arriving data by the sensors installed in the building. The huge quantity of data generated by these sensors is sent to the "Transmission Control Protocol" port where it is stored. It could be useful in the automatic management of luminosity and smoke detection. Furthermore, [46] proposed an energy-aware communication system for the IoT environment where the data was processed through Hadoop systems for efficient balancing of load during peak time.

Apart from this, autonomous connected electric vehicles in smart cities are possible with advances in wireless and vehicular communications, which can help cities go greener by cutting down emissions. BDA can be utilized for the development of policies for the location of charging stations, development of smart charging, and energy solutions for improved efficacy [47]. New-generation electric vehicles could produce gigabytes of data during mobility due to the variety of sensors used for autonomy.

Thus, a government's choice to adopt such technological change is difficult without proper proof-of-concept and data security assurance. Critics also point out the limited capacity to overcome siloed data storage structure and management of diverse stakeholders. Such challenges are thoroughly investigated in [48]. As we have seen already, energy efficiency is the end goal of energy BDA in smart cities. To extract consumption patterns in time series data, a model was proposed in [49] so that valuable conclusions can be made. The integration of important features and operations offers decision-making capabilities for the building's energy efficiency. An architecture based on service orientation to support BDA for energy management was proposed in [50].

A systematic review of BDA in smart cities has been conducted in [51]. Nowadays, several tools like SVMs, decision trees, Bayesian inference, different types of NNs, long short-term memory (LSTM), generative adversarial networks (GANs), etc. are apparent in applications. One clear thing is to provide a universal platform by integrating with the web. Thus, BDA-embedded smart city architecture was proposed in [52], which was integrated with the web via a smart gateway.

While the smart city is still a concept far from reality, smart meters, which are the enhanced meters, have the capability of not only measuring the energy but also communicating it [53]. The downside of all this is the enormous load on the data transmission lines and the high costs of data storage. A study on compression methods like singular value decomposition, principal component analysis, symbolic aggregation approximation, linear regression-based dimension reduction, etc. for Big Data can be found in [54]. After this step, the data also needs to be analyzed which is explored in [55] using AMI.

However, most of the Big Data research stagnated on concepts, design, and issues and lacked potential further development. A core–broker–client system for BDA was presented in [56] and was coined smart meter analytics scaled by Hadoop (SMASH). This could perform on data sets at 20-terabyte scale. Cloud computing has always provided an ideal solution for both data processing and communication-associated problems.

We have till now discussed many ways for energy saving in BDA of smart meters. Another aspect of achieving this goal is to explore the relationship between the depth of submetering and the level of savings in energy, as given in [57] using the data gathered regarding energy savings, depth of submetering, and the involved costs for 21 cases.

6.5 MISCELLANEOUS

Energy management systems (EMS) are complex and expensive systems used to monitor/control/optimize the industrial consumption of energy. Big Data emerged as a solution, but with limitations such as handling large data due to a higher sampling rate. To combat the same, [58] proposed an industrial EMS with cloud computing abilities. Challenges that they highlighted for their solution were concerned with the security of the collected data and overall performance as there is a delay between the data collection and its analysis. They eliminated latency issues that helped them analyze the collected data at a faster rate so that companies could optimize their efficiency by implementing changes faster. They evaluated their work using five use cases and parameters, like the ease of obtaining results among other parameters. Their research paves way for other EMS implementations for cost- and energy-saving goals.

Another contribution to this field was made in [59], where a general methodology was proposed to integrate IoT and data analytics with the existing EMS. In their study, they highlighted the importance of collecting data from a variety of sources and geographically dispersed repositories to truly analyze a company's energy management. Their model took care of this specific challenge as well as improving equipment effectiveness of machine tools. All these factors combined would, in theory, result in great productivity and promote smart factory development. The fundamental difference is that, in IoT platforms, collection and analysis are done parallelly, whereas Big Data is about the collection first and, later, the processing.

While there is an undeniable involvement of Big Data in the industrial realm owing to the predominant role of IoT sensors/devices, there exist challenges in Big Data analytics. The limited computational and storage resources are the main problems that are faced by experts, which are systematically presented in [60] that classified various analytic tools and techniques from an industrial usage perspective. They also gave a detailed discussion of Big Data adoption in IIoT (also known as Industry 4.0). They pointed out seven principles that affect the design of IIoT. Thus, the influence of Big Data on the industry has not let the automation go untouched.

6.6 CONCLUSIONS

The Big Data handling aspect for smart energy systems is discussed in this chapter. The chapter also deliberates upon the numerous techniques for large-scale data analysis involved in price and energy estimation of loads in the context of the intelligent grid environment. Besides this, a host of other issues pertaining to smart grids, including building an intelligent cloud-based grid platform, connecting intelligent grids with the IoT, automated demand response, real-time intelligent grid pricing, etc, are taken up for the study. The chapter also supports energy management techniques involving large data sets. It also addresses the importance of BDA in smart cities and advanced measurement systems. It highlights the involvement of Big Data in different industrial applications as well. Finally, several different uses for Big Data are now being taken into accounts, such as the energy internet, maintenance systems, and social and environmental sustainability. Thus, this chapter provides a detailed insight into the intelligent energy systems in the Big Data sense. Some potential applications including the networked control system, blockchain technique for cybersecurity, and internet of energy in the smart grid scenario are upcoming areas for research and development in a Big Data environment.

REFERENCES

1. Diamantoulakis, P. D., Kapinas, V. M., & Karagiannidis, G. K. (2015). Big data analytics for dynamic energy management in smart grids. *Big Data Research*, 2(3), 94–101.
2. Barrile, V., Bonfa, S., & Bilotta, G. (2016). Big data analytics for a smart green infrastructure strategy. In *International Conference on Advanced Material Technologies (ICAMT)* (Vol. 2016, No. 27th), Visakhapatnam.
3. He, Y., Yu, F. R., Zhao, N., Yin, H., Yao, H., & Qiu, R. C. (2016). Big data analytics in mobile cellular networks. *IEEE Access*, 4, 1985–1996.
4. Gabbar, H. (2016). *Smart Energy Grid Engineering*. Academic Press.
5. Jaradat, M., Jarrah, M., Bousselham, A., Jararweh, Y., & Al-Ayyoub, M. (2015). The internet of energy: smart sensor networks and big data management for smart grid. *Procedia Computer Science, 56*, 592–597.
6. Zhou, K., Fu, C., & Yang, S. (2016). Big data driven smart energy management: From big data to big insights. *Renewable and Sustainable Energy Reviews*, 56, 215–225.
7. Zhang, X., Cheng, M., Liu, Y., Li, D. H., & Wu, R. M. (2014). Short-term load forecasting based on big data technologies. In *Applied Mechanics and Materials* (Vol. 687, pp. 1186–1192). Trans Tech Publications Ltd, Switzerland.
8. Wang, D., & Sun, Z. (2015). Big data analysis and parallel load forecasting of electric power user side. *Proceedings of the CSEE*, 35(3), 527–537.
9. Zhang, P., Wu, X., Wang, X., & Bi, S. (2015). Short-term load forecasting based on big data technologies. *CSEE Journal of Power and Energy Systems*, 1(3), 59–67.
10. Coelho, V. N., Coelho, I. M., Coelho, B. N., Reis, A. J., Enayatifar, R., Souza, M. J., & Guimarães, F. G. (2016). A self-adaptive evolutionary fuzzy model for load forecasting problems on smart grid environment. *Applied Energy, 169*, 567–584.
11. Grolinger, K., L'Heureux, A., Capretz, M. A., & Seewald, L. (2016). Energy forecasting for event venues: Big data and prediction accuracy. *Energy and Buildings, 112*, 222–233.
12. Rahman, M. N., Esmailpour, A., & Zhao, J. (2016). Machine learning with big data an efficient electricity generation forecasting system. *Big Data Research*, 5, 9–15.
13. Li, L., Ota, K., & Dong, M. (2017). When weather matters: IoT-based electrical load forecasting for smart grid. *IEEE Communications Magazine*, 55(10), 46–51.
14. Wang, K., Xu, C., Zhang, Y., Guo, S., & Zomaya, A. Y. (2017). Robust big data analytics for electricity price forecasting in the smart grid. *IEEE Transactions on Big Data*, 5(1), 34–45.

15. Zhou, Z., Xiong, F., Huang, B., Xu, C., Jiao, R., Liao, B.,.... Li, J. (2017). Game-theoretical energy management for energy Internet with big data-based renewable power forecasting. *IEEE Access*, *5*, 5731–5746.
16. Dong, X., Qian, L., & Huang, L. (2017, February). Short-term load forecasting in smart grid: A combined CNN and K-means clustering approach. In *2017 IEEE International Conference on Big Data and Smart Computing (BigComp)* (pp. 119–125). IEEE, Jeju Island.
17. Zhao, H., Tang, Z., Shi, W., & Wang, Z. (2017, May). Study of short-term load forecasting in big data environment. In *2017 29th Chinese Control And Decision Conference (CCDC)* (pp. 6673–6678). IEEE, Chongqing.
18. Dong, X., Qian, L., & Huang, L. (2017, August). A CNN based bagging learning approach to short-term load forecasting in smart grid. In *2017 IEEE SmartWorld, Ubiquitous Intelligence & Computing, Advanced & Trusted Computed, Scalable Computing & Communications, Cloud & Big Data Computing, Internet of People and Smart City Innovation (SmartWorld/SCALCOM/UIC/ATC/CBDCom/IOP/SCI)* (pp. 1–6). IEEE, San Francisco, CA.
19. Fu, Y., Sun, D., Wang, Y., Feng, L., & Zhao, W. (2017, October). Multi-level load forecasting system based on power grid planning platform with integrated information. In *2017 Chinese Automation Congress (CAC)* (pp. 933–938). IEEE, Jinan.
20. Saber, A. Y., & Alam, A. R. (2017, November). Short term load forecasting using multiple linear regression for big data. In *2017 IEEE Symposium Series on Computational Intelligence (SSCI)* (pp. 1–6). IEEE, Honolulu, HI.
21. Singh, S., & Yassine, A. (2018). Big data mining of energy time series for behavioral analytics and energy consumption forecasting. *Energies*, *11*(2), 452.
22. Li, Y., Han, D., & Yan, Z. (2018). Long-term system load forecasting based on data-driven linear clustering method. *Journal of Modern Power Systems and Clean Energy*, *6*(2), 306–316.
23. Vrablecová, P., Ezzeddine, A. B., Rozinajová, V., Šárik, S., & Sangaiah, A. K. (2018). Smart grid load forecasting using online support vector regression. *Computers & Electrical Engineering*, *65*, 102–117.
24. Perçuku, A., Minkovska, D., & Stoyanova, L. (2018). Big data and time series use in short term load forecasting in power transmission system. *Procedia Computer Science*, *141*, 167–174.
25. Preda, S., Oprea, S. V., & Bâra, A. (2018). PV forecasting using support vector machine learning in a big data analytics context. *Symmetry*, *10*(12), 748.
26. Moon, J., Kim, K. H., Kim, Y., & Hwang, E. (2018, January). A short-term electric load forecasting scheme using 2-stage predictive analytics. In *2018 IEEE International Conference on Big Data and Smart Computing (BigComp)* (pp. 219–226). IEEE, Shanghai.
27. Chen, P., Cheng, H., Yao, Y., Li, X., Zhang, J., & Yang, Z. (2018, June). Research on medium-long term power load forecasting method based on load decomposition and big data technology. In *2018 International Conference on Smart Grid and Electrical Automation (ICSGEA)* (pp. 50–54). IEEE, Changsha.
28. Xu, M., Huang, G., Zhang, M., Cui, P., & Wang, C. (2018, July). Load forecasting research based on high performance intelligent data processing of power big data. In *Proceedings of the 2018 2nd International Conference on Algorithms, Computing and Systems* (pp. 55–60), Beijing.
29. Ali, M., Adnan, M., Tariq, M., & Poor, H. V. (2020). Load forecasting through estimated parametrized based fuzzy inference system in smart grids. *IEEE Transactions on Fuzzy Systems*.
30. Hafeez, G., Alimgeer, K. S., & Khan, I. (2020). Electric load forecasting based on deep learning and optimized by heuristic algorithm in smart grid. *Applied Energy*, *269*, 114915.

31. Pérez-Chacón, R., Asencio-Cortés, G., Martínez-Álvarez, F., & Troncoso, A. (2020). Big data time series forecasting based on pattern sequence similarity and its application to the electricity demand. *Information Sciences*.

32. Simmhan, Y., Aman, S., Kumbhare, A., Liu, R., Stevens, S., Zhou, Q., & Prasanna, V. (2013). Cloud-based software platform for big data analytics in smart grids. *Computing in Science & Engineering*, *15*(4), 38–47.

33. Mayilvaganan, M., & Sabitha, M. (2013, December). A cloud-based architecture for Big-Data analytics in smart grid: A proposal. In 2013 *IEEE International Conference on Computational Intelligence and Computing Research* (pp. 1–4). IEEE.

34. Shyam, R., HB, B. G., Kumar, S., Poornachandran, P., & Soman, K. P. (2015). Apache spark a big data analytics platform for smart grid. *Procedia Technology*, *21*, 171–178.

35. Yu, N., Shah, S., Johnson, R., Sherick, R., Hong, M., & Loparo, K. (2015, February). Big data analytics in power distribution systems. In 2015 *IEEE Power & Energy Society Innovative Smart Grid Technologies Conference (ISGT)* (pp. 1–5). IEEE, Columbia.

36. Hu, J., & Vasilakos, A. V. (2016). Energy big data analytics and security: Challenges and opportunities. *IEEE Transactions on Smart Grid*, *7*(5), 2423–2436.

37. Chin, W. L., Li, W., & Chen, H. H. (2017). Energy big data security threats in IoT-based smart grid communications. *IEEE Communications Magazine*, *55*(10), 70–75.

38. Tu, C., He, X., Shuai, Z., & Jiang, F. (2017). Big data issues in smart grid—A review. *Renewable and Sustainable Energy Reviews*, *79*, 1099–1107.

39. Akhavan-Hejazi, H., & Mohsenian-Rad, H. (2018). Power systems big data analytics: An assessment of paradigm shift barriers and prospects. *Energy Reports*, *4*, 91–100.

40. Bhattarai, B. P., Paudyal, S., Luo, Y., Mohanpurkar, M., Cheung, K., Tonkoski, R., ... Manic, M. (2019). Big data analytics in smart grids: state-of-the-art, challenges, opportunities, and future directions. *IET Smart Grid*, *2*(2), 141–154.

41. Marlen, A., Maxim, A., Ukaegbu, I. A., & Nunna, H. K. (2019, February). Application of big data in smart grids: Energy analytics. In 2019 *21st International Conference on Advanced Communication Technology (ICACT)* (pp. 402–407). IEEE, PyeongChang.

42. Khan, Z., Anjum, A., & Kiani, S. L. (2013, December). Cloud based big data analytics for smart future cities. In 2013 *IEEE/ACM 6th International Conference on Utility and Cloud Computing* (pp. 381–386). IEEE, Washington, DC.

43. Kumar, A., & Prakash, A. (2014). Role of big data and analytics in smart cities. *International Journal of Scientific Research (IJSR)*, *6*(14), 12–23.

44. Horban, V. (2016, August). A multifaceted approach to smart energy city concept through using big data analytics. In 2016 *IEEE First International Conference on Data Stream Mining & Processing (DSMP)* (pp. 392–396). IEEE, Ukraine.

45. Bashir, M. R., & Gill, A. Q. (2016, December). Towards an IoT big data analytics framework: smart buildings systems. In 2016 *IEEE 18th International Conference on High Performance Computing and Communications; IEEE 14th International Conference on Smart City; IEEE 2nd International Conference on Data Science and Systems (HPCC/SmartCity/DSS)* (pp. 1325–1332). IEEE, Sydney, NSW.

46. Khan, M., Babar, M., Ahmed, S. H., Shah, S. C., & Han, K. (2017). Smart city designing and planning based on big data analytics. *Sustainable Cities and Society*, *35*, 271–279.

47. Li, B., Kisacikoglu, M. C., Liu, C., Singh, N., & Erol-Kantarci, M. (2017). Big data analytics for electric vehicle integration in green smart cities. *IEEE Communications Magazine*, *55*(11), 19–25.

48. Giest, S. (2017). Big data analytics for mitigating carbon emissions in smart cities: opportunities and challenges. *European Planning Studies*, *25*(6), 941–957.

49. Pérez-Chacón, R., Luna-Romera, J. M., Troncoso, A., Martínez-Álvarez, F., & Riquelme, J. C. (2018). Big data analytics for discovering electricity consumption patterns in smart cities. *Energies*, *11*(3), 683.

50. Mohamed, N., Al-Jaroodi, J., & Jawhar, I. (2018, June). Service-oriented big data analytics for improving buildings energy management in smart cities. In *2018 14th International Wireless Communications & Mobile Computing Conference (IWCMC)* (pp. 1243–1248). IEEE, Cyprus.

51. Soomro, K., Bhutta, M. N. M., Khan, Z., & Tahir, M. A. (2019). Smart city big data analytics: An advanced review. *Wiley Interdisciplinary Reviews: Data Mining and Knowledge Discovery*, *9*(5), e1319.

52. Silva, B. N., Khan, M., & Han, K. (2020). Integration of Big Data analytics embedded smart city architecture with RESTful web of things for efficient service provision and energy management. *Future Generation Computer Systems*, *107*, 975–987.

53. Liu, X., & Nielsen, P. S. (2016). A hybrid ICT-solution for smart meter data analytics. *Energy*, *115*, 1710–1722.

54. Wen, L., Zhou, K., Yang, S., & Li, L. (2018). Compression of smart meter big data: A survey. *Renewable and Sustainable Energy Reviews*, *91*, 59–69.

55. Mohammad, R. (2018, August). AMI smart meter big data analytics for time series of electricity consumption. In *2018 17th IEEE International Conference On Trust, Security And Privacy In Computing And Communications/12th IEEE International Conference On Big Data Science And Engineering (TrustCom/BigDataSE)* (pp. 1771–1776). IEEE.

56. Wilcox, T., Jin, N., Flach, P., & Thumim, J. (2019). A Big Data platform for smart meter ata analytics. *Computers in Industry*, *105*, 250–259.

57. Zhai, Z. J., & Salazar, A. (2020). Assessing the implications of submetering with energy analytics to building energy savings. *Energy and Built Environment*, *1*(1), 27–35.

58. Sequeira, H., Carreira, P., Goldschmidt, T., & Vorst, P. (2014, December). Energy cloud: Real-time cloud-native energy management system to monitor and analyze energy consumption in multiple industrial sites. In *2014 IEEE/ACM 7th International Conference on Utility and Cloud Computing* (pp. 529–534). IEEE.

59. Bevilacqua, M., Ciarapica, F. E., Diamantini, C., & Potena, D. (2017). Big data analytics methodologies applied at energy management in industrial sector: A case study. *International Journal of RF Technologies*, *8*(3), 105–122.

60. ur Rehman, M. H., Yaqoob, I., Salah, K., Imran, M., Jayaraman, P. P., & Perera, C. (2019). The role of big data analytics in industrial Internet of Things. *Future Generation Computer Systems*, *99*, 247–259.

7 Optimum Placement of Multiple Distributed Generators in Distribution Systems for Loss Mitigation Considering Load Growth

D. Kavitha, B. Ashok Kumar, R. Divya
Thiagarajar College of Engineering

S. Senthilrani
Velammal College of Engineering and Technology

CONTENTS

7.1   INTRODUCTION

Nowadays, the usage of electricity has increasing day by day. The electricity delivered to the consumers must meet their requirement. The power system network consists of three major components: Generation, Transmission, and Distribution. On average, the majority of the losses in the power systems are due to the distribution network. Because the distribution network consists of passive components, it dissipate more energy. Moreover, reactance to resistance ratio in the system is more compared with that of the transmission system. Considering these facts, the distribution network losses have to be minimized to make the system efficient. One of the best ways to achieve this is by installing Distribution Generators (DGs) in the distribution system.

Regarding interfacing the DG, it is classified into inverter- and non-inverter-based DG. Inverter-based DGs include solar Photo Voltaic (PV), wind generators, micro-turbines, and fuel cells, since the generation is DC in those cases and we need inverters to convert DC to AC to synchronize the power to distribution grid. The non-inverter-based DGs are small hydro-induction generators and synchronous generators. Usually, the DG component is installed in the load side. So, the power can be easily injected to the load. Therefore, the power transmission and subsequently the power losses are reduced. Moreover, the reactive and active power injection through the DG improves the voltage profile and system's power factor.

Power demand has been continuously increasing and represented as the load growth, and it is considered as an important factor. In the present scenario, it is considered a main criterion while placing DGs in optimal locations. These installed DGs must meet the load growth. So, various techniques were proposed for optimal position of DGs in distribution networks in previous literatures. Reference [1] discusses optimal placement of multiple DGs for reducing the loss. These losses are estimated by the load flow algorithm, which has been studied in [2]. Optimal placement of DGs with multi objectives are employed in [3] and [4]. The multi objectives are loss reduction and voltage profile improvement [5]. Proposed equivalent current injection in the branches need to be done with DG placement to account loss sensitivity factor.. At the same time, multiple DG placements should be done in an economical manner. So, the cost minimization of overall network is discussed in literatures [6–9]. Various methodologies are employed. Loss reduction of distribution network is also achieved by shunt capacitor placement and static VAr components such as D-STATCOM. Shunt capacitors injects reactive power, which reduces the transmission losses. The inductive loads are satisfied by this method [10].In [11], Voltage Stability and Optimization Package were employed that use the graphical user interfaceability of MATLAB® for improvement of Voltage Stability. Most of the literatures discuss the best location of DG for active and reactive power losses. In few papers, voltage profile improvement is discussed. To examine the stability of the position of DG, a parallel small signal model was developed in [12]. Diesel generator placement in the distribution network also creates pollution by the discharge of impure gases including CO_2, NO, and NO_2. As the proposed idea should favor the environment, researchers start working with renewable resources. There are two types of DG: renewable and nonrenewable. Renewable DG placement is introduced to reduce the emissions. Reference [13] discussed the renewable DG placement for loss reduction. Mixed Integer Nonlinear Programming (MINLP) technique has been utilized for reducing the loss. Both improvements in stability index and loss reduction in optimal renewable DG placement are studied in [14]. Particle Swarm Optimization (PSO) technique for loss reduction is employed in [15–18]. In [18], the voltage profile improvement and line loading effects were discussed. Load growth is an important criterion, but most of the literatures did not discuss it. Reference [19] uses Genetic Algorithm (GA) with load model for loss reduction. The increasing load demand and the objectives to meet the loads were studied in [20].A different type of DG instalment using PSO algorithm is proposed in [21]. It also discussed power factor improvement to meet the load. So, the losses can be reduced by operating the DGs at nearer to the loading power factor. Shuffled Frog Leaping Algorithm (SFLA) and PSO were used in [22] for cost minimization during DG placement. It also discusses reliability index. All of these literatures used various techniques for optimal positioning of DG. In all these literatures, the capability of DG to meet the load growth is not discussed. The research questions addressed in this work areas follows:

1. What will be the reduction in losses if the number of DGs installed in a distribution network increases?
2. How long will the DG installation plan hold good as the load is continuously growing?
3. Is there any appreciable contribution of voltage stability index (VSI) in planning of DG?

In this work, SFLA is proposed for multiple DG placement. This technique gives the best solution when compared with the other techniques. The results were compared with other techniques as well. Moreover, the load growths is also employed in this chapter with multiple DG placements. It gives the details about accommodation of DG for years which can minimize losses.

7.2   PROBLEM FORMULATION

The objective function minimizes the active power losses of the radial distribution network by injecting real power by multiple DGs. The total loss of real power in the system is estimated by summing all line losses (branch), as given below:

$$\text{Min } P_{\text{Loss}} = \sum_{i=1}^{n} P_L = \sum_{i=1}^{n} I^2 R \tag{7.1}$$

I = Line current between two buses
R = Resistance of the line
n = Number of lines
Operational Constraints
Bus Voltage Limit
The bus voltage should be in the expected range and maintained within bearable functioning range all through the optimization process as follows:

$$V_{\text{min}} \leq V \leq V_{\text{max}} \tag{7.2}$$

V_{min}= Minimum voltage value
V_{max}= Maximum voltage value
The value of maximum and minimum voltage is considered as 1.05 and 0.90 p.u, respectively.
Distributed Generation Capacity Limit:

$$P_{DG,min} \leq P_{DG} \leq P_{DG,max} \tag{7.3}$$

Effect of Load Growth:

$$PL(new)=PL(0)(1+growth) \tag{7.4}$$

$$QL(new)=QL(0)(1+growth) \tag{7.5}$$

$PL(0)$ = Base case real power load
$QL(0)$ = Base case reactive power load
growth= Load growth rate
growth = 0.1 if load growth is considered as 10%
Voltage Stability Index (VSI):
Stability index, SI, is calculated using the following formula taken from the reference

$$SI = V_s^4 - 4^*(P_r^* x_{ij} - Q_r^* x_{ij})^2 - 4^*(P_r^* x_{ij} + Q_r^* x_{ij})^2 \left| V_s^2 \right| \tag{7.6}$$

The SI value has to be greater than 0 always
V_S=Sending end bus voltage
P_r=Receiving end active load
Q_r=Receiving end reactive load
r_{ij} = line i–j resistance
x_{ij}=line i–j reactance

7.2.1 Assumption Made for Multiple DG Placement

1. The DG considered for this proposed work is synchronous generator-based DG.
2. DG has been assumed to be working at power factor closer to load power factor.
3. The highest extent of the multiple DGs has been considered to be the same as the system's total active load.
4. The location of the DGs are at the load side.

Assumption 2 is based on the proof that the best power factor of DG units operated seemed to be nearer to the power factor of the collective load of respective systems for loss reduction purpose [1].

7.3 PROPOSED METHODOLOGY

The main aim of the proposed method is to acquire the optimum DG size and location for minimizing the power loss in the distribution network, and this can be done using both analytical and optimization techniques. This type of problem is solved using various optimization algorithms as reported in the literature. In this work, a heuristic optimization technique known as SFLA is used as a solving tool to minimize active power loss. The results were compared with other optimization techniques.

7.3.1 SFLA (Shuffled Frog Leaping Algorithm)

It is necessary to resolve the optimization issues with a very accurate algorithm to obtain globally optimum results at the proper time rather than being trapped in local optima. Shuffled frog leaping is a kind of meta-heuristic search optimization mechanism. The aim of this algorithm is to solve complicated optimization problems to obtain globally optimum results with less computational time. SFLA is a combination of PSO and memetic algorithms. This has been evolved from the memetic evolution of frog groups as they seek places that have more amount of food available for them to eat. The SFLA algorithm is focused on global information exchange and local search technique. Frogs within each memeplex improve their position to reach the best solution through local search. If the local search is not able to determine an optimal solution, then some virtual frogs are generated at random and utilized in the population. These shuffling and local search processes will be carried till reaching the convergence condition. The SFLA parameters are

- Frog Numbers (i.e. population)
- Memeplex Numbers
- Frog Numbers in each memeplex
- Iteration Numbers in each memeplex ahead of shuffling

The steps of SFLA areas follows:

1. Initialize an arbitrary population of p solution (frogs).
2. Estimate the fitness function for every frog.
3. Arrange the frogs in ascending sequence based on the strength values.
4. m groups (memeplexes): Partition of the population as follows: The frogs are allocated to the groups according to the fitness values. The frog with the lowest value occupies the first group; the next frog having the next lower value moves to the second group; and the mth lowest frog occupies the last group. Then, the $m+1$th frog with $m+1$th lower value again occupies the first group. These process are carried until the final frog is allocated to a group. Finally, each group contains n frogs. Thus, $p=n \times m$, as shown in Figure 7.1.
5. From every memeplex, select the frogs with the bad fitness value y_w and good fitness value y_b. Also, assign global fitness y_b to the frog. At predetermined times, an evolutionary process has to be applied to advance the frog with the bad fitness in every iteration. The worst frog is adjusted as follows according to its previous position.

$$\text{Change in fog position } y_i = \text{rand}()(y_a - y_w)y_i \qquad (7.7)$$

$i = 1,2,3,\ldots,m$

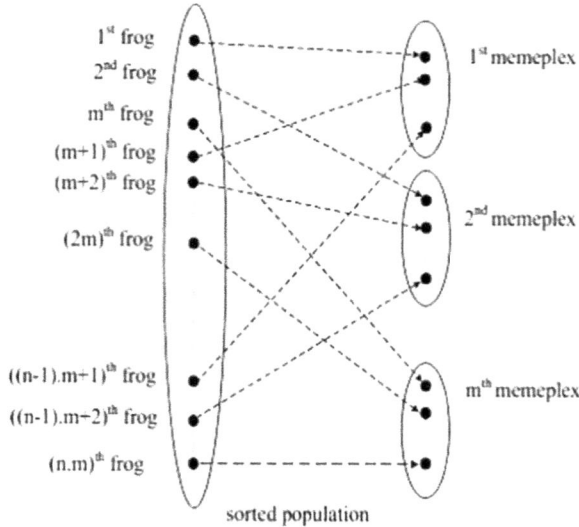

FIGURE 7.1 Memeplex partitioning process.

$$\text{New position } y_i' = y_i + \text{rand}()(y_a - y_w) \tag{7.8}$$

$i=1,2,3,...,m$

If $y_i'>y_i$, global search is carried out using the following equation

$$\text{Change in fog position } y_i = \text{rand}()(y_g - y_w)\text{dd} \tag{7.9}$$

$i=1,2,3,...,m$

$$\text{New position } y_i' = y_i + \text{rand}()(y_g - y_w) \tag{7.10}$$

$i=1,2,3,...,m$

If $y_i'>y_i$, a fresh frog y_i' is arbitrarily created to swap the bad frog (Figure 7.2).

7.3.2 LOAD FLOW ALGORITHM

Step 1: Examine load data and line data of the considered test system.

Step 2: Voltage at all buses is initialized as 1.0 p.u. After first iteration, voltage at each bus will change.

Step 3: Calculate the power supplied from the distribution system using the equation

$$S_i = P_i + jQ_i = V_s I^* \text{ where}, i = 1 \text{ to } N \tag{7.11}$$

Step 4: After finding the complex power, calculate the current using the equation

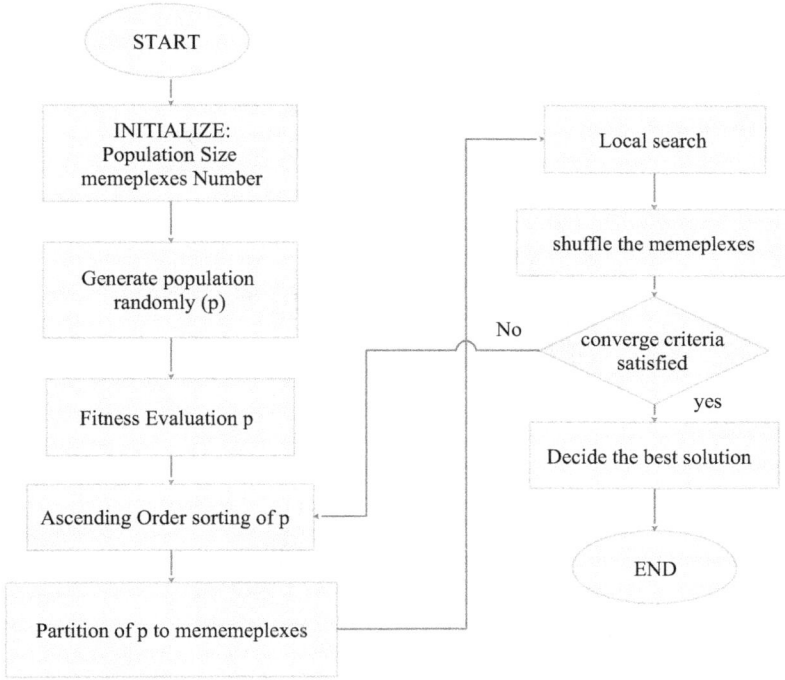

FIGURE 7.2   Flow chart for SFLA.

$$I_i^k = I_i^k V_i^k + j I_i^k \left(V_i^k \right) = \left(\frac{P_i + jQ_i}{V_i^k} \right) \qquad (7.12)$$

where V_i and I_i are the voltage and injected current of bus i in the kth iteration, respectively.

Step 5: Formulate BIBC matrix by applying KCL to the test system.

Step 6: After formulating BIBC matrix, the branch current(B) can be calculated using the equation

$$[B]=[BIBC][I] \qquad (7.13)$$

where [BIBC] = Bus injection / branch current matrix

Step 7: This branch current is used to estimate new voltage value at each bus,

$$V_R = V_S - B^*(R + jX) \qquad (7.14)$$

Step 8: The difference between the previous voltage and new value of voltage has to be found. If it is lower than 0.0001, algorithm converges and program proceeds to step 9. If it not so, again start at Step 3.

Step 9: Power loss in the line section can be found out using the equation

$$P_{loss} = I^2 * R \tag{7.15}$$

where I=Branch current
R=Resistance of the line
Equation 7.16 gives the reactive power injection at bus i

$$Q_{DG} = \frac{P_{DG}\sqrt{1 - PF_{DG}^2}}{PF_{DG}} \tag{7.16}$$

The whole system power loss can be obtained by means of summing all line losses, which is given by Eq. (7.1).

Equation 7.17 is used to find the optimal DG size at bus i,

$$P_i = P_{DGi} - P_{Di} \tag{7.17}$$

7.3.3 ALGORITHM FOR SFLA APPROACH

Number of control variables=3(location, size, and power factor for single DG placement)

Number of considered DG =3

Step1: Read load data and bus data.

Step2: Generate initial population and make size of DG as discrete and evaluate the fitness function using Eq. (7.1); sort the population in ascending order.

Step3: Identify the global best frog (i.e. size, power factor, DG, and location for the minimum loss).

Step4: After evaluating the fitness function, divide the population into five memeplexes. Every memeplex consists of 10 frogs.

Step5: Identify the worst and best frogs in each memeplexes.

Step6: Update the worst frog in each memeplex using Eq. (7.7).

Step7: If the newly obtained frog is enhanced, then the worst frog is replaced with newly obtained frog; load flow is run again. Otherwise, update the worst frog using Eq. (7.9).

Step8: If no improvement is obtained, replace the worst frog by creating a frog randomly and run the load flow.

Step9: The SFLA will stop when maximum user-specified iteration is reached.

Step10: The location, size, and power factor with minimum real power loss has been considered as a best solution.

7.3.4 COMPUTATIONAL PROCEDURE FOR MULTIPLE DG PLACEMENT

Step 1: All the system parameters are input and the base case load flow is done. The total active power loss of the system P_{loss} is found.

Step 2: Initial population is generated with the number of DGs. If the number of DGs to be installed = 3, then each solution in the population will be having 6 variables with first 3 as different bus numbers, except the source node bus, and the last 3 as the capacity of the DG, which is governed by minimum and maximum constraints. The initial population is random in nature.

Step 3: Place DG at generated node and according to the DG capacity, load flow is run with new power data and P_{loss} is determined.

Step 4: The population is updated in each iteration through an optimization algorithm (Frog Leap in this work), and the program is stopped on reaching convergence criteria.

Step 5: The solution corresponding to minimum loss is taken and the DG locations and their sizes are recorded.

7.3.5 Voltage Stability Index and Load Growth Analysis

Once the solution is obtained, VSI of all buses are calculated using [15]. Using this SI, the vulnerable bus is obtained. The system is considered as a strong system once the lowest rate of SI is higher than 0.75, which is considered as threshold value. Load is assumed to be growing 1% every year; for each year's load growth, SI is determined after the installation of DG(s). The year in which lowest value of SI falls below the threshold is obtained and reported as the number of years in which the installed DG(s) can accommodate the load growth.

7.4 RESULTS AND DISCUSSIONS

7.4.1 Test System

The SFLA and other approaches such as IA, MINLP, and PSO, proposed in the present investigation is tested in IEEE 33-bus radial distribution systems with 3.7 MW and 2.3 MVAr loads, respectively. The system is shown in Figure 7.3. The combined load factor of the system is 0.85. These optimization techniques have been carried out in MATLAB for the load flow and to find out the optimal size of bus and multiple DGs. The algorithm is tested with iterations or generations varying from 200 to 300

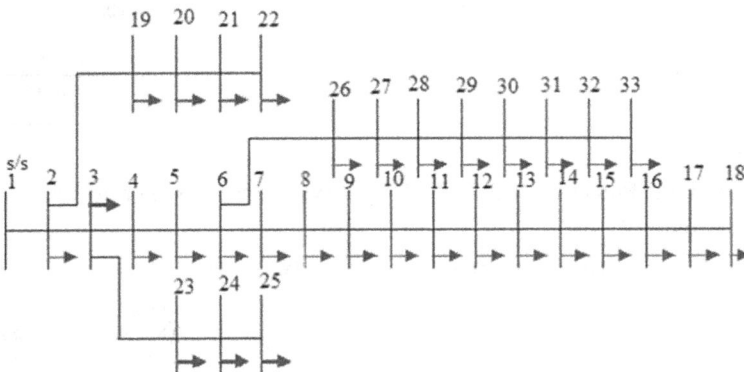

FIGURE 7.3   Test system.

with population varying from 25 to 50. Each case is tested with a minimum of 30 runs to check the stability of the algorithm to determine the best solution.

7.4.2 ASSUMPTIONS

In this chapter, four different cases were considered for this proposed work.

Case 1: System with base case (with no DG).
Case 2: System by means of one installed DG.
Case 3: System by means of two installed DGs.
Case 4: System by means of three installed DGs.

The load flow was carried out for base case. Table 7.1 shows that in the directive approach, the real power loss is 210.82 kW, which is less than the loss in IA method (211 kW) for the base case condition.

Table 7.2 discusses the results of case2 where one DG is installed. The loss reduction of one DG placement is achieved by the optimal location of bus 6. The DG size and power factor are represented in support of different techniques in the table. SFLA algorithm gives the best solution when compared with other techniques.

Table 7.3 discusses the results of case3 where two DGs are installed. The loss reduction of two DGs placement is achieved by the optimal location of bus 13 and 30 for SFLA technique. The DG size and power factor in the table are also shown for different techniques. SFLA algorithm gives the best solution rather than the other techniques.

Table 7.4 discusses the results of case4 where three DGs are installed. The loss reduction of three DGs placement is achieved by the optimal location of bus 13, 24, and 30 for SFLA technique. The DG size and power factor in the table are given for

TABLE 7.1
Comparison of Real Power Loss Obtained by Different Load Flow Analysis Methods for the Base Case

Different Load Flow Analysis Method	Real Power Loss in kW
Directive approach for load flow solution method	210.82
IA method	211

TABLE 7.2
One DG Unit Placement in IEEE 33-Bus System with Real and Reactive Power Capability

Cases	Techniques	Location	Size in MW	Power Factor	P_{loss} in kW
One DG	SFLA	6	2.600	0.82	67.884
	IA	6	2.637	0.85	68.157
	MINLP	6	2.558	0.823	67.854
	PSO	6	2.557	0.826	67.857

TABLE 7.3

Two DG Units Placement in IEEE 33-Bus System with Real and Reactive Power Capability

Cases	Techniques	Location	Size in MW	Power Factor	P_{loss} in kW
Two DG	SFLA	13	0.8	0.82	32.013
		30	1.4	0.82	
	IA	6	1.8	0.85	44.84
		30	0.9	0.85	
	MINLP	13	0.819	0.88	29.31
		30	1.550	0.80	
	PSO	12	0.818	0.822	39.10
		30	1.669	0.822	

TABLE 7.4

Three DG Units Placement in IEEE 33-Bus System with Real and Reactive Power Capability

Cases	Techniques	Location	Size in MW	Power Factor	P_{loss} in kW
Three DG	SFLA	13	1.100	0.82	17.16
		24	1.000		
		30	0.800		
	IA	6	0.900	0.85	23.05
		14	0.630		
		30	0.900		
	MINLP	13	0.766	0.87	12.74
		24	1.044	0.88	
		30	1.146	0.80	
	PSO	13	0.764	0.82	15.0
		24	1.068	0.87	
		30	1.016	0.83	

different techniques. SFLA algorithm gives the best solution when comparing the other techniques.

The VSI for different cases as reported by SFLA algorithm is represented in Table 7.5. VSI for base case condition is 0.6716, and it is found in bus number 18. The optimal sizing and location of multiple DGs in the distribution network improve the VSI. It is understood by considering case4 where three DGs are installed. The VSI value for three DGs placement is represented as 0.9762. Figure 7.4 shows the comparison of voltage magnitudes for different cases in various buses. The voltage profile is improved in three DG placement. Branch current for different cases are represented in Figure 7.5, which shows the branch current has reduced by placement of three DGs. So, the real power losses can be reduced.

TABLE 7.5

Voltage Stability Analysis

Different Cases	Minimum VSI Value
Base case	0.6716
One DG	0.8522
Two DG	0.9269
Three DG	0.9762

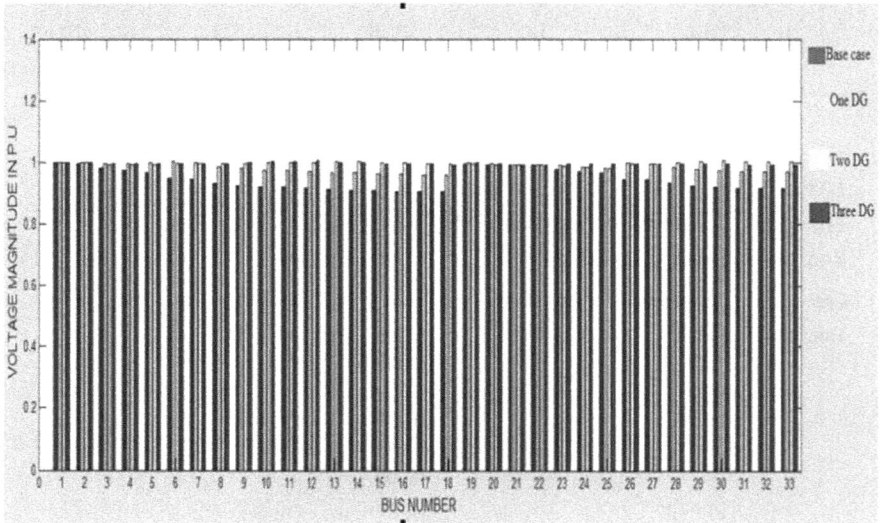

FIGURE 7.4　Comparison of voltage magnitude for different cases.

For this different cases, the load demand requirements are also proposed in this chapter. And so, the optimal placement of multiple DGs must meet the load demand every year. Only then can the losses be reduced by the optimal way.

In this work, it is assumed that load grows by 1% during the first year, 2% during the second year, and so on from the base case value. At different loading factors, the values of real power, reactive power, minimum voltage value, and VSI Index are represented in Table 7.6. Thus, loads are varied year by year. The loading factor in Table 7.6 represents the increasing loads in the distribution network. Thus, we have to meet the load demand. In order to achieve this criterion, loading capacity of DGs are studied in this work. After multiple DG placements, the capability of DG to withstand the loading for a number of years is mentioned in Table 7.7. Accommodation of load growth for one DG, two DGs, and three DGs is 8, 13, and 15years, respectively.

It is obvious that if one DG is placed, the DG can withstand the increase in load without compromising the SI below threshold for up to 8 years. A similar trend holds good for other cases also.

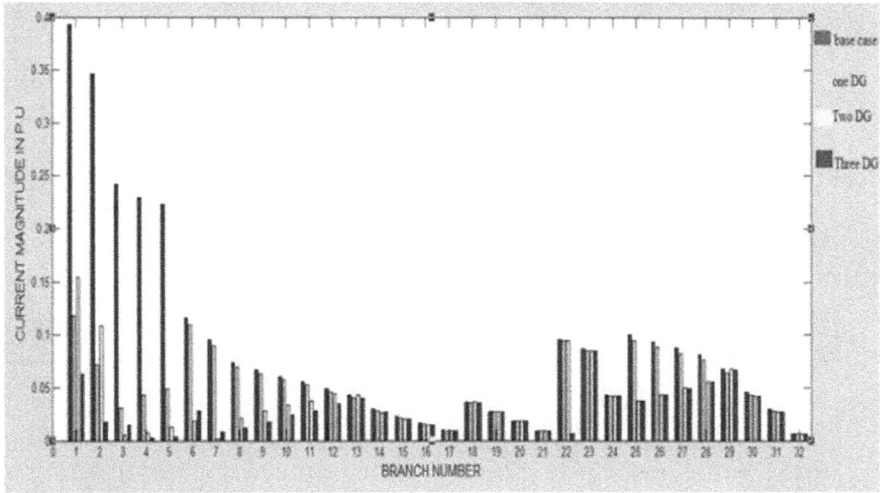

FIGURE 7.5 Comparison of branch currents for different cases.

TABLE 7.6
Load Growth Without DG Placement

	Base Case	Loading Factor 0.01	Loading Factor 0.02	Loading Factor 0.03	Loading Factor 0.04	Loading Factor 0.05	Loading Factor 0.06
Active power (kW)	3715	3752	3789	3826	3863	3900	3937
Reactive power (kVAr)	2300	2323	2346	2369	2392	2415	2438
V_{min} (p.u.)	0.09052 (18)	0.09042 (18)	0.9031	0.9021	0.9011	0.9000	0.8990
VSI_{min} (p.u.)	0.6716	0.6685	0.6654	0.6624	0.6593	0.6562	0.6532

Figures 7.6–7.8 represent the convergence characteristics of fitness function using SFLA in IEEE 33-bus test system for one DG, two DG, and three DG placement. This shows the relationship between real power loss in p.u and number of iterations. In one DG placement, after 60 iterations the real power losses are condensed to optimal, as revealed in Figure 7.5. Similarly, after 54 iterations, the real power losses are reduced to optimal in two DGs placement, as revealed in Figure 7.6. Figure 7.7 shows that after 162 iterations, real power losses are reduced to optimal for three DGs placement. When comparing these three, Figures 7.6–7.8, an increased amount of reduction in real power losses is achieved by three DGs placement. So, multiple DGs being placed improve the loss reduction efficiently.

TABLE 7.7

Accommodation of Load Growth (in Years) after Multiple DG Placement

	Location	Size in MW	Accommodation of Load Growth in Years
One DG	6	2.600	8
Two DG	13	0.8	13
	30	1.4	
Three DG	13	1.100	15
	24	1.000	
	30	0.800	

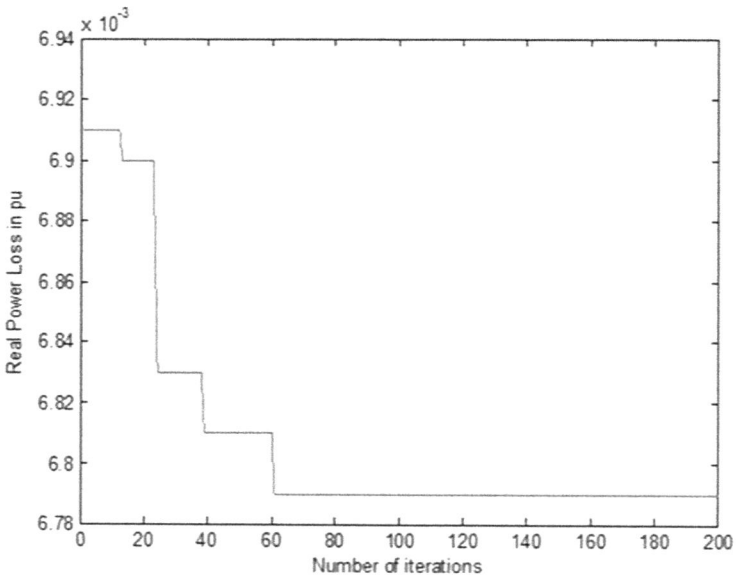

FIGURE 7.6 Convergence characteristics of fitness function using SFLA in IEEE 33-bus system for one DG placement.

7.5 CONCLUSION

This chapter examined the best placement of multiple DGs in a distribution network, leading to reduced power loss reduction and VSI improvement. SFLA was used to resolve the problem, and an IEEE 33-bus system was used for result demonstration. The outcomes were compared with other methodologies. It also showed that the best power factor of DG units operated seemed to be nearer to the power factor of the shared load of respective systems for loss reduction purpose. The increase in system power loss and increase in voltage drop due to rise in load growth can be overcome by placing multiple DGs at suitable locations in the distribution network. So, multiple

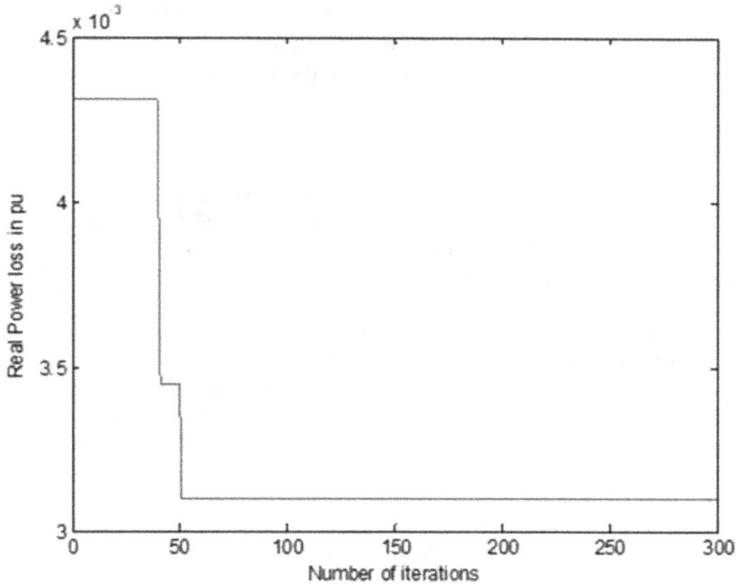

FIGURE 7.7　Convergence characteristics of fitness function using SFLA in IEEE 33-bus system for two DGs placement.

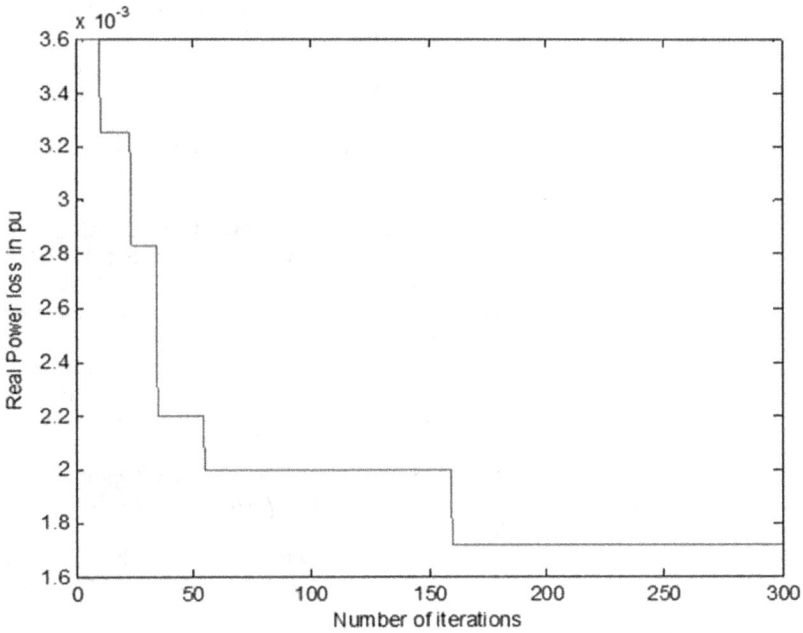

FIGURE 7.8　Convergence characteristics of fitness function using SFLA in IEEE 33-bus system for three DGs placement.

DG placement will improve voltage profile, reduce branch current due to the injection of local generating current, and also reduce loss in real power to a great extent. Future expansion planning may also be addressed with the VSI along with the DG placement.

REFERENCES

1. Hung, D.Q., and N. Mithulananthan. "Multiple distributed generator placement in primary distribution networks for loss reduction." *IEEE Transactions on Industrial Electronics* 60(4) (2011): 1700–1708.
2. Teng, J.-H. "A direct approach for distribution system load flow solutions." *IEEE Transactions on Power Delivery* 18(3) (2003): 882–887.
3. Aman, M.M., G.B. Jasmon, A.H.A. Bakar, and H. Mokhlis. "A new approach for optimum DG placement and sizing based on voltage stability maximization and minimization of power losses." *Energy Conversion and Management* 70(2013):202–210.
4. A. Barin, L.F. Pozzatti, L.N. Canha, R.Q. Machado, A.R. Abaide, and G. Arend. "Multi-objective analysis of impacts of distributed generation placement on the operational characteristics of networks for distribution system planning." *International Journal of Electrical Power & Energy Systems* 32(10) (2010):1157–1164.
5. Gözel, T., and M.H.Hocaoglu. "An analytical method for the sizing and siting of distributed generators in radial systems." *Electric Power Systems Research* 79(6) (2009): 912–918.
6. Naderi, E., H. Seifi, and M.S. Sepasian. "A dynamic approach for distribution system planning considering distributed generation." *IEEE Transactions on Power Delivery* 27(3) (2012): 1313–1322.
7. Ebrahimi, R., M. Ehsan, and H. Nouri. "A profit-centric strategy for distributed generation planning considering time varying voltage dependent load demand." *International Journal of Electrical Power & Energy Systems* 44(1) (2013): 168–178.
8. Gomez-Gonzalez, M., A. López, and F. Jurado. "Optimization of distributed generation systems using a new discrete PSO and OPF." *Electric Power Systems Research* 84(1) (2012): 174–180.
9. Borges, C.L.T., and D.M. Falcao. "Optimal distributed generation allocation for reliability, losses, and voltage improvement." *International Journal of Electrical Power & Energy Systems* 28(6) (2006): 413–420.
10. Roy, N.K., H.R. Pota, M.J. Hossain, and D. Cornforth "An effective VAR planning to improve dynamic voltage profile of distribution networks with distributed wind generation." *2012 IEEE Power and Energy Society General Meeting.* IEEE, (2012).
11. Gözel, T., U. Eminoglu, and M.H. Hocaoglu. "A tool for voltage stability and optimization (VS&OP) in radial distribution systems using MATLAB graphical user interface (GUI)." *Simulation Modelling Practice and Theory* 16(5) (2008): 505–518.
12. Marwali, M.N., J.-W. Jung, and A. Keyhani. "Stability analysis of load sharing control for distributed generation systems." *IEEE Transactions on Energy Conversion* 22(3) (2007): 737–745.
13. Atwa, Y.M., E.F. El-Saadany, M.M.A. Salama, and R. Seethapathy "Optimal renewable resources mix for distribution system energy loss minimization." *IEEE Transactions on Power Systems* 25(1) (2009): 360–370.
14. Roy, N.K., H.R. Pota, and A. Anwar. "A new approach for wind and solar type DG placement in power distribution networks to enhance systems stability." *2012 IEEE International Power Engineering and Optimization Conference* Melaka, Malaysia. IEEE (2012).
15. Moradi, M.H., and M. Abedini. "A combination of genetic algorithm and particle swarm optimization for optimal DG location and sizing in distribution systems." *International Journal of Electrical Power & Energy Systems* 34(1) (2012): 66–74.

16. AlRashidi, M.R., and M.F. AlHajri. "Optimal planning of multiple distributed generation sources in distribution networks: A new approach." *Energy Conversion and Management* 52(11) (2011): 3301–3308.

17. Ishak, R., A. Mohamed, A.N. Abdalla, and M.Z.C. Wanik. "Optimal placement and sizing of distributed generators based on a novel MPSI index." *International Journal of Electrical Power & Energy Systems* 60(2014): 389–398.

18. El-Zonkoly, A.M. "Optimal placement of multi-distributed generation units including different load models using particle swarm optimization." *Swarm and Evolutionary Computation* 1(1) (2011):50–59.

19. Singh, D., D.Singh, and K. S. Verma. "Multiobjective optimization for DG planning with load models." *IEEE Transactions on Power Systems* 24(1) (2009): 427–436.

20. Mistry, K.D., and R. Roy. "Enhancement of loading capacity of distribution system through distributed generator placement considering techno-economic benefits with load growth." *International Journal of Electrical Power & Energy Systems* 54 (2014): 505–515.

21. Kansal, S., V. Kumar, and B. Tyagi. "Optimal placement of different type of DG sources in distribution networks." *International Journal of Electrical Power & Energy Systems* 53 (2013):752–760.

22. Gitizadeh, M., A.A. Vahed, and J. Aghaei. "Multistage distribution system expansion planning considering distributed generation using hybrid evolutionary algorithms." *Applied Energy* 101 (2013): 655–666.

8 Big Data for Smart Energy

Hare Ram Sah and Yash Negi
SAGE University

CONTENTS

8.1 INTRODUCTION

The term of Big Data was coined due to the explosive increase of global data and was mainly used to describe these enormous datasets. Over the past 20 years, data has increased in a large scale in various fields. According to a report from International Data Corporation, in 2011, the overall created and copied data volume in the world was 1.8ZB, which has increased by nearly nine times within 5 years. This figure will double at least every 2 years in the near future. Compared with traditional datasets, Big Data generally includes masses of unstructured data that need more real-time analysis. The rapid growth of Big Data mainly comes from people's daily life, especially related to the service of Internet companies. For example, Google processes data of hundreds of PB and Facebook generates log data of over 10 PB per month; like this, there are many other companies that produce even more vast amounts of data per month. And this will only increase over time [1].

While the amount of large datasets is drastically rising, it also brings about many challenging problems demanding prompt solutions.

8.2 SMART ENERGY

A smart energy system is a cost-effective energy system combining the efficient use of energy and the use of renewable sources. It is a system in which energy production, distribution, and consumption are linked together intelligently in an integrated and flexible way. Using energy efficiently in smart homes saves money, enhances sustainability, and reduces carbon footprint. Consequently, the need for smart energy management is on the rise for smart homes and for smart cities in general. However, the lack of low-cost, easy-to-deploy, and low-maintenance technology has somewhat limited the large-scale deployment. Business Intelligence plays an essential role in energy management decisions for home owners and the utility alike. The data can be monitored, collected, and analyzed using predictive analysis and advanced methods to obtain actionable information in the form of reports, graphs, and charts. Thus, this analyzed data in real-time can aid homeowners and, utilities and utility eco-systems providers to gain significant insights on energy consumption of smart homes. The energy service providers can use the power consumption data available with analytics engine to provide flexible and on-demand supply with appropriate energy marketing strategies. The consumers, being aware of their consumption behavior and having a close interaction with the electricity utilities, can adjust and optimize their power consumption and reduce their electricity bills [2]. For a cost-efficient system, it is very important to monitor and control peak power consumption and identify where the energy is being used more and identify ways to minimize this. This is where the combination of IoT technology, Big Data analytics, and BI comes into play for implementing energy management solutions. Big Data can help us to manage and identify the cause of where the energy is being used extensively as it will generate data from multiple sensors and lights, and everything associated with Smart Energy.

> Energy doesn't communicate in English, French, Chinese or Swahili, but it does speak clearly.

With the rapid development of sensor technology, wireless transmission technology, network communication technology, cloud computing, and smart mobile devices, large amounts of data have been accumulated in almost every aspect of our lives. Moreover, the volume of data is growing rapidly, with increasingly complex structures and forms. The innovation brought by Big Data is changing the landscape of the industry, but everyone knows everything is not perfect; it comes with its own set of issues and challenges [3].

> It really matters whether people are working on generating clean energy or improving transportation or making the Internet work better and all those things. And small groups of people can have a really huge impact.

The energy sector is currently facing various challenges, such as the following:

- Operational efficiency and cost control
- System stability and reliability
- Renewable energy management
- Energy efficiency and environmental issues
- Service improvement

This is where Smart Grid comes into play. So, what exactly is Smart Grid?

The "Grid" refers to the electric grid, a network in short, the digital technology that allows for two-way communication between the utility and its customers, and the sensing along the transmission lines is what makes the grid smart. Like the Internet, the Smart Grid will consist of controls, computers, automation, and new technologies and equipment working together, but in this case, these technologies will work with the electrical grid to respond digitally to our quickly changing electric demand of transmission lines, substations, transformers and, more than that, deliver electricity from the power plant to the computer. An "electricity grid" is not a single entity but an aggregate of multiple networks and multiple power generation companies with multiple operators employing varying levels of communication and coordination, most of which is manually controlled [4].

8.2.1 What Makes a Grid "Smart?"

In short, the digital technology that allows for two-way communication between the utility and its customers, and the sensing along the transmission lines is what makes the grid smart. Like the Internet, the Smart Grid will consist of controls, computers, automation, and new technologies and equipment working together, but in this case, these technologies will work with the electrical grid to respond digitally to our quickly changing electric demand.

> On climate and clean energy, government sets the international framework, and the private sector uses that framework to do what it does best: innovate, create, and drive global progress.

8.2.2 What does a Smart Grid do?

The above diagram tells us about the various components in a Smart Grid (Figure 8.1). The components of a Smart Grid are a combination of intelligent appliances and heavy equipment that play an important role in the production of electricity. These appliances work in a predefined manner they are smart enough to understand the incoming power supply and how to utilize it.

Some basic components are discussed in what follows.

8.2.2.1 Smart Appliances

These appliances are set to consumer's predefined preference level, and they have an idea on when to consume energy on what level. These tech appliances have an important impact on the grid generators since they help in understanding the power position and reduce the peak load factors.

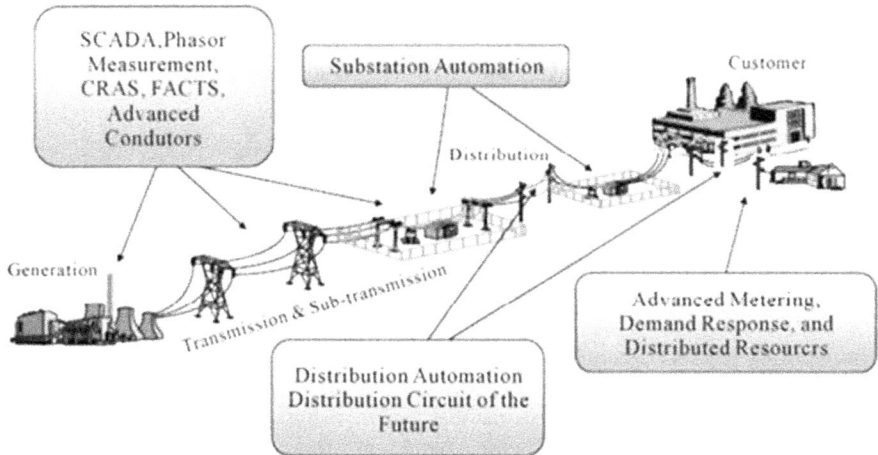

FIGURE 8.1 Components of Smart Grid.

8.2.2.2 Smart Meters

The smart meters are a two-way communicator that helps create a bridge between the power providers and the end consumer. It automates the billing data collection in every convenient manner, detects system failures, and sends repairing teams much faster than before because as soon as a system or a unit fails, the service providers are notified immediately [5].

8.2.2.3 Synchrophasors

The recent advancement in synchrophasor technology has played a key role in the supervisory control and data acquisition (SCADA) and energy management systems (EMS). The most common advantages of phasor measurement technology include:

- Dynamic monitoring of the whole interconnected system
- Post-event analysis
- Oscillation detection
- Island detection

Synchrophasors gather data from various locations of the grid to get a coherent picture of the whole network using GPS and transmit the data for analysis to central locations (Figure 8.2).

Aims of the Smart Grid:

- Provide a user-centric approach and allow new services to enter the market.
- Establish innovation as an economical driver for the electricity network renewal.
- Maintain security of supply, ensure integration, and aid in interoperability.
- Provide accessibility to a liberalized market and foster competition.
- Enable distributed generation and utilization of renewable energy sources.

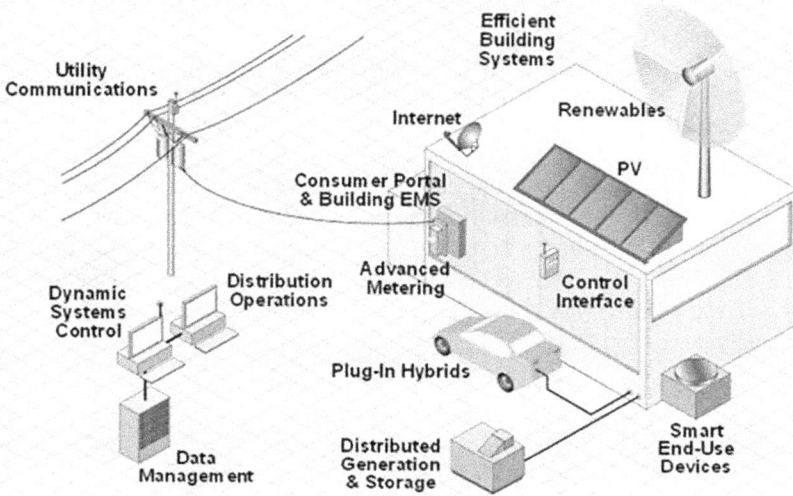

FIGURE 8.2 Aims of the Smart Grid.

- Ensure best use of central generation.
- Consider appropriately the impact of environmental limitations.
- Inform the political and regulatory aspects.
- Consider the societal aspects.

8.2.3 SMART GRID APPLICATIONS

Smart Grid has not only transformed the way power stations used to work but it has also provided an upper hand to the consumer market in reliability, efficiency, and accessibility regarding electricity. Some of its applications are as follows.

- Improving the working of transmission lines.
- Quick repairing system after any sort of breakage or misfire.
- Reducing the cost of electricity.
- Reducing the peak demand.

Benefits:
- The Smart Grid has been able to provide better power management technologies through its integrated systems, providing a better user interface.
- It has also provided a protective management system in case of emergency crisis or setbacks.
- It has been noted that Smart Grids provide better supply and demand management.
- The Smart Grid has not only provided longer battery timing but also better power quality to its consumers.

- Previously, grid stations used to emit a large amount of carbon dioxide, leading to climate change.
- It has also provided the convenience of reading meters remotely. Meter readers will not have to physically appear at the property and check for meter readings. It will all be done through IT resources.

We have seen how Smart Grid can help us minimize the cost and improve the efficiency of the electricity, thus resulting in low-cost bills and saving the environment. Big Data alone cannot do anything. There are many things that contribute to Big Data; Smart Grid is one of the technologies that helps us to better understand the cost and where the electricity is being used. These technologies generate data, and that data contributes to Big Data.

> Smart development invests in insulation, efficient cars, and ever-renewed sources of energy. Dumb growth crashes around looking for more oil.

8.3 INTERNET OF THINGS (IoT)

The Internet of Things (IoT) refers to the use of intelligently connected devices and systems to leverage data gathered by embedded sensors and actuators in machines and other physical objects. IoT is also the key player in Smart Energy because of its working capability with sensors and the ability to generate more accurate data or information, which helps us to better understand the things. Always being connected to the internet meaning that wherever you go, data comes with you so that you are always connected. Let us see IoT in more detail to help better understand its capability and how it is transforming the world (Figure 8.3).

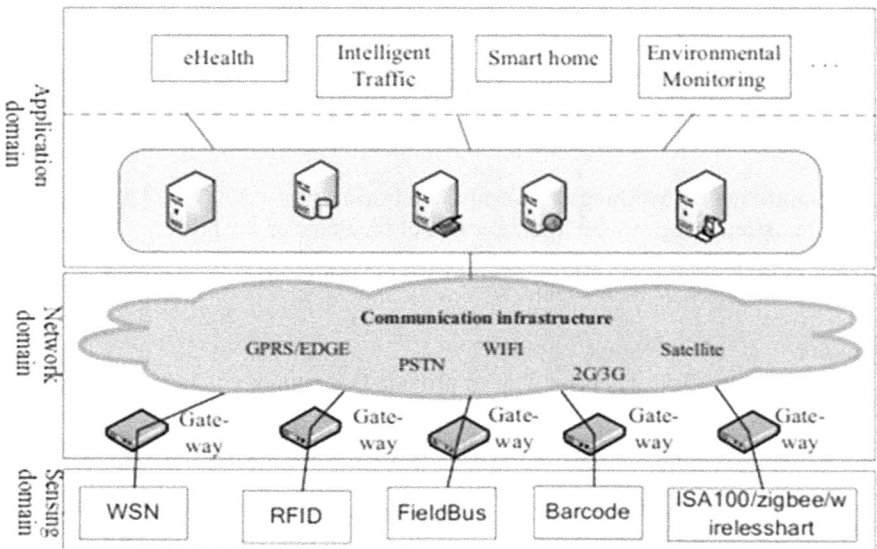

FIGURE 8.3 Three domains of IoT architecture.

The typical IoT architecture can be divided into three domains: sensing domain, network domain, and application domain, as shown in the above diagram. The sensing domain plays an important role in IoT, it enables "things" to interact and communicate among them and with existing or evolving communication infrastructure. It consists of many "'Smart Things'," Smart Things make our world smarter. Smart Things are a group of devices which can be monitored and controlled via a hub device (central processors) and web services. Smart Things adding support for popular connected products such as the Belkin WeMo family of devices, Philips Hue color-changing bulbs, and the Sonos home music system. Network domain builds on existing or evolving communication infrastructure, such as PSTN, 2G, 3G, LTE, and Satellite. The main aim of this domain is to transfer the data collected from the sensing domain to a remote destination. Application domain is responsible for data processing and services providing. The data from the transmission layer is handled by corresponding management systems, and then various services will be provided to all kinds of users.

Note: There is no single consensus on architecture for IoT, agreed upon universally. Different architectures have been proposed by different researchers.

If you invent a breakthrough in artificial intelligence, so machines can learn, that is worth 10 Microsofts.

- Bill Gates

8.3.1   IoT GATEWAY

An IoT Gateway is a solution for enabling IoT communication, usually device-to-device communications or device-to-cloud communications. The gateway is typically A hardware device housing application software that performs essential tasks. IoT gateways can be essential in making this connection possible **because** gateways act as bridges between sensors/devices and the cloud. Many sensors/devices will "talk" to a gateway, and the gateway will then take all that information and "talk" to the cloud (Figure 8.4).

The above diagram shows how IoT Gateways work together in a meaningful way. From collecting data from sensors, processing it, pushing it to the cloud via internet, and syncing that data across devices, a simple way to conceive of an IoT Gateway is to compare it to your home or office network router or gateway. Such a gateway facilitates communication between your devices, maintains security, and provides an administrative interface where you can perform basic functions. An IoT Gateway does this and much more [5].

8.4   IoT GATEWAY FEATURE SET

A versatile IoT Gateway may perform any of the following:

- Facilitating communication with legacy or non-internet connected devices.
- Data caching, buffering, and streaming
- Data pre-processing, cleansing, filtering, and optimization

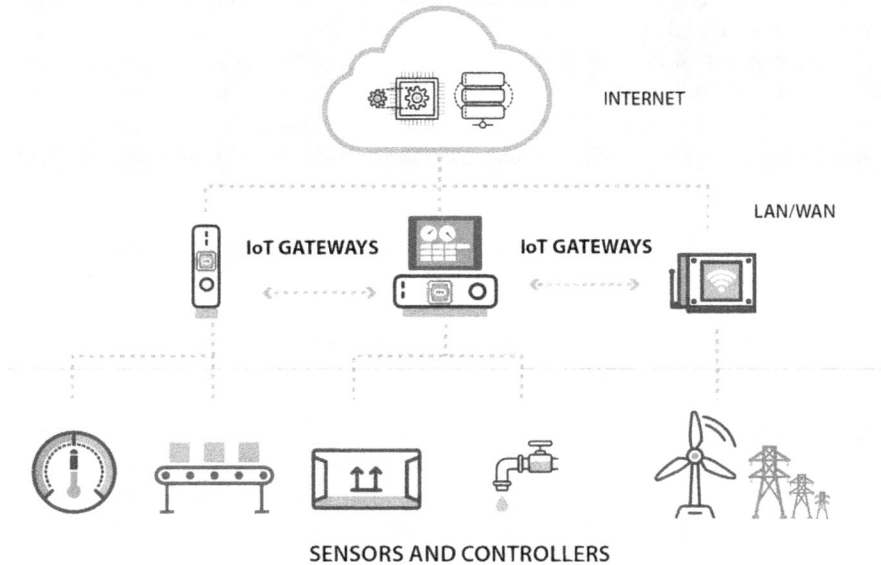

INTERNET

LAN/WAN

IoT GATEWAYS   IoT GATEWAYS

SENSORS AND CONTROLLERS

FIGURE 8.4   Sensors and controllers.

- Some data aggregation
- Device to device communications/M2M
- Networking features and hosting live data
- Data visualization and basic data analytics via IoT Gateway applications
- Short-term data historian features
- Security – manage user access and network security features.
- Device configuration management
- System diagnostics

Let us see an example of IoT-based Smart Energy Meter.

Smart meter is a next-generation meter that is highly efficient and user friendly and provides a great way to save and control the usage of energy. The smart meter is wirelessly connected to users by means of IoT. This means users can easily have control on the meter as per their needs. Smart energy meter has a unit measuring meter. This device will be connected to the main server with the help of an IoT. On smart phones, the user will get the bill on a monthly basis. What is the advantage of this?

The advantage of using smart meter is that on the smartphone the user will get the bill on a monthly basis; he/she can set the limits of consumption per their requirements. Smart Meter application will also have the option to optimize the bill based on the usage, and the user can also control the connected appliances with the help of smartphone, irrespective of whether it is Android or iOS (Figure 8.5) [6].

The above block diagram is a Smart Meter that tells us about how data is being used from various stages to have a better understanding of how it works.

So, the main question how much will it cost?

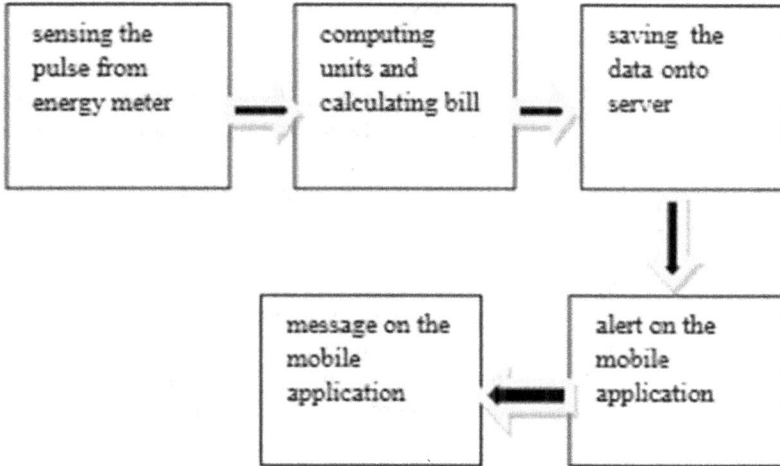

FIGURE 8.5   A Smart Meter which tells us about how data is used.

There is no extra charge for installation. Your smart meter will be installed by your energy supplier at no charge. The cost of the roll out is covered already in your energy bill in the same way as that for the installation and maintenance of traditional meters.

IoT has opened many dimensions of possibilities to act and develop new and smart applications to help us work smart. IoT can help us find to solutions to the most complex problems if used correctly.

What the Internet of Things is really about is information technology that can gather its own information. Often what it does with that information is not tell a human being something, it [just] does something.

We have seen how IoT can help us to minimize our electricity bills and how smartly it works and how beautifully and seamlessly data is transferred across all devices. But here is the catch: how data is being pushed to all devices via internet there is surely another technology involved here in it and that technology is known as Cloud Computing, which helps IoT and Big Data to send their data to all the devices which a user can access. Cloud Computing plays a vital role in every field as it is because of this that the data can travel wirelessly to unlimited devices across the internet.

So, what exactly is Cloud Computing?

8.5   CLOUD COMPUTING

It is the storing and accessing of data and programs over the internet instead of your computer's hard drive. Cloud computing is a virtualized pool of resources, from raw computer power to application functionality, available on demand. Cloud plays a vital role in Big Data as it pushes and syncs data across multiple devices on multiple platforms.

The advantages and essential characteristics of using cloud computing are as follows:

- *On-Demand Self-Service*: A consumer can individually provision computing capabilities as needed automatically without requiring human interaction with each service provider.
- *Broad Network Access*: Capabilities are available over the network. It can be accessed through standard mechanisms, to be used by heterogeneous thin or thick client platforms.
- *Resource Pooling*: A multi-tenant model is used to serve multiple consumers from a pool of computing resources. The customer has no control over the exact location of the provided resources.
- *Rapid Elasticity*: Cloud computing supports elastic nature of storage and memory devices. It can expand and reduce itself according to the demand from the users, as needed.
- *Measured Service*: Cloud computing offers metering infrastructure to customers. Cost optimization mechanisms are offered to users, enabling them to provision and pay for their consumed resources only (Figure 8.6).

The above diagram is a model of cloud computing definition.

Virtualization technology can be used in cloud computing that can take a variety of different types of computing resources as abstracted services to users. These cloud services are divided into Infrastructure-as-a-Service (IaaS), Platform-as-a-Service (PaaS), and Software-as-a-Service (SaaS).

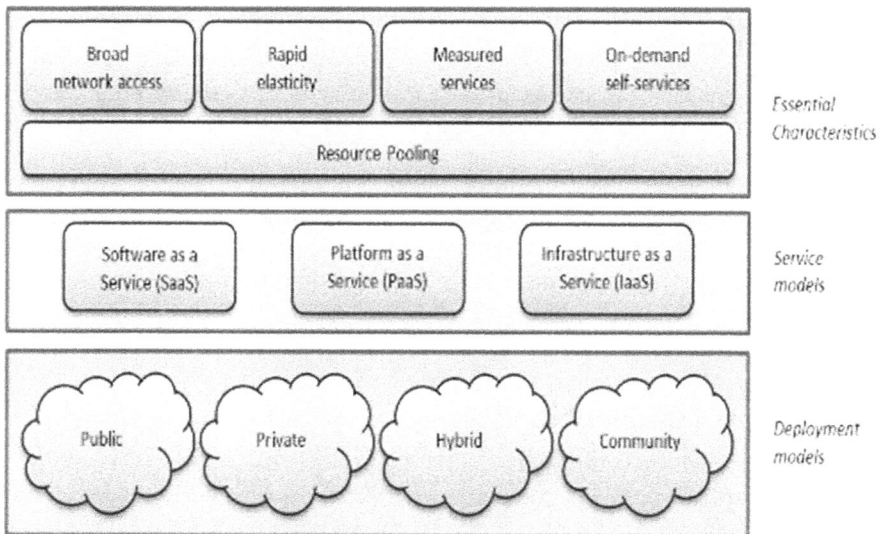

FIGURE 8.6   A model of cloud computing definition.

- *Infrastructure-as-a-Service*: IaaS provides scalable infrastructure, e.g., servers, network devices, and storage disks, to consumers as services on demand. The access to the cloud is provided through various user interfaces, such as web service application programming interface (API), command-line interfaces (CLI), and graphical user interfaces (GUI), each of which provides different levels of abstraction. The consumer has control over operating systems, storage, and deployed applications, but they are not required to manage or control the cloud infrastructure.
- *Platform-as-a-Service*: PaaS provides a platform where users or customers can create and run their applications or programs. The users can build and deliver web applications without downloading and installing the required software, as PaaS service completes the requirements. It is responsible for the runtime execution of users' given task. The most important customers for this layer are the developers.
- *Software-as-a-Service*: SaaS is responsible for delivering various kinds of applications plus the interfaces for the end users. This feature of cloud computing is accessible through web browsers. The SaaS provides the modeling of software deployment where users can run their applications without installing software on his/her own computer. However, this service is limited to the users, i.e., only the existing set of services is available to the customers.

According to the deployment model, a cloud can be classified as public, private, community, and hybrid cloud.

- Public Clouds

  In a public cloud, individual businesses share the premises and access to basic computer infrastructure (servers, storage, networks, development platforms, etc.) provided by a CSP (Cloud Service Providers). Each company shares the CSP's infrastructure with the other companies that have subscribed to the cloud. Payment is usually pay-as-you-go, with no minimum time requirements. Some CSPs derive revenue from advertising and offer free public clouds.
- Private Clouds

  In a private cloud, a business has access to infrastructure in the cloud that is not shared with anyone else. The business typically deploys its own platforms and software applications on the cloud infrastructure. The business's infrastructure usually lies behind a firewall that is accessed through the company intranet over encrypted connections. Payment is often based on a fee-per-unit-time model.
- Hybrid Cloud

  In a hybrid cloud, a company's cloud deployment is split between public and private cloud infrastructure. Sensitive data remains within the private cloud where high security standards can be maintained. Operations that do not make use of sensitive data are carried out in the public cloud where infrastructure can be scaled to meet demands and costs are reduced.

*   Community Clouds
    Community clouds are a recent variation on the private cloud model
    that provides a complete cloud solution for specific business communities.
    Businesses share infrastructure provided by the CSP for software and devel-
    opment tools that are designed to meet community needs. In addition, each
    business has its own private cloud space that is built to meet the security,
    privacy, and compliance needs that are common in the community.

I don't need a hard disk in my computer if I can get to the server faster... carrying
around these non-connected computers is byzantine by comparison.

Steve Jobs

8.5.1 CLOUD APPLICATIONS FOR ENERGY MANAGEMENT

Energy management is the process of monitoring, controlling, and conserving
energy. In Smart Grid, energy management is a major concern. It is needed for
resource conservation, climate protection, and cost saving without compromising work processes by optimally coordinating several energy sources. Building
Energy Management System (BEMS) and Home Energy Management System
(HEMS), dynamic pricing, and load shifting are different applications that were
implemented by researchers in the past to address energy management. The growing importance of energy security and sustainability is resulting in an increased
need to monitor and control energy assets for their optimal use. Smart power grids,
which leverage large scale deployment of Smart Meters and sensors connected to
pervasive communication infrastructure, are being deployed globally. These support real-time, two-way communication between utilities and consumers and allow
software systems at both ends to control and manage power use. Software architecture to support Smart Grid applications will need Cloud platforms as an intrinsic
component due to the many benefits they offer. Information sources from smart
energy grids will approach internet scales, with millions of consumers acting as
data sources. Cloud's data centers were built for such scales of data interactions-
and can scale better than centralized systems at utilities. See the diagram below
(Figure 8.7) for further details:

Information on real-time energy usage and power pricing will need to be shared
with consumers through online portals. The web presence of Cloud platforms again
is well suited for this. In addition, data collected and integrated from various sources
will need to be accessible to third-party applications, after meeting data privacy
concerns. Cloud works seamlessly across devices and transfers data with speed and
security.

Cloud is about how you do computing, not where you do computing.

-Paul Maritz, CEO of VMware

The rapid growth of cloud computing and IoT further promote the sharp growth
of data. Cloud Computing provides safety, access sites, and channels for data assets.
In the paradigm of IoT sensors all over the world are collecting and transmitting data
that will be stored and processed in the cloud. Such data in both quantity and mutual

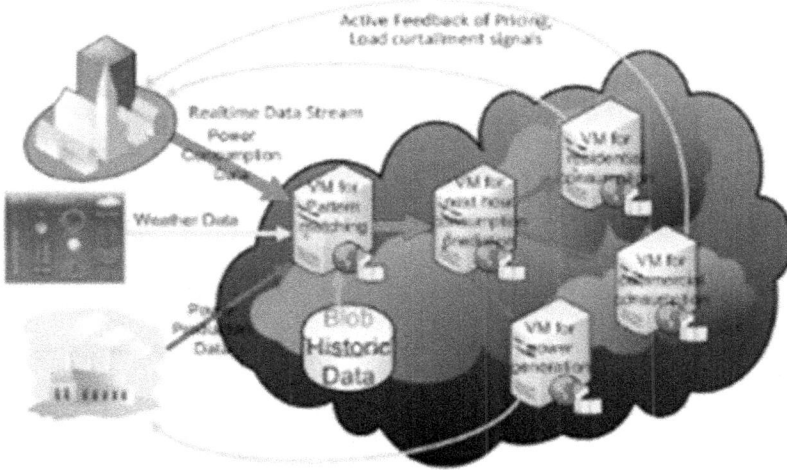

FIGURE 8.7   Smart meters and consumers interacting with demand-response applications on the cloud.

relations will far surpass the capacities of the IT architectures and infrastructure of existing enterprises, and its real-time requirement will greatly stress the available computing capacity. [7]

> Data are becoming the new raw material of business: Economic input is almost equivalent to capital and labor.
>
> *-Said by Economist in 2010.*

8.6   BIG DATA

Big Data is an abstract concept. Apart from masses of data, it also has some other features, which determine the difference between itself and "massive data" or "very Big Data." In 2010, Apache Hadoop defined Big Data as "datasets which could not be captured, managed and processed by general computers within an acceptable scope-" (Figures 8.8 and 8.9) [8].

1. Volume of Big Data
     The volume of data refers to the size of the data sets that need to be analyzed and processed, which are now frequently larger than TB and PB in size. The sheer volume of the data requires distinct and different processing technologies than traditional storage and processing capabilities. For example, all credit card transactions on a day within Europe.
2. Velocity of Big Data
     It refers to the speed with which data is generated. High-velocity data is generated with such a pace that it requires distinct processing techniques. An example, of a data that is generated with high velocity would be the number of tweets in a day.

The Phenomenon of Big Data

1.8ZB

Data generated during 2 days in 2011
(larger than the accumulated amount of data
generated from the origin of civilization to 2003)

750 million

The amount of pictures
uploaded to Facebook

966PB

In 2009, the storage capcity of American
manufacturing industry

209 billion

The number of RFID tags in 2021
(12 million in 2011)

200+TB

Data downloaded during a computer
geek's 2450 thousand hours

200PB

The amount of data generated by
a smart urban project in China

800 billion dollars

Personal location data
in 10 years

300 billion dollars

Medical expense saving by
big data analysis in America

$32+B

The purchase amount of the
4 big companies since 2010

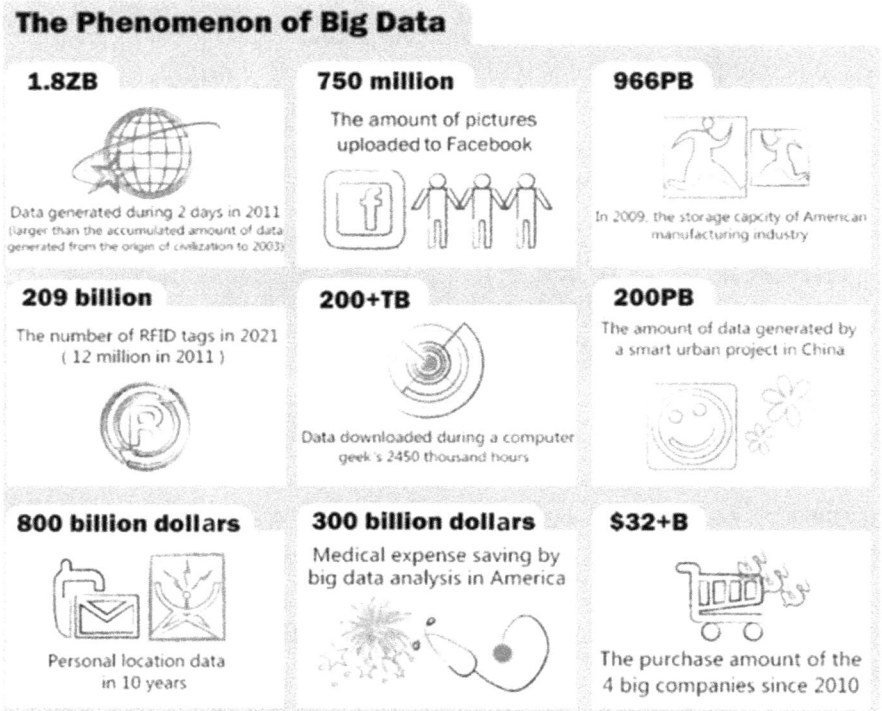

FIGURE 8.8 The 4Vs of Big Data.

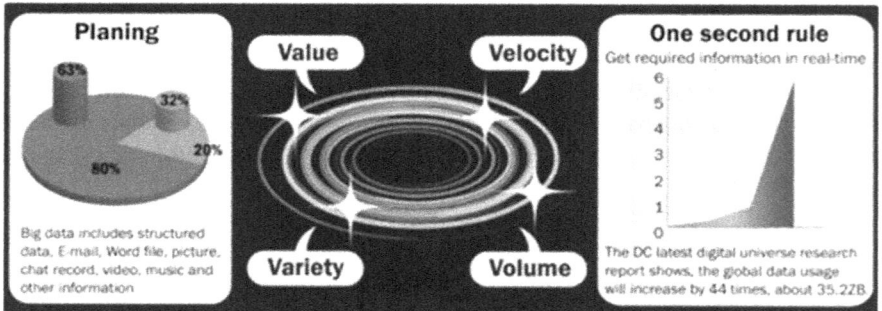

Planing

63%
32%
20%
80%

Big data includes structured
data, E-mail, Word file, picture,
chat record, video, music and
other information

Value

Velocity

Variety

Volume

One second rule

Get required information in real-time

6
5
4
3
2
1
0

The DC latest digital universe research
report shows, the global data usage
will increase by 44 times, about 35.2ZB.

FIGURE 8.9 Volume of Big Data.

3. Variety of Big Data

Variety makes Big Data really big. Big Data comes from a great variety of sources, generally one of three types: structured, semi-structured, and unstructured data. The variety in data types frequently requires distinct processing capabilities and specialist algorithms. An example of high-variety data sets would be the CCTV audio and video files that are generated at various locations in a city.

4. Veracity of Big Data

Veracity refers to the quality of the data that is being analyzed. High-veracity data has many records that are valuable to analyze and that contribute in a meaningful way to the overall results. Low-veracity data, on the other hand, contains a high percentage of meaningless data. The non-valuable information in these data sets is referred to as noise. An example of high-veracity data set would be data from a medical experiment or trial.

You can have data without information, but you cannot have information without data.

-Daniel Keys Moran, American Computer Programmer and
Science Fiction Writer.

8.6.1 RELATIONSHIP WITH CLOUD COMPUTING AND BIG DATA

Cloud Computing is closely related to Big Data. The key components of cloud computing are:

- Traditional applications and services
- Big Data applications and services
- Virtual Resource pool
- Flexible resource scheduling management
- Virtualization
- Distributed storage
- Parallel computing
- Inquiry, analysis, and excavate parallel algorithm.

Big Data is the object of the computation operation and stresses the storage capacity and computing capacity of a cloud server. The main objective of cloud computing is to use huge computing resources and computing capacities under concentrated management, so as to provide applications with resource sharing at the granular level and provide Big Data applications with computing capacity. The development of cloud computing provides solutions for the storage and processing of Big Data. There are many overlapping concepts and technologies in cloud computing, but they differ in the following two major aspects.

1. The concepts are different in the sense that cloud computing transforms the IT architecture while Big Data influences business decision-making; Big Data depends on cloud computing as the fundamental infrastructure for smooth operation.
2. Big Data and cloud computing have different target customers. Cloud computing is a technology and product targeting Chief Information Officers as advanced IT solutions. Big Data is a product targeting Chief Executive Officers focusing on business operations.

8.6.2 RELATIONSHIP BETWEEN IoT AND BIG DATA

The IoT and Big Data are massive, complex ideas. While interrelated, they are also distinct. The IoT consists of millions of devices that collect and communicate information, but Big Data encompasses a much wider landscape. In one sense, the IoT is a series of creeks and rivers that feed into an ocean of Big Data. The enormous collection of connected sensors, devices, and other "things" that represent the IoT. The tools created for Big Data and analytics are useful for corralling the influx of data streaming in from IoT devices. While both Big Data and the IoT refer to collecting large sets of data, only the IoT seeks to run analytics simultaneously to support real-time decisions. While the focus of IoT is more on the immediate analysis and use of incoming data, Big Data tools can still support those functions. Big Data and IoT are distinctive ideas, but they depend on each other for ultimate success. Both emphasize the need for converting data into tangible forms that can be acted upon. IoT and Big Data have an important relationship that will develop as technology advances.

Torture the data, and it will confess to anything.

-Ronald Coase

8.6.3 DATA-DRIVEN SMART ENERGY MANAGEMENT

In the energy sector, large amounts of energy production and consumption data are being generated and the energy systems are being digitized, with the increasing penetration of emerging information technologies. In a certain sense, smart energy system can be regarded as the convergence of the Internet, and the various intelligent devices and sensors spread throughout the energy system. The smart meters usually collect customer's '-electricity consumption information every 15 minutes, and the meter readings alone have created and accumulated massive amount5 of data. It is estimated the number of readings will surge from 24 million a year to 220 million per day for a large-utility company when the AMI (Advanced Metric Infrastructure) is adopted and implemented (Figure 8.10).

The above diagram shows the 4V's and 3E's in Big Data. It means that energy savings can be achieved by Big Data analytics. Data exchange indicates that the Big Data in energy systems need to exchange and integrate with the Big Data from other sources to better realize its value. Empathy means that better energy services can

FIGURE 8.10 "4V' and "3E" characteristics of energy Big Data.

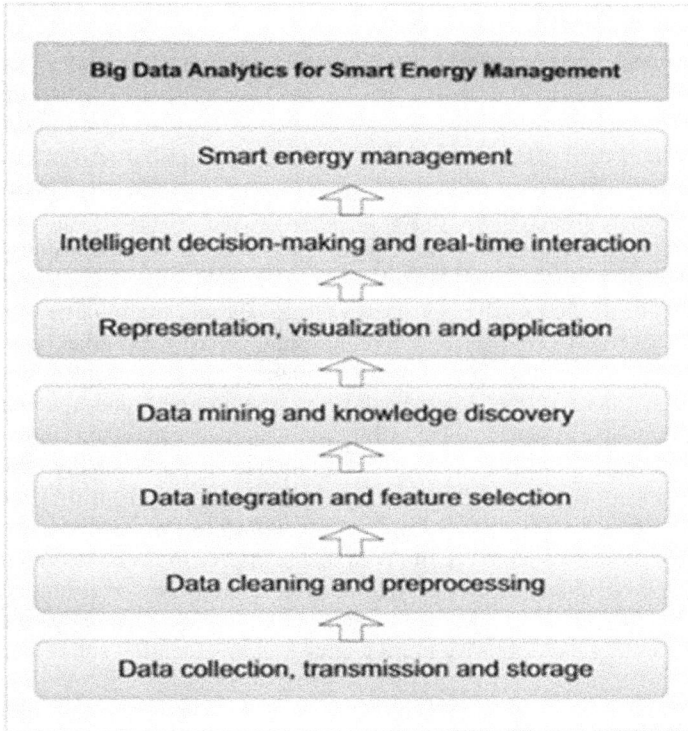

FIGURE 8.11   "4V" and "3E" Characteristics of energy.

be provided to the user in order to better understand energy services and improve it based on energy Big Data analytics (Figure 8.11) [9].

The above diagrams indicates that there are seven major steps for Big Data-driven smart energy management tasks. In the process model, data collection, transmission, storage, cleaning, preprocessing, integration, and feature selection are important preparation phases for Big Data mining. Then, data mining and knowledge discovery are the key steps and the core content of Big Data driven smart energy management. Afterwards, the knowledge extracted from energy Big Data should be represented, visualized, and applied, thus supporting the decision making and control throughout the energy system. Finally, the various smart energy management objectives, including energy efficiency, consumer engagement, real-time monitoring, demand response, intelligent control, and dynamic pricing, can be achieved. Big Data analytics play an important role in the whole process of Smart Grid management, including power generation, power transmission, and power distribution and transformation, as well as demand-side management.

War is 90% information.

-Napoleon Bonaparte

8.7   CONCLUSION

As we have seen in this chapter, Big Data cannot contribute alone to the Smart Energy Sector; it relies on different mechanisms and processes which that all will give the information or data to process it further. IoT + Cloud Computing + Smart Grid contributes to Big Data in order to process, push, pull, and store data across multiple devices. We have seen how IoT works with sensors to cost-efficiently manage the electricity, thus resulting in less lower electricity bills. Large amounts of data are increasingly accumulated in the energy sector with the continuous application of sensors, wireless transmission, network communication, and cloud computing technologies. To fulfill the potential of energy, Big Data and the obtained insights are considered to achieve smart energy management. Smart Grid takes this approach very smartly and differently; it uses smart appliances and IoT- enabled Smart meters to put users in control of the electricity flowing in house and, with the help of cloud computing, he/she can see the cost, billing and pay their bills at on time using only their smart phones. Cloud platforms are well suited to support such data and compute intensive, always-on applications. Big Data combines data from different sensors and, the latest trends so that they can save more money by implementing smart ideas. Technologies generate a meaning full data that is pushed to the cloud and pull by the user through their smart phones. Living in this world where everything is digitized, and smart people should be aware, and they should update themselves regularly on latest trends so that they can save more money by implementing smart ideas.

REFERENCES

1. M.A. Beyer and D. Laney. *"The Importance of Big Data"*, Gartner 2012.
2. R. Papa, and R. Fistola, *"Smart Energy in the Smart City"*, Springer, 2016.
3. G. Mingming, S. Liangshan, H. Xiaowei, and S. Qingwei, "The System of wireless Smart House Based on GSM and Zigbee", *2010 International Conference on Intelligent Computation Technology and Automation*, Changsha, 2010, pp. 1017–1020.
4. H.-P. Chao, "Efficient pricing and investment in electricity markets with intermittent resources", *Energy Policy*, Vol 39, 2011, pp. 3945–3953.
5. H. Chen, X. Jia, and H. Li. *"A Brief Introduction to IoT Gateway"*, ICCTA 2011.
6. Q. Zhu, R. Wang, Q. Chen, Y. Liu, and W. Qin, "IoT gateway: Bridging wireless sensor networks into internet of things", *2010 IEEE/IFIP International Conference on Embedded and Ubiquitous Computing*.
7. A.R. Al-Ali, I.A. Zualkernan, M. Rashid, R. Gupta, and M. AliKarar, "A smart home energy management system using IoT and big data analytics approach" *IEEE Transactions on Consumer Electronics*, Vol 63, No 4, November 2017, pp. 426–434.
8. B. Franks. *"Taming the Big Data Tidal Wave"*, Wiley 2012.
9. K. Zhou, C. Fu, and S. Yang, "Big Data Driven Smart Energy Management: From Big Data to Big Insights", Elsevier, 2015.

9 An Intelligent Security Framework for Cyber-Physical Systems in Smart City

Dukka Karun Kumar Reddy,
H.S. Behera, and Bighnaraj Naik
Veer Surendra Sai University of Technology

CONTENTS

9.1 INTRODUCTION

In modern days, cities are viewed as the reflecting face of a nation. A smart city is a multifaceted and a modernistic urban area that addresses and serves the needs of inhabitants. With technological progress in computing and communication, people are moving to urban areas for improved amenities, healthcare, employment, and education. The metropolitan populace of the world has grown swiftly from 751 million in 1950 to 4.2 billion in 2018. As of 2018 urbanization and human settlements values, 55% of the global population resides in metropolitan areas, and the percentage is projected to reach 68% by 2050 [1]. The latest projections indicate that in developing regions by 2030, the world is expected to have 43 megacities with over 10 million inhabitants. For the successful development of any country, sustainable urbanization is also a key factor. As the world proceeds to urbanize, sustainable advancement relies progressively upon the effective management of urban development. Thereby,

167

urban areas are turning out to be "smart" to guarantee economically feasible and comfortable. Urbanization becomes a global phenomenon.

Smart cities are enabled by CPS, which includes connecting devices and systems through IoT technologies. IoT is a disruptive technology that provides the potential for significant improvements and innovations in business and societal environments [2]. The advancement of IoT implementations is drawn from the synergy of physical components and computational power, specifically by the use of CPSs. CPSs are physical and engineered systems whose operations are integrated, coordinated, controlled, and monitored by computing and communication networks. Just as the internet transformed how people interact with each other, similarly CPS will transmute how we communicate with the physical world around us. CPSs enable a more accurate actuation through monitoring cyber- and physical indicators, sensors, and actuators [3]. The architecture of CPS is shown in Figure 9.1. The progress of emerging technologies leads to the formation of smart cities. A smart city is

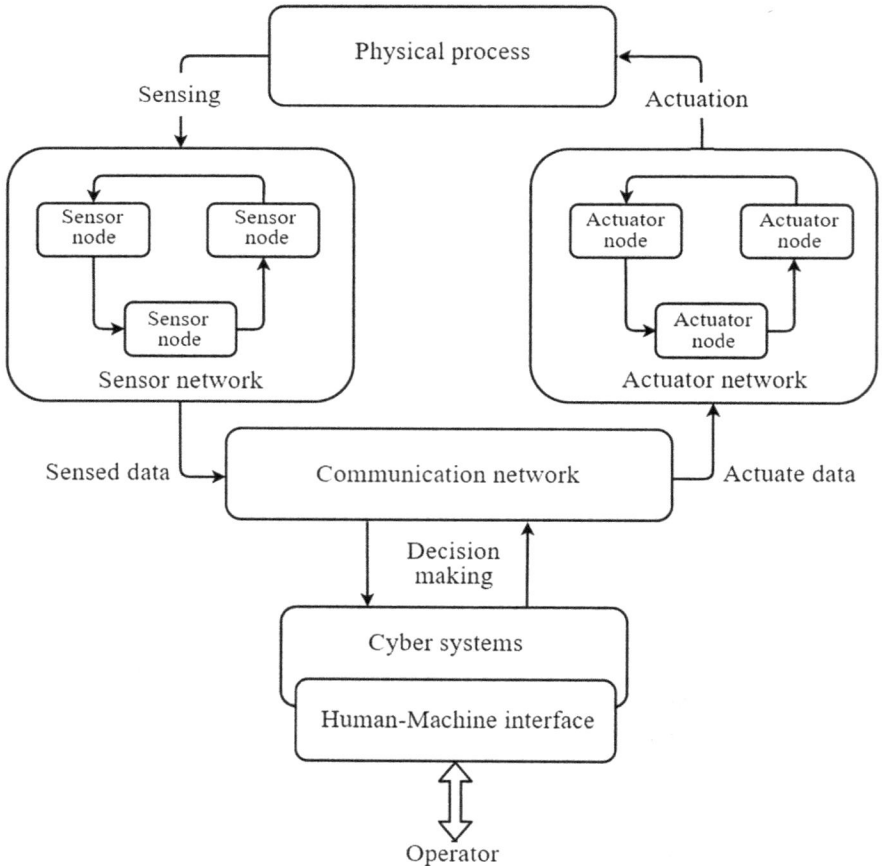

FIGURE 9.1 Cyber physical system architecture.

an amalgamation of the technological platform to address the following challenges like public safety, health care, municipal corporation services, traffic and mobility, and transportation [4]. Designing and organizing smart cities needs specialists from different fields, including sociology, economics, Information and Communications Technology, engineering, and policy regulations. Numerous outlines describing the framework of smart cities have been proposed by industry and scholastic sources. The most generally modified and appropriate reference model is projected by the U.S. National Institute of Standards and Technology [5]. Smart cities are composite systems, called "systems of systems" including process components, infrastructure, and people. Figure 9.2 shows the smart cities model with essential components of environment, mobility, economy, government, people, and living.

CPSs acquire a profound knowledge of the environment by the use of sensors and actuators connected to distributed intelligence in the environment. CPSs are intended to assist applications that can supervise a wide variety of data and generate huge amounts of data from the environment. CPSs require improved design tools that enable design methodology that supports scalability and complexity management. CPSs are intelligent real-time feedback systems with lead to few significant protection worries for the end clients as it is progressively covering the social life aspect. It is a combination of assorted networks that not just includes security of sensor network problems, internet, and communication of mobile networks but specifically privacy-related protection problems, access control, network authentication

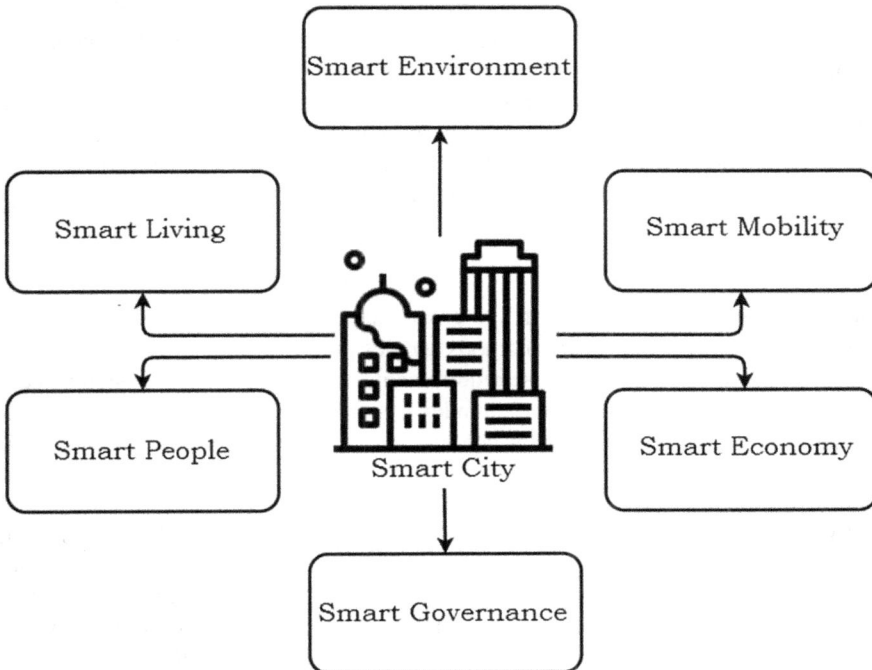

FIGURE 9.2 Essential components of smart city.

problems, information management, etc. [6]. Due to the mounting complexity of the infrastructure framework and security breach have become general phenomena that need to be solved to improve the undesirable weaknesses of the systems. The crucial role of CPSs is trust management for proficient context-aware intelligence services, reliable data mining with improved user confidentiality, and information security. To gain the confidence of people, we must conquer the perception of risk and ambiguity in user utilization and acceptance on IoT applications and services [7]. Thus, the vision of a smart city can be boosted by CPSs using communication and information technologies for more effective and efficient resource management. The term smart cities emphasizes delivering better-quality and innovative services to its citizens by enhancing the infrastructure of the city and reducing the overall costs.

The main contributions of this chapter are as follows:

  i. Design and implement gradient boosting approach with various ML classifiers for fine tuning of parameters.
 ii. Understand the influencing characteristics of DS2OS data set for CPS to recognize the feasible consequences and make it competent to detect various categorical attacks
iii. Evaluate the performance using accuracy, precision, recall, F1-score values, ROC, and precision-recall curve.

The proposed model has an enduring effect on the classification of various attacks and the accuracy of the solution compared to existing ML classifiers. This chapter is organized as follows: Section 9.2 provides a brief study of major related work in this area. Section 9.3 summarizes the proposed gradient boosting technique with various ML algorithms for the data analysis part. Section 9.4 provides the description and preprocessing of the data set. Section 9.5 illustrates the experimental setup and result in analysis. Section 9.6 concludes the chapter.

9.2 LITERATURE REVIEW

The CPSs vary in their levels of operations, applications, and characteristics for smart cities' operation. Different methods and designs have been proposed by several researchers to address these diverse requirements. CPS is a system of systems due to the presence of heterogeneous and complex systems with physical components and the series of networked systems for communication and computation interaction in a continuous manner. The novel prototype for a smart city in an urban environment deals with new innovative services and amenities like environmental monitoring, emergency response, healthcare, business, transportation, energy distribution, and social activities. This augmented use of CPS brings to the fore various attacks and vulnerabilities that could lead to major consequences. So, the security problem is the global issue in this area has prompted researchers to design a secure, efficient, and robust CPS for a smart city. Here, we are presenting some of the methodologies proposed by various researchers in characterizing anomaly detection techniques in various application fields related to a smart city.

Orojloo and Azgomi [8] proposed state-based Semi-Markov Chain model for evaluating the damage to physical components from cyber-attack. The integrity of quantitative data evaluation for CPS security is designed to overcome DoS passive attacks through eavesdropping. The system functionality is composed of cyber and physical components susceptible to confidentiality failure, availability failure, and cyber integrity failure through cyber-attacks. The physical components parameters that are taken into consideration for evaluating the attacks are interdependencies of the component, kinetic knowledge of the attacker, ease of access, cost, and reward.

Sargolzaei et al. [9] proposed a connected vehicle CPS to interact with the surrounding environment to improve mobility, sustainability, and transportation safety. A fault detection technique with a neural network-based control strategy is applied for detection and tracking of fault data injection attacks on the control layer of connected vehicles in real time. The decision-making system is designed to decrease the severity and probability of any consequent accident. The decision-making algorithm structure is constructed on a fuzzy decision-making system that preserves a safe gap among the vehicles.

Junejo and Goh [10] discuss the behavior-based ML approach for intrusion detection. The ML algorithms model the physical process of the CPS to detect anomalous behavior change. The secure water and treatment testbed is demonstrated with a complete replica of the control and physical components of a real-time modern water treatment facility. To study the effectiveness of the simulation model, various types of attacks are inserted at different stages of waterflow like inflow into the process, outflow from the process, and water level of the tank in the system. The effectiveness of the water system is validated by nine ML algorithms, with the Best-first tree exhibiting accuracy (99.72), precision (99.7), recall (99.7), and low false-positive rate.

Mitchell and Chen [11] presented a specification-based methodology with behavior rules for intrusion detection of medical devices embedded in a medical CPS. This methodological behavior transforms the rules into the state machine. The medical CPS device is monitored for its behavior and verified against the deviation from its behavior specification provided for the transformed state machine. The state machine identifies attacks performed by a compromised sensor/actuator that will drive the medical CPS into a certain attack. The state machine distinguishes this through examining the specification-based behavior rules. This specification-based technique performs anomaly-based detection by sensing abnormal patient behaviors in healthcare applications. The vital sign monitor fetches the detection of false alarm rate probability with less than 25% for random attackers and less than 5% for reckless attackers over a varied series of environmental noise levels.

Yang et al. [12] proposed a security model for industrial system integration of cyber and physical domains to promote the efficiency and flexibility of management. A Zone Partition and Conflicts Disposition algorithmic technique is designed for an anomaly-based detection approach for industrial CPS. The modeling knowledge is performed by a backpropagation neural network approach for zone function approximation. The Zone Partition algorithm is used to partition the crucial state, and the Conflicts Disposition algorithm used to resolve magnitude zones with intersections and contradictions among the zones with other physical states. The performance of the model is tested on a testbed with correlation and errors under the spoofing attack. The model performed well on 2,000 instances data set with 5 false-positives, 0 false-negatives, FPR of 1.16%, and accuracy of 99.18%.

Maleh [13] proposed an analysis study of ML algorithms for intrusions detection in aerospace CPS. The proposed study is based on the Cooja IoT simulator is chosen to generate high-fidelity attack data in IoT 6LoWPAN networks. Various machine models with efficient network architecture are selected based on the performance of various network scenarios and network topologies, namely, hello flood, wormhole, and sinkhole IoT intrusion data sets are taken into consideration. The experimental results show that k-NN and RF achieved better results for the Hello flood data set. Similarly, RF, Naïve Bayes, and MLP achieved 100% accurate results for the wormhole and sinkhole data set.

9.3 PROPOSED METHOD

Ensemble learning is the ML concept with the progression of multiple models, predominantly used to perk up the function approximation, classification, prediction, etc. These multiple models, like experts or classifiers, are generated and combined through a strategic perspective for solving a specific computational intelligence problem by producing one predictive optimal model [14]. Boosting is one of the most effectively utilized algorithmic calculations in data science, which comes under the branch of ensemble learning techniques. Boosting is another effective way to enhance the predictive control of the regression tree and classical decision models, wherein a set of base learners (weak learners) alludes and is combined together to craft a strong learner that acquires better execution than a single one. The boosting techniques pull together base learners to shape them into a strong rule. The base learning algorithm of ML is used with different distributions in a sequential order to identify the weak rule and correct its predecessor, which then produces a new rule for the weak prediction. The integration of all these weak rules is used to build a strong single prediction rule by applying the base learning algorithm every time. This entire iterative procedure followed by choosing the right distribution for each round with copious iterations is known as the general boosting technique, shown in Figure 9.3.

Boosting Algorithm

Step 1:   The base learner receives all the distributions and allots them with equivalent weight to each observation.

Step 2:   In case of any misclassification in the prediction, through the initial base ML algorithm, at that point we give higher consideration to the observations having prediction error. At that point, the next base ML algorithm is applied.

Step 3:   Repeat step 2 until the limit of the base learning algorithm with high accuracy is accomplished.

Gradient boosting algorithm (GBA) is the most efficient and popular implementations of the Adaptive Boosting algorithm. The derivation pursues a similar idea of the existing AdaBoost. It comes under the method of supervised learning, which is based on function approximation by optimization of a specific loss function. GBA trains the models in a steady, stabilized, and sequential manner. The main transformation between AdaBoost and GBA is how these algorithms recognize the inadequacies of

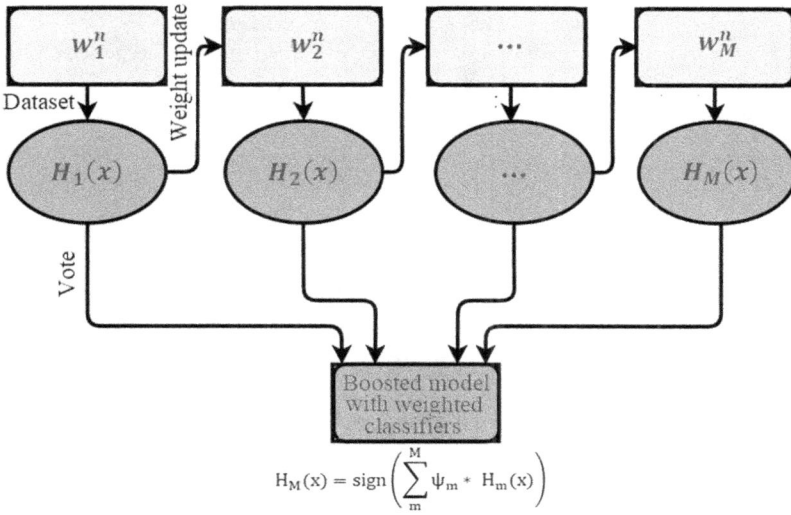

FIGURE 9.3 General boosting technique.

weak learners (e.g. decision trees). Whereas the AdaBoost algorithm recognizes the inadequacies using high-weight data points, GBA achieves the same by gradients in the loss function. GBA trains the model into various sub-models in a regular, additive, and sequential mode, where every individual sub-model progressively minimizes its loss function by using the gradient descent method [15,16]. This course of action successively fits all the sub-models through learning and makes them endowed with a highly accurate estimation for the variables. The GBA integrates all the predictions from various base learners to generate the final output predictions. In GBA, decision trees are the base learners. These decision trees confine every decision tree's nodes that take various subset features for choosing the best split. Where every sub-model (new tree) reflects on the previous tree's errors and taking these considerations into account, a consecutive decision tree is built from previous tree errors, by ensemble all the previous predecessors of the decision trees model..

Gradient Boosting Algorithm

Step 1:   Initialization of weights to $^c w : w_1, w_2, \ldots w_n = \dfrac{1}{n}$ where n is the total instances of the data set i.e. x is the data set and y is corresponds labels. Here, $\{x_1, x_2 \ldots x_n\}$ is the recorded IoT network activity data with various attacks, $x_n = \{x_{n,1}, x_{n,2} \ldots x_{n,k}, y_n\}$ denotes instance of past recorded network data, k is the number of features in the data set, and $y_n \in y$ is any one of the attack types. $y = \{y^1, y^2, y^3, y^4, y^5, y^6, y^7, y^8\}$, i.e.

$$= \begin{cases} \text{DoS attack,Scan,Malicious Control,Malicious Operation,} \\ \text{Spying,Data probing,Wrong setup,Normal} \end{cases}$$

Step 2:   Initialize f with some constant value, where f is the ensemble function.

Step 3: While $^c M < M$ do: where M is the number of models to be grown

Step 3.1 A model is created and gets the hypothesis as $H_m(x_n)$ for all data points, where x_n is the data set and y_n corresponds labels.

Step 3.2 Compute the negative gradient $g_m(H_m(x_n))_{(n=1)}^T$

$$g_m\left(H_m\left(x_n\right)\right) = \left(\frac{\partial \psi\left(y_n, f\left(H_m\left(x_n\right)\right)\right)}{\partial f\left(H_m\left(x_n\right)\right)}\bigg| H_m\left(x_n\right)\right)_{f(H_m(x_n))=f^{m-1}(H_m(x_n))}$$

where T is the number of leaf nodes of tree

Step 3.3 Fitting a new base learner function $h(H(x), \theta_m)$

Step 3.4 Calculating the best gradient descent step size ρ_m

$$(\rho_m, \theta_m) = \arg \min_{\rho, \theta} \sum_{n=1}^T \psi\left(\left(y_n, f^{m-1}\left(H_m\left(x_n\right)\right)\right) \rho h\left(H_m\left(x_n\right), \theta_m\right)\right)$$

Step 3.5 Computing ψ_m, which is a specific loss function $\psi(y, \underline{f})$

Step 3.6 Updating the function estimate for model

$$f_m = f_{m-1} + \rho_m h\left(H_m\left(x_n\right), \theta_m\right)$$

Step 4: Following $^c M$ iterations, the final regression functional output is:

$$f(x) = \sum_M^m f_m(x)$$

9.4 DATA SET DESCRIPTION

Exploratory data observation and analysis are the major criteria required for ML research. Feeding data with an appropriate methodology is the most important task for a classification model. The Distributed Smart Space Orchestration System (DS2OS) was designed by [17] for generating synthetic data using a virtual IoT environment. Their design is an assortment of micro-services communication by Message Queuing Telemetry Transport (MQTT) protocol. The traffic traces gathered from the IoT in the DS2OS data set are taken from different simulated IoT sites with various type of services like smartphones, thermometers, light controllers, washing machines, movement sensors, batteries, smart doors, and thermostats. The DS2OS data set comprises of different IoT nodal communications. The IoT nodes are accessed by using an address like "/agent4/lightcontrol4." The nodes have a type like "/sensorservice" and also a location like "dinningroom." The traffic traces gathered from the IoT in the DS2OS data set are collected from the Kaggle website.

The data set contains 357,952 data samples of categorical data of 13 features with Source ID, Source Address, Source Type, Source Location, Destination Service Address, Destination Service Type, Destination Location, Access Node Address, Access Node Type, Operation, Value, Timestamp, and Normality, as given in Table 9.1. The Normality feature is the target feature classified into Normal and Anomalous categories. The frequency distribution of Normal and Anomalous attacks is given in Table 9.2, where the Anomalous attacks are further classified into seven types of categorical attacks, as DoS attack, 5,780; scan, 1,547; malicious control, 889; malicious operation, 805; spying, 532; data probing, 342; and wrong setup, 122.

TABLE 9.1

Features Description of DS2OS Data Set

S. No.	Features	Data Type
1	Source ID	Nominal
2	Source address	Nominal
3	Source type	Nominal
4	Source location	Nominal
5	Destination service address	Nominal
6	Destination service type	Nominal
7	Destination location	Nominal
8	Access node address	Nominal
9	Access node type	Nominal
10	Operation	Nominal
11	Value	Continuous
12	Timestamp	Discrete
13	Normality	Nominal

TABLE 9.2

Normality Feature Distribution in DS2OS Data Set

Label Feature	Frequency Count	% of Total Data	Ratio between Normal and Anomaly Class
Normal	347,935	97.20	1: 0.03
Anomaly	10,017	2.80	

The frequency distribution of attacks is given in Table 9.3. A brief explanation of these categorical attacks is given in Table 9.4.

9.4.1 DATA PREPROCESSING

Data preprocessing is an important task for addressing and prioritizing the features because data processing increases productivity with reliable and accurate results. Exploratory data observation and analysis are the major criteria required for ML research. Feeding the data with an appropriate classifier is the principal task for a classification model. Therefore, data cleaning is an initial step to deal with in an ML project. The DS2OS data set does not consist of features with a single value. The Timestamp feature is discrete by nature; thus, the feature is not considered during the process. The data set attributes with the features Value and Accessed Node Type are preprocessed accordingly because missing data is the initial step to deal with. The features Accessed Node Type and Value hold missing values and noisy data due to anomaly raised during data transferring. The Accessed Node Type feature consists of 148 instances containing NaN (Not

TABLE 9.3
Distribution of Anomaly in DS2OS Data Set

Target Feature	Frequency Count	Percentage of Total Data	Percentage of Anomalous Data
Anomaly (DoS attack)	5,780	1.61	57.7
Anomaly (Scan)	1,547	0.43	15.45
Anomaly (Malicious control)	889	0.25	8.85
Anomaly (Malicious operation)	805	0.22	8.04
Anomaly (Spying)	532	0.15	5.31
Anomaly (Data probing)	342	0.09	3.41
Anomaly (Wrong setup)	122	0.03	1.22

TABLE 9.4
Anomaly Classes of DS2OS Data set

Attacks	Definition
DoS	In Denial of Service attack, the perpetrator seeks to make the network resource engaged to its proposed users by indefinitely distracting services of a host connected to the Internet.
Scan	To identify the network users, establish the devices and state of systems and take a list of network elements while acquiring data through scanning, the attacker violates the applications and target systems.
Malicious control	Malicious Control is the information controlling mechanism to execute by detecting the system's entry and exit points in an unauthorized manner
Malicious operation	Unauthorized modifications are made to the files, folders, registry entries, other software and operating system by the attacker's malicious code.
Spying	The act of monitoring/obtaining information without the knowledge of the holder of the information using different methods over the Internet.
Data probing	Data probing is the practice of acquiring control over a data network or telecommunication without distracting the network being monitored or altering the structure of data.
Wrong setup	If the system framework is not legitimately set up, it leads to risk posed by compromised credentials varying with the level of problems and the access it provides.

a Number), and the corresponding Normality feature of that row is found to be as anomalous (Malicious Operation). Eliminating these 148 instances may cause loss of significant information; so, the NaN of Accessed Node Type is filled with "/sensorService." Accessed Node Type for anomalous (Malicious Operation) consists of the majority of "/sensorService." Figure 9.4 illustrates the attribute Values, which contain NaN and noisy data. The Value feature contains 25966 instances of "False" that are replaced with 0, 14,460 instances of "True" that are replaced with 1, and 200 instances of "Twenty" that are replaced with 20. The Value features

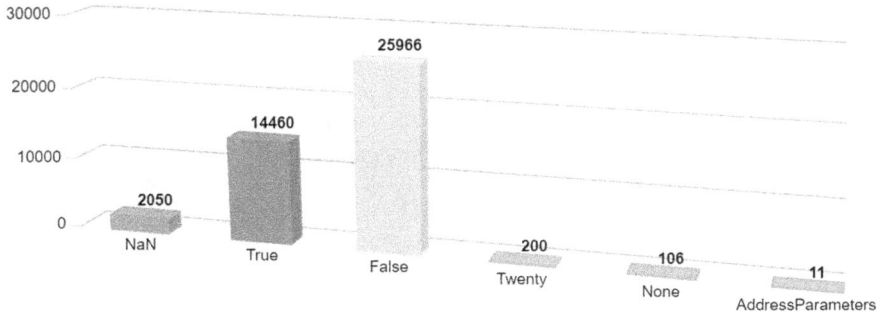

FIGURE 9.4   Value feature with NaN and noisy data.

with "none," "org.ds2os.vsl.core.utils.AddressParameters," and NaN values are
also replaced with 1 to facilitate accuracy. During the process, categorical values
are encoded to numeric data. The preprocessed data set values are further partitioned into training and testing data sets in the ratio of 80%: 20%, i.e., 80% data,
i.e., 286,361 instances, are considered for training and the other 20% data, i.e.,
71,591 instances, are considered for testing.

9.5   EXPERIMENTAL SETUP AND RESULT ANALYSIS

To test the effectiveness of the study and visualization for the data set, it is impera-
tive that the process is tested, since any data set would not demonstrate the appro-
priateness under variability and realistic noise structures. The experimental setup
and results analysis disclose undetected phenomenon cases from the data set. The
experimental setup was done on a LENOVO (IdeaPad 330) laptop where the operating system was Windows 10 Enterprise 64-bit and processor was Intel(R) Core (TM)
i5–8250U CPU @ 3.10GHz (4 CPUs). The system memory contains 8GB RAM. For
data preprocessing, NumPy framework and Pandas framework are used. For data
analysis and for information visualization, Scikit-learn framework and Matplotlib
framework, respectively, were utilized through Spyder-integrated development
environment.

9.5.1   EVALUATION CRITERIA

To study the efficiency of the projected method in comparisons with other models
after efficient learning, performance evaluation aims to estimate the metrics of a
model on future data. The metrics are a statistical determination of decision-making
for the most appropriate model for this work. The confusion matrix enlightens the
classification problem through prediction results. It gives us errors obtained by a
classifier and the types of errors analysis through classification by true positive (TP),
false positive (FP), false negative (FN), and true negative (TN), and correctness of
the classifier through accuracy (Eq. 9.1), true positive rate (TPR or also known as
sensitivity or recall) (Eq. 9.2), false-positive rate (FPR) (Eq. 9.3), precision (Eq. 9.4),

specificity (Eq. 9.5), F1-score (Eq. 9.6), receiver operating characteristic (ROC) curve that is the plot of true positive rate concerning the false positive rate at different threshold situations, and precision-recall curve.

$$\text{Accuracy} = \frac{\text{TP+TN}}{\text{TP+TN+FP+FN}} \tag{9.1}$$

$$\text{TRP} = \frac{\text{TP}}{\text{TP+FN}} \tag{9.2}$$

$$\text{FRP} = \frac{\text{FP}}{\text{TN+FP}} \tag{9.3}$$

$$\text{Precison} = \frac{\text{TP}}{\text{TP+FP}} \tag{9.4}$$

$$\text{Specificity} = \frac{\text{TN}}{\text{TN+FP}} \tag{9.5}$$

$$\text{F1} - \text{score} = \frac{2*\text{TP}}{2*\text{TP+FP+FN}} \tag{9.6}$$

9.5.2 PARAMETER SETTING

This section represents the result analysis of the proposed ML algorithm as MLP, DT, ETC, K-NN, and RF classifiers. Table 9.5 shows the fine-tuned parameter setting of classifiers for better prediction results. The IoT traffic traces from DS2OS consist of 357,952 instances with multi-classification, where 286,361 instances are

TABLE 9.5
Parameter Setting for ML Classifiers

Data set	Technique	Parameter Setting
DS2OS data set	ETC	criterion: gini
		n_estimators: 20
		max_depth: 15
	DT	criterion: entropy
		splitter: best
		max_depth: 10
		n_estimators: 30
	RF	max_depth: 10
		n_estimators: 40
	k-NN	n_neighbors: 2
		algorithm: kd_tree
		Activation = 'relu',
	MLP	batch_size = 200

used for training and 71,591 instances are used for testing purposes. To validate the performance of our proposed method, we considered performance measures such as confusion matrix, TP, TN, FP, FN, TPR/Recall, FPR, F1-score, precision, individual class accuracy, and overall accuracy.

9.5.3 RESULT ANALYSIS

Tables 9.6–9.11 show the evaluation factors of ML classifier with Multi-layer Perceptron (MLP), Random Forest (RF), k-Nearest Neighbor (k-NN), Extra Trees (ET), and Decision Tree (DT), compared with GBA. From Table 9.6, Malicious Control and Wrong Setup classes are precisely predicted, as the False Positive and False Negative of these classes are evaluated as 0. The overall accuracy of the MLP comes to 99.04. Table 9.7 shows the RF classifier evaluated metrics. The TPR value for Malicious Control, Spying, and Wrong Setup classes is 0, and FPR for Normal class is 0.24. The RF classifier produces an overall accuracy of 99.28%. Table 9.8 shows k-NN classifier's overall accuracy of 99.3%, where the TPR for classes DoS Attack and Scan are 0.64 and 0.93. The FPR of Normal class is 0.22. The evaluated results of ET and DT classifiers are given in Tables 9.9 and 9.10, and both classifiers produced the same results. The False-Positive value for Normal and False-Negative value for DoS Attack had 477 instances each. The TPR for all the classes is evaluated as 1.00, except for DoS Attack, with 0.64, and FPR, with all classes as 0 except for Normal, with 0.21. The ET and DT classifiers produced an overall accuracy of 99.37%. The GB classifier produces an enhanced result compared to ET and DT classifiers in classifying the DoS Attack. The proposed classifier gives an overall accuracy of 99.44%; FPR value for Normal is 0.19; and the F1-score for DoS Attack is 0.81, as given in Table 9.11.

Figure 9.5 illustrates the confusion matrix of various ML classifiers. Figure 9.5a of MLP shows that the instances with Normal (528) and Scan (4) are predicted as DoS Attack. The Malicious Operation (25), Scan (23), and Spying (50) are predicted

TABLE 9.6
Evaluation Factors Using MLP Classifier

	Data Probing	DoS Attack	Malicious Control	Malicious Operation	Normal	Scan	Spying	Wrong Setup
True positive	20	1,255	176	140	68,938	297	56	28
True negative	71,521	69,804	71,415	71,426	1,976	71,265	71,485	71,563
False positive	0	532	0	0	148	1	0	0
False negative	50	0	0	25	529	27	50	0
TPR/Recall	0.29	1.0	1.0	0.84	0.99	0.91	0.53	1.0
FPR	0.0	0.007	0.0	0.0	0.07	0.0	0.0	0.0
F1 score	0.44	0.83	1.00	0.92	1.00	0.95	0.69	1.00
Precision	1.00	0.70	1.00	1.00	1.00	1.00	1.00	1.00
Accuracy	1.00	0.99	1.00	1.00	0.99	1.00	1.00	1.00
Overall accuracy	99.04							

TABLE 9.7
Evaluation Factors Using RF Classifier

	Data Probing	DoS Attack	Malicious Control	Malicious Operation	Normal	Scan	Spying	Wrong Setup
True positive	61	808	176	122	69,467	308	106	28
True negative	71,521	70,336	71,415	71,426	1,609	71,267	71,485	71,563
False positive	0	0	0	0	515	0	0	0
False negative	9	447	0	43	0	16	0	0
TPR/Recall	0.87	0.64	1.00	0.74	1.0	0.95	1.0	1.0
FPR	0.0	0.0	0.0	0.0	0.24	0.0	0.0	0.0
F1 score	0.93	0.78	1.00	0.85	1.00	0.97	1.00	1.00
Precision	1.00	1.00	1.00	1.00	0.99	1.00	1.00	1.00
Accuracy	1.00	0.99	1.00	1.00	0.99	1.00	1.00	1.00
Overall accuracy	99.28							

TABLE 9.8
Evaluation Factors Using k-NN Classifier

	Data Probing	DoS Attack	Malicious Control	Malicious Operation	Normal	Scan	Spying	Wrong Setup
True positive	70	808	174	165	69,444	302	106	28
True negative	71,515	70,336	71,412	71,426	1,656	71,256	71,479	71,563
False positive	6	0	3	0	468	11	6	0
False negative	0	447	2	0	23	22	0	0
TPR/Recall	1.00	0.64	0.99	1.00	1.00	0.93	1.00	1.00
FPR	0.0	0.0	0.0	0.0	0.22	0.0	0.0	0.0
F1 score	0.96	0.78	0.99	1.00	1.00	0.95	0.97	1.00
Precision	0.92	1.00	0.98	1.00	0.99	0.96	0.95	1.00
Accuracy	1.00	0.99	1.00	1.00	0.99	1.00	1.00	1.00
Overall accuracy	99.3							

TABLE 9.9
Evaluation Factors Using ET Classifier

	Data Probing	DoS Attack	Malicious Control	Malicious Operation	Normal	Scan	Spying	Wrong Setup
True positive	70	808	176	165	69,467	324	106	28
True negative	71,521	70,336	71,415	71,426	1,677	71,267	71,485	71,563
False positive	0	0	0	0	447	0	0	0
False negative	0	447	0	0	0	0	0	0
TPR/Recall	1.00	0.64	1.00	1.00	1.00	1.00	1.00	1.00
FPR	0.0	0.0	0.0	0.0	0.21	0.0	0.0	0.0
F1 score	1.00	0.78	1.00	1.00	1.00	1.00	1.00	1.00
Precision	1.00	1.00	1.00	1.00	0.99	1.00	1.00	1.00
Accuracy	1.00	0.99	1.00	1.0	0.99	1.00	1.00	1.00
Overall accuracy	99.37							

TABLE 9.10

Evaluation Factors Using DT Classifier

	Data Probing	DoS Attack	Malicious Control	Malicious Operation	Normal	Scan	Spying	Wrong Setup
True positive	70	808	176	165	69,467	324	106	28
True negative	71,521	70,336	71,415	71,426	1,677	71,267	71,485	71,563
False positive	0	0	0	0	447	0	0	0
False negative	0	447	0	0	0	0	0	0
TPR/Recall	1.00	0.64	1.00	1.00	1.00	1.00	1.00	1.00
FPR	0.0	0.0	0.0	0.0	0.21	0.0	0.0	0.0
F1 score	1.00	0.78	1.00	1.00	1.00	1.00	1.00	1.00
Precision	1.00	1.00	1.00	1.00	0.99	1.00	1.00	1.00
Accuracy	1.00	0.99	1.00	1.0	0.99	1.00	1.00	1.00
Overall accuracy	99.37							

TABLE 9.11

Evaluation Factors Using GB Classifier

	Data Probing	DoS Attack	Malicious Control	Malicious Operation	Normal	Scan	Spying	Wrong Setup
True positive	70	855	176	165	69,467	324	106	28
True negative	71,521	70,336	71,415	71,426	1,724	71,267	71,485	71,563
False positive	0	0	0	0	400	0	0	0
False negative	0	400	0	0	0	0	0	0
TPR/Recall	1.00	0.68	1.00	1.00	1.00	1.00	1.00	1.00
FPR	0.0	0.0	0.0	0.0	0.19	0.0	0.0	0.0
F1 score	1.00	0.81	1.00	1.00	1.00	1.00	1.00	1.00
Precision	1.00	1.00	1.00	1.00	0.99	1.00	1.00	1.00
Accuracy	1.00	0.99	1.00	1.00	0.99	1.00	1.00	1.00
Overall accuracy	99.44							

as Normal, and one Normal instance is predicted as Scan. The RF classifier in Figure 9.5b shows that Data Probing (9), DoS Attack (447), Malicious Operation (43), and Scan (16) instances are predicted as Normal, with all remaining classes classified correctly. The confusion matrix of k-NN as shown in Figure 9.5c illustrates that Normal (6) instances are predicted as Data Probing class, and Scan (3) instances are predicted as Malicious Control. The DoS Attack (447), Malicious Control (2), and Scan (19) instances are predicted as Normal classes. The confusion matrices of ET and DT classifiers are shown in Figure 9.5d–e and illustrate that DoS Attack (447) instances are classified as Normal class, with all remaining classes classified correctly. Figure 9.5f illustrates the proposed model confusion matrix. The GBA classifies 855 DoS Attack instances correctly compared to other ML classifiers. The remaining DoS Attack (400) instances are classified as Normal. The confusion matrix clearly shows that GBA classifies all the target classes accurately. Figures 9.6

Confusion Matrix — A:MLP

	Data Probing	DoS Attack	Malicious Control	Malicious Operation	Normal	Scan	Spying	Wrong Setup
Data Probing	20	0	0	0	50	0	0	0
DoS Attack	0	1255	0	0	0	0	0	0
Malicious Control	0	0	176	0	0	0	0	0
Malicious Operation	0	0	0	140	25	0	0	0
Normal	0	528	0	0	68938	1	0	0
Scan	0	4	0	0	23	297	0	0
Spying	0	0	0	0	50	0	56	0
Wrong Setup	0	0	0	0	0	0	0	28

Confusion Matrix — B:RF

	Data Probing	DoS Attack	Malicious Control	Malicious Operation	Normal	Scan	Spying	Wrong Setup
Data Probing	61	0	0	0	9	0	0	0
DoS Attack	0	808	0	0	447	0	0	0
Malicious Control	0	0	176	0	0	0	0	0
Malicious Operation	0	0	0	122	43	0	0	0
Normal	0	0	0	0	69467	0	0	0
Scan	0	0	0	0	16	308	0	0
Spying	0	0	0	0	0	0	106	0
Wrong Setup	0	0	0	0	0	0	0	28

Confusion Matrix — C: k-NN

	Data Probing	DoS Attack	Malicious Control	Malicious Operation	Normal	Scan	Spying	Wrong Setup
Data Probing	70	0	0	0	0	0	0	0
DoS Attack	0	808	0	0	447	0	0	0
Malicious Control	0	0	174	0	2	0	0	0
Malicious Operation	0	0	0	165	0	0	0	0
Normal	6	0	0	0	69444	11	6	0
Scan	0	0	3	0	19	302	0	0
Spying	0	0	0	0	0	0	106	0
Wrong Setup	0	0	0	0	0	0	0	28

Confusion Matrix — D:ET

	Data Probing	DoS Attack	Malicious Control	Malicious Operation	Normal	Scan	Spying	Wrong Setup
Data Probing	70	0	0	0	0	0	0	0
DoS Attack	0	808	0	0	447	0	0	0
Malicious Control	0	0	176	0	0	0	0	0
Malicious Operation	0	0	0	165	0	0	0	0
Normal	0	0	0	0	69467	0	0	0
Scan	0	0	0	0	0	324	0	0
Spying	0	0	0	0	0	0	106	0
Wrong Setup	0	0	0	0	0	0	0	28

Confusion Matrix — E:DT

	Data Probing	DoS Attack	Malicious Control	Malicious Operation	Normal	Scan	Spying	Wrong Setup
Data Probing	70	0	0	0	0	0	0	0
DoS Attack	0	808	0	0	447	0	0	0
Malicious Control	0	0	176	0	0	0	0	0
Malicious Operation	0	0	0	165	0	0	0	0
Normal	0	0	0	0	69467	0	0	0
Scan	0	0	0	0	0	324	0	0
Spying	0	0	0	0	0	0	106	0
Wrong Setup	0	0	0	0	0	0	0	28

Confusion Matrix — F:GBA

	Data Probing	DoS Attack	Malicious Control	Malicious Operation	Normal	Scan	Spying	Wrong Setup
Data Probing	70	0	0	0	0	0	0	0
DoS Attack	0	855	0	0	400	0	0	0
Malicious Control	0	0	176	0	0	0	0	0
Malicious Operation	0	0	0	165	0	0	0	0
Normal	0	0	0	0	69467	0	0	0
Scan	0	0	0	0	0	324	0	0
Spying	0	0	0	0	0	0	106	0
Wrong Setup	0	0	0	0	0	0	0	28

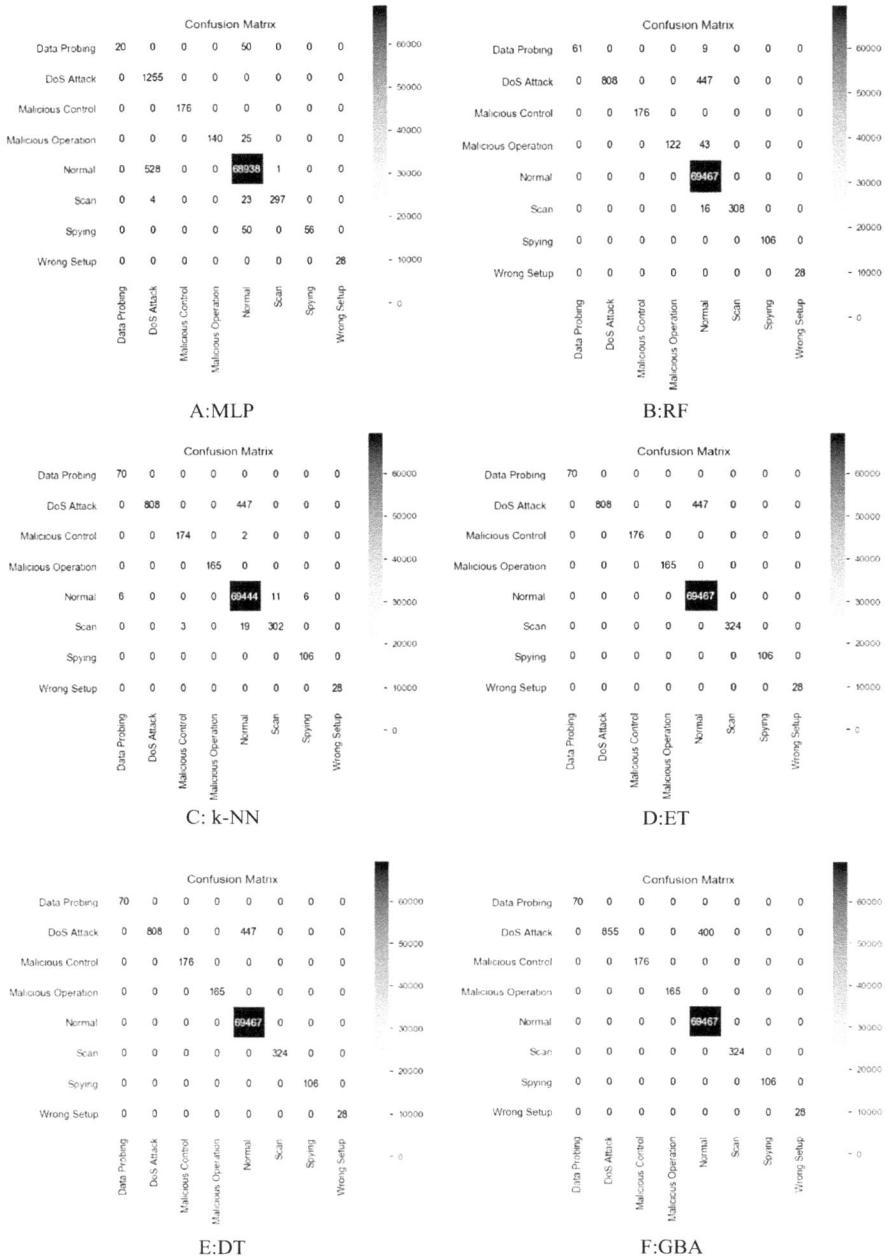

FIGURE 9.5   Confusion matrix of (a) MLP, (b) RF, (c) k-NN, (d) ET, (e) DT, and (f) GBA.

and 9.7 depict the ROC and precision-recall curves of MLP, RF, k-NN, ET, DT, and GBC, respectively, whereas the ROC curve in the case of MLP the Data Probing, Malicious Operation, and Spying is equivalent to 0.95,0.95, and 0.93. The remaining models with all the classes produce area under the curve of 1.00. Similarly, the precision-recall curve in the case of MLP for Datatype Probing, Spying, DoS Attack, and Malicious Operation area under the curve is equivalent to 0.32, 0.53, 0.88, and 0.85. The remaining models with all the class have area under the precision-recall curves of 1.00, except the DoS Attack class, which gives an area of 0.903. The precision-recall curve is taken into consideration in view of imbalanced data set. Finally, Figure 9.8 represents the various graph plots for TPR, FPR, F1-score, precision, accuracy, and overall accuracies of the models concerning the data set classes.

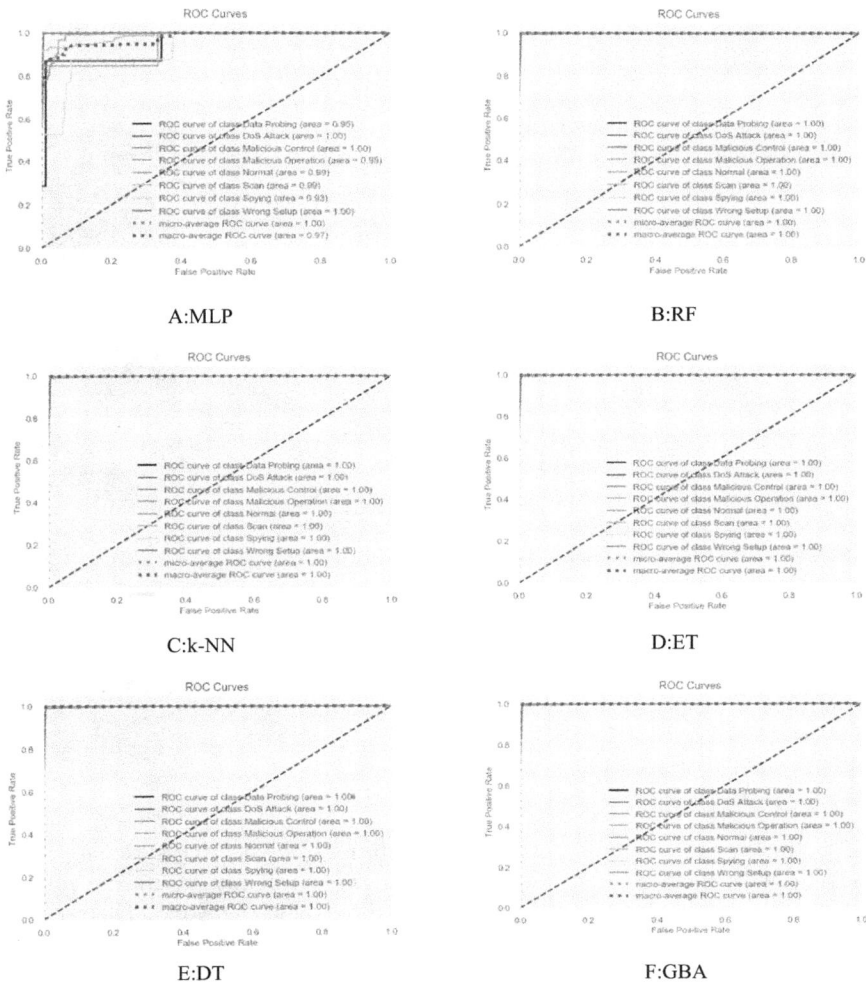

FIGURE 9.6   ROC of (a) MLP, (b) RF, (c) k-NN, (d) ET, (e) DT, and (f) GBA.

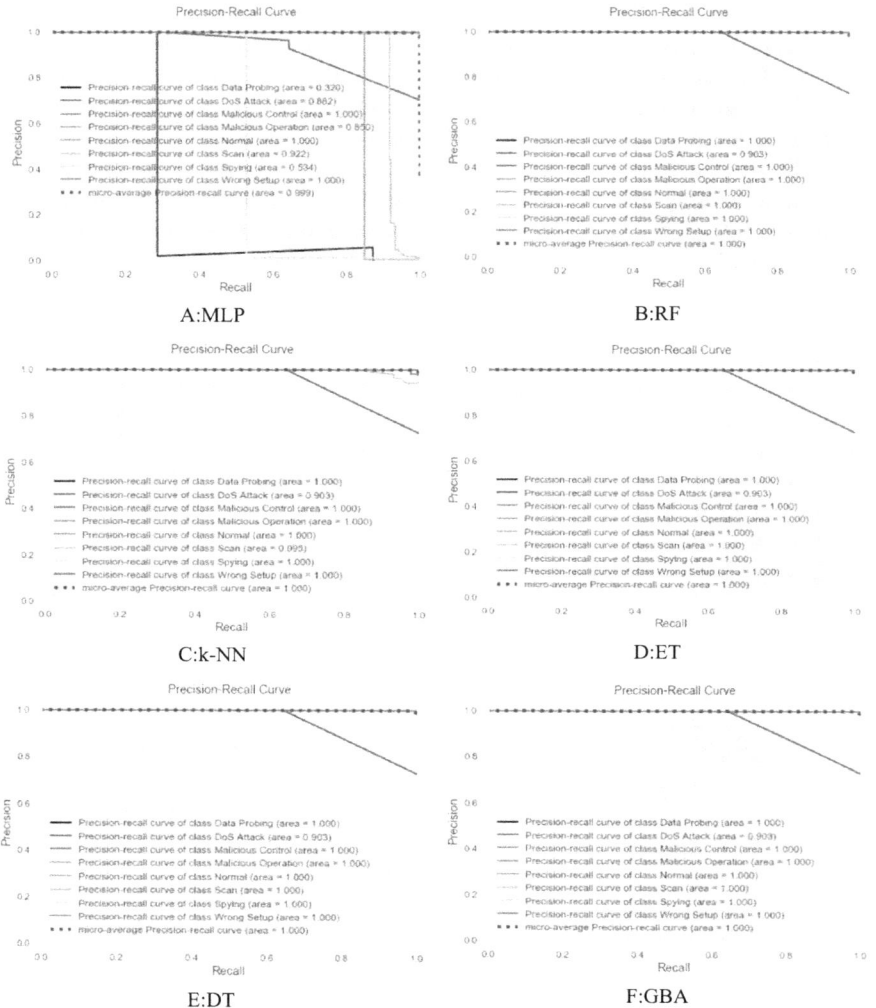

FIGURE 9.7 Precision-recall curve of (a) MLP, (b) RF, (c) k-NN, (d) ET, (e) DT, and (f) GBA.

9.6 CONCLUSION

The technological evolution of CPS is the future of engineering systems. In particular, it will enhance the quality of services and eventually benefit the environment as they are implemented in smart cities throughout the world. The proposed GB algorithm framework is used for analyzing the security of a CPS with sophisticated ML techniques. In conjunction with the objective of finding various anomalies and attack classifications from DS2OS data set, this chapter proposes an intelligent

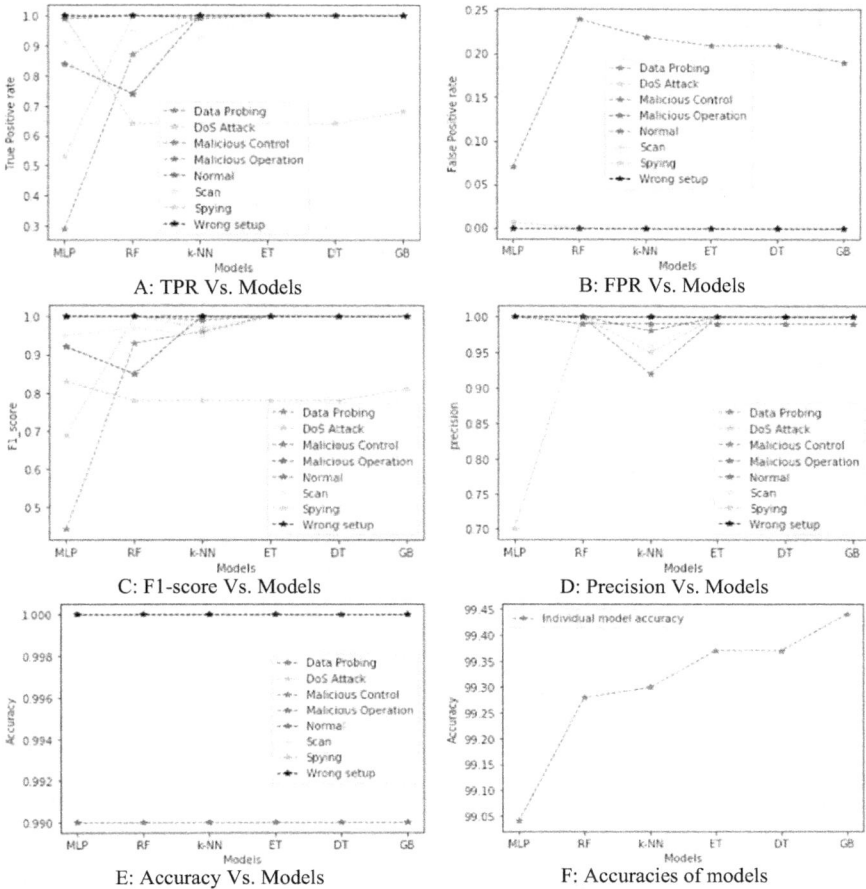

FIGURE 9.8   (a): TPR Vs. Models, (b) FPR Vs. Models, (c) F1-score Vs. Models, (d) Precision Vs. Models, (e) Accuracy Vs. Models, and (f) Accuracies of models

security framework for an anomaly exposure system by analyzing the attack and anomaly classification in the network. We conducted a distinct combination search to assess the best learning parameters for ML classifiers on the DS2OS data set. The ET and DT classifiers achieved an overall accuracy of 99.37%. The experimental study and result analysis from modeling knowledge of GBA shows a promising approach for anomaly detection in categorizing Normal and DoS Attack comparative with ML techniques . The FPR for all the attacks is evaluated as zero and the F1-score achieved by all the classes is 100%, except for DoS Attack (81%). From the experimental examination, it can be concluded that GBA is a viable approach in detecting anomalies and attacks in smart home devices when compared with the different ML algorithms that were used in this study.

REFERENCES

1. "UN," Notes and Queries, 2018. [Online]. Available: https://www.un.org/development/desa/en/news/population/2018-revision-of-world-urbanization-prospects.html.
2. P. B. Dash, J. Nayak, B. Naik, E. Oram, and S. H. Islam, "Model based IoT security framework using multiclass adaptive boosting with SMOTE," *Secur. Priv.*, no. April, pp. 1–15, Jun. 2020.
3. L. Cao, X. Jiang, Y. Zhao, S. Wang, D. You, and X. Xu, "A survey of network attacks on cyber-physical systems," *IEEE Access*, vol. 8, pp. 44219–44227, 2020.
4. M. A. Jabbar, S. Samreen, R. Aluvalu, and D. K. Kiran Reddy, "Cyber physical systems for smart cities development," *Int. J. Eng. Technol.*, vol. 7, no. 4.6, p. 36, Sep. 2018.
5. R. Khatoun and S. Zeadally, "Smart cities," *Commun. ACM*, vol. 59, no. 8, pp. 46–57, Jul. 2016.
6. A. Singh and A. Jain, "Study of cyber attacks on cyber-physical system," *SSRN Electron. J.*, pp. 686–690, 2018.
7. B. Naik, M. S. Obaidat, J. Nayak, D. Pelusi, P. Vijayakumar, and S. H. Islam, "Intelligent secure ecosystem based on metaheuristic and functional link neural network for edge of things," *IEEE Trans. Ind. Informatics*, vol. 16, no. 3, pp. 1947–1956, Mar. 2020.
8. H. Orojloo and M. A. Azgomi, "A method for modeling and evaluation of the security of cyber-physical systems," *2014 11th International ISC Conference on Information Security and Cryptology*, Tehran, 2014, pp. 131–136.
9. A. Sargolzaei, C. D. Crane, A. Abbaspour, and S. Noei, "A machine learning approach for fault detection in vehicular cyber-physical systems," *Proceeding of 2016 15th IEEE International Conference on Machine Learning and Applications*, Anaheim, CA. ICMLA 2016, pp. 636–640, 2017.
10. K. N. Junejo and J. Goh, "Behaviour-based attack detection and classification in cyber physical systems using machine learning," *Proceedings of the 2nd ACM International Workshop on Cyber-Physical System Security - CPSS '16*, New York, no. Ml, pp. 34–43, 2016.
11. R. Mitchell and I.-R. Chen, "Behavior rule specification-based intrusion detection for safety critical medical cyber physical systems," *IEEE Trans. Dependable Secur. Comput.*, vol. 12, no. 1, pp. 16–30, Jan. 2015.
12. J. Yang, C. Zhou, S. Yang, H. Xu, and B. Hu, "Anomaly detection based on zone partition for security protection of industrial cyber-physical systems," *IEEE Trans. Ind. Electron.*, vol. 65, no. 5, pp. 4257–4267, May 2018.
13. Y. Maleh, "Machine learning techniques for IoT intrusions detection in aerospace cyber-physical systems," in *Studies in Computational Intelligence*, vol. 836, Springer International Publishing, pp. 205–232, 2020.
14. R. Polikar, "Ensemble learning," *Scholarpedia*, vol. 4, no. 1, p. 2776, 2009.
15. J. Friedman, T. Hastie, and R. Tibshirani, "Additive logistic regression: a statistical view of boosting (With discussion and a rejoinder by the authors)," *Ann. Stat.*, vol. 28, no. 2, pp. 337–407, Apr. 2000.
16. J. H. Friedman, "Machine," *Ann. Stat.*, vol. 29, no. 5, pp. 1189–1232, Oct. 2001.
17. M. O. Pahl and F. X. Aubet, "All eyes on you: Distributed multi-dimensional iot microservice anomaly detection," *14th International Conference on Network and Service Management, CNSM 2018 and Workshops, 1st International Workshop on High-Precision Networks Operations and Control, HiPNet 2018 and 1st Workshop on Segment Routing and Service Function Chaining, SR+SFC 2*, Rome, 2018, no. CNSM, pp. 72–80.

10 Big Data and Its Application in Smart Education during the COVID-19 Pandemic Situation

Saumyadip Hazra, Souvik Ganguli
Thapar Institute of Engineering and Technology

CONTENTS

10.1 INTRODUCTION

From 2020, the world has been witnessing a pandemic situation that has put a stop to almost all the sectors. Since then, the situation has not been brought under the control despite the efforts put in by the government. The global pandemic situation has prevented students from studying in colleges and schools, all the technological meets and conferences have been called off, and the ongoing research has also been affected. The education industry has experienced a huge loss, forcing the authorities to look for an alternative solution. The concept of smart education through online mode has become more popular in recent months.

Schools and colleges that were shut down due to the impact of the pandemic started conducting online lectures. A proper time-table was prepared and the online lectures were being engaged according to that. Here, the teacher started delivering a lecture from a presentation virtually. Many platforms like Zoom, Google meet, CISCO Webex etc gained popularity due to this, and became one of the widely used software or applications. The doubts and queries of the students were put to the teacher either by typing it or by asking it with the help of the enabled microphone option. For small kids, the teacher started teaching them through activities that they performed while keeping their camera on.

Conduction of laboratory experiments seemed to be a challenge for the school or college authorities as students had to be present physically. Some of the laboratories which taught the students about some software or programming started virtual lessons without much difficulty and the students were able to execute the code based on what was shown to them on the screen. Thus, the simulation software was being used and a new category of laboratories known as virtual laboratory gained much popularity. Through virtual laboratories or simulation software, students were able to place various components required by them and perform the experiment virtually to generate the result. Virtual laboratories also have an added advantage of proper monitoring of students to check whether they have performed the experiment or not.

Many competitions also started happening online that were previously scheduled to be conducted in offline mode. These games or competitions included the joining of students in time. Then the host asked the students to perform their respective parts. The seminars or talk shows that were supposed to happen in institutions, schools, or colleges started taking place online and are known as webinars. It carries on like any seminar where first of all the participants are addressed and then the main speaker of the seminar delivers the content. At the end of the webinar, the participants are allowed to ask their queries through a microphone option or chat. All the candidates also get awarded an e-certificate for attending the webinar.

A large number of research scholars were affected due to the pandemic as all of the conferences were called off. It caused a huge amount of problem to the organizers as well as to the participants as their research papers were also on hold. As a solution to it, the organizers decided to conduct an e-conference. A schedule was prepared for it as it was done for any conference where, first, the guests and participants were welcomed. The e-conference was then addressed by the keynote speaker and, after that, the process of presentation of paper started. Each presentation was allotted a specific amount of time at the end of which questions were raised by the panel and suggestions were made on it. The conference ended with a valedictory ceremony where the various awards as decided by the organizers were presented. For attending the e-conference, the participants also received a certificate.

Big Data analytics is used for analyzing a very large amount of data quickly and to perform some operations on it. The importance of Big Data analytics rose as the conventional data analyzing processes were not able to provide satisfactory results for large data. The size of data in Big Data analytics may reach up to zettabytes. When a school or college holds an online lecture, a large number of students take part in that. The same is the case with the webinars and e-conferences. For handling all the students or participants at the same time, Big Data analytics is used by the online platform that organizes and structures the available data. It also handles all the students or participants based on their camera or microphone feature. The operation of a Big Data algorithm can be understood using the 6V concept, which has been represented in Figure 10.1. Earlier there were only 3 Vs, but as the time passed, more Vs have been added. Volume represents the amount of data it can handle, which can rise to several zettabytes. Variety

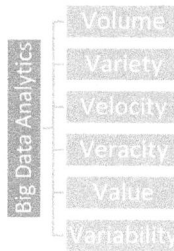

FIGURE 10.1   The 6 Vs in Big Data analytics.

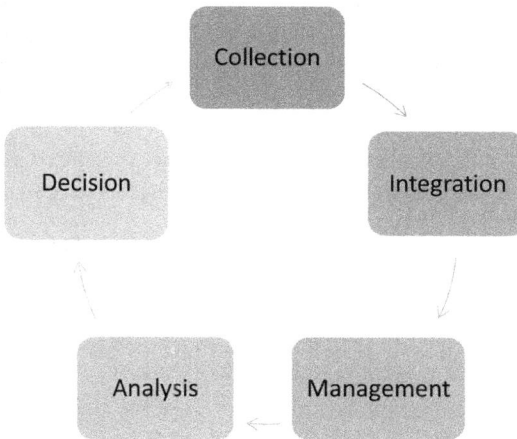

FIGURE 10.2   The cycle of operation of Big Data analytics.

represents the different types of data that are to be processed. The data may be structured, semi-structured, or unstructured. Velocity is the speed with which the data is being processed and is probably the most important V. Veracity represents the correctness of the data as the processing of wrong data may result in some irregularities. Value is the identification of data that has some value. Not all data contain value, but they may be quite important. Variability is used for the processing of data in multiple ways [1,2].

The proper working of Big Data is based on five steps that follow a continuous cycle, as shown in Figure 10.2. The first step is the collection of data, wherein the data is collected from the sources available. As the data starts accumulating, the integration process is done where the information from all the sources is integrated. After the integration, the data is managed, wherein it is segregated and is then stored. The stored data is then analyzed based on the type of operation desired, and, finally, the decision is taken [3,4].

The COVID-19 pandemic presented a large number of obstacles in front of schools, colleges, and educational events. Keeping everything in mind, they all adapted to the situation and brought changes to their existing system, hence depicting flexibility.

The remainder of the chapter is organized in the following way. The effect of COVID-19 on the education sector is discussed in Section 10.2. Conduction of online lectures is addressed in Section 10.3. In Section 10.4, the working of the learning management system is deliberated, while in Section 10.5 the concept of virtual laboratories is taken up. Discussions on webinars and e-conferences have been considered in Sections 10.6 and 10.7, respectively. Section 10.8 presents the conclusion of the chapter.

10.2 EFFECT OF COVID-19 ON THE EDUCATION SECTOR

Right from the beginning of the COVID-19 pandemic, every aspect of human life has been affected. The education sector is no different and has been affected critically. It all started around March 2020 when the educational institutes started taking the strict decision for the suspension of classes and closure of the institute till further notice. COVID-19 was declared as a global pandemic by the World Health Organization (WHO) on 12 March 2020; since then, many countries across the world have taken rapid decisions to order the closure of various educational institutes, such as schools, colleges, coaching institutes, etc. It can be considered as a major step and, hence, a turning point in the education sector as the authorities were not prepared for the pandemic and had no clue on how to stop it.

The decision to close schools and colleges was taken keeping in mind the health of children as they possess a comparatively weaker immune system; also, this would reduce the number of people gathering at a particular place. There is a positive impact of the closure of the school that the children could be kept safe at their homes, and hence the chances of transmission of disease would be comparatively less. The negative impact of closure was that the academics of the children got badly affected. Since then, the schools and colleges could not open due to the pandemic, and the effect on academics of children became more adverse. In colleges, the students hail from different parts of the country, and it was not even possible for them to travel due to travel restrictions. These prolonged conditions affected the children not only from the academic point of view but also mentally and physically. Further, the laboratory sessions, which are considered as crucial from the learning perspective as it includes the application of theoretical learning into practical implementation, was no longer taking place. Even if the students were gaining knowledge from any source, such as the internet or scanned document of the reference book, the knowledge is still considered incomplete due to the unavailability of laboratory sessions. To keep the pace of education going as the pandemic was not ending, a new perspective of education was demanded, and that was smart education. Smart education for the students of schools and colleges was intended to teach the students virtually from their home, and that was only possible through online virtual teaching modes. Smart education became a necessity for schools and colleges to keep the flow of education going. The online mode of study or smart education only eased up the learning process a bit, but the laboratory sessions still could not take place as it required the students to perform certain tasks. For the completion of laboratory sessions, smart laboratories were being organized or virtual laboratories were being implemented. The details about these laboratory sessions have been discussed in the upcoming sections.

Seminars or the previously scheduled talks were also affected. The main motive behind the organization of a seminar is either to obtain an outcome for a particular set of the problem through mutual discussion and presentation or to make the audience aware about some topic. Due to the pandemic situation, all these activities could not take place anymore in schools, colleges, or any other institutes. This had a major effect on the students who were to gain knowledge due to these seminars and proved to be a negative point for not only from their academic point of view but also for their overall development. Since there was no way that these seminars could be organized, a new smart way of organizing them was required, and the only feasible option was to conduct the session online [5–7].

Apart from these aspects, the pandemic also affected the conferences that were previously scheduled for research paper presentation. All the conferences were either called off or were postponed for the next few days, as it would have led to the gathering of a large number of people at a single place and would also have necessitated the traveling of participants from different places to attend the conference. As a result of this step, particularly the research scholars were badly affected as their research work was not going to get published due to the pandemic situation. A large number of conferences take place throughout all the states in India, which all of a sudden stopped due to the COVID-19 situation. The conferences not held were piling up, and some of the forthcoming conferences were even canceled. As a large number of conferences were piling up, a solution was required to conduct them so that the research work can continue. The term "e-conference" was introduced, which acted as a solution for the pending conferences.

10.3   CONDUCTION OF ONLINE LECTURES

The pandemic situation presented a great challenge as well as an opportunity in front of the educational authorities, such as schools and colleges, to keep the educational pace going. The pandemic situation shunned all the schools and colleges which can be considered as a great decision as it reduced the contact of children from each other and the staff members of schools or colleges. Further, it reduced the number of people gathering in a particular place. Thus it protected the children who are considered to have a comparatively weaker immune system. The challenge was gradually overcome by the organization of classes and lectures through online mode, hence shifting towards smart education. The opportunity was presented as it would shift the paradigm from conventional classes towards smart classes. India has been constantly working towards the development of smart classes, and this period has proved to be of great importance in terms of that shift. All the schools and colleges, no matter which grade they are teaching, started delivering the classes and lectures through the online mode.

The online lectures are being organized with the help of an application or software which can be accessed through smartphones or computers. An active internet connection is required, and the student needs to create an account on that platform. The instructor acts as a host for the session, which has to be scheduled by him. The link for joining that session is then provided by the instructor to the students who need to join the lecture at the scheduled timing. After joining the session, the student can find the option of microphone and web camera, and it is dependent on the instructor or the host to decide whether the students are allowed to use their microphone and web camera or not. If it is decided by the instructor that all the students need to remain muted and their web camera should be off, then forcefully all the microphones would be muted and web cameras would be kept off for the students. The instructor then can start delivering the lecture using the presentation prepared by him through online mode. There is also a chatbox feature in it where the students can put up their queries or can discuss doubts. If any student has any doubt and feels like he or she needs to ask it, then he or she can use the raise the hand feature, which enables notification on the screen of the instructor and the instructor can immediately get to know the query of the student. Then he or she can let the student ask his doubt which is cleared by him afterward. There is also a pen feature that can be used by the instructor for marking or locating or highlighting any part of the presentation.

The question now arises about how the maximum can be gained from the online lecture as most of the students have not attended any online lectures before. Figure 10.3 describes the steps that need to be followed by the students when they

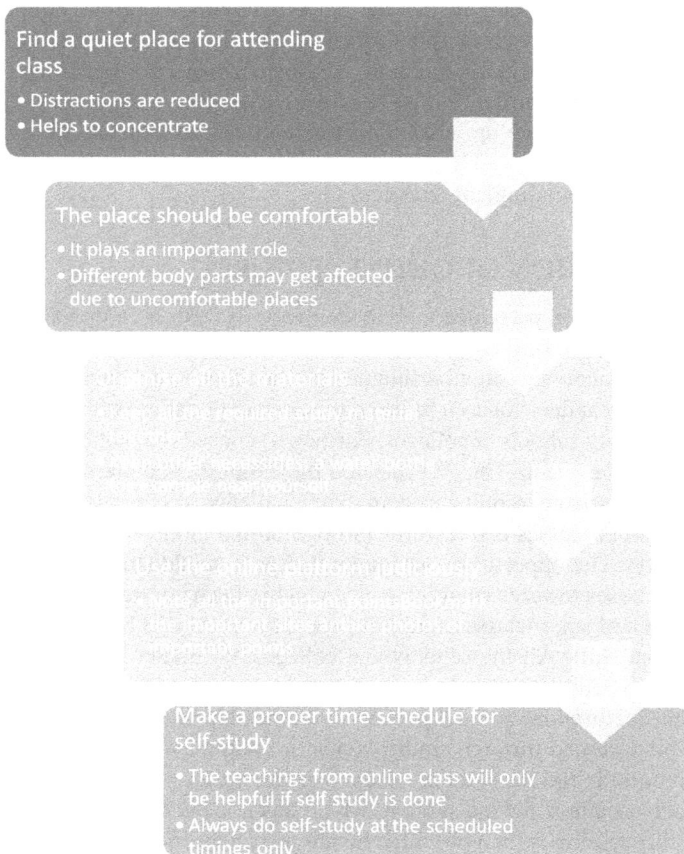

FIGURE 10.3 How to study effectively from online lectures.

are studying for online lectures. The students should find a perfect place where they should study every day while the lectures are going on. The place should be comfortable and should be free from all distractions. If the place is not comfortable, then the concentration will be diverted, and prolonged usage of an uncomfortable environment result in various body ailments. All the study material should be arranged in one place and, if possible, it should have all the other necessities such as a water bottle. The important points told by the teacher during the lecture should be noted down and any material should be downloaded from the internet whenever required. The most important part is self-study, which should also follow a proper schedule.

There are many problems that are also associated with these events. In the case of a poor internet connection, the session cannot be started. During the online lecture, if the internet connection is interrupted due to some reason, then it may result in getting disconnected from the session. In case the participants or the instructor is not familiar with the software, this may create a lot of nuisance. Besides this, most of the instructors have spent most of their career teaching the students through conventional classrooms, and it becomes difficult for them to suddenly switch to online mode. The instructors are also not generally habituated in teaching through presentations. The same is the case with the students also who have not studied through online mode, and this would create difficulties for them.

The volume of data is large to handle, and hence certain algorithms are incorporated by the software or the application for ensuring the smooth operation of the whole online lecture. After the collection of data from the users, in this case, the teacher and the students, the algorithm first sorts the data. It may include various types of sorting based on the categories on which the algorithm has been built. The algorithm also ensures that whatever is being delivered by the teacher is delivered to the student screens without any delay and vice versa. When the number of students increases, it makes sure that the website or application or software does not crash due to its inability to handle the volume of data efficiently. There may be assignments or chats that have happened among the students, which are also saved, so that it can be used further, and this is also performed with the help of Big Data analytics [8–10].

10.4   LEARNING MANAGEMENT SYSTEM (LMS)

When the schools and colleges started conducting the lectures through online mode, a system was required that could manage the delivery of materials to the students, provide the links for lectures to the students, take the submission of assignments from students, etc. Hence, most of the educational institutes initiated the usage of a learning management system (LMS). LMS is a customizable software for the management, tracking, delivery, reporting, etc. of educational programs. LMS can be considered as a direct link between an educational institute and a smart education system. With the incorporation of LMS, various activities based on the actions of schools and colleges can directly be controlled, and the direct contact can be ensured with the students remotely. As has been mentioned earlier, it is a customizable software that is nowadays developed by most of the educational institutes themselves. It is generally a web-based or application-based platform where the teacher can conduct any distance classes, circulate the course-related materials and videos, and can also forward the assignments along with their due dates. In some advanced LMS, many other customizable features are integrated, such as the progress report, online tracking, etc. One of the key features of LMS is that it identifies all the participants differently. There is a provision for creating many profiles such as host, teacher, students, parents, and visitors. The host, who is mostly the instructor or teacher, is given access by the administrator, and the teacher can add and edit all the materials. The LMS can also be used as an interface for conducting various evaluations where the monitoring can be done by the teacher. Besides, after the completion of the unit, there can also be a self-evaluative test added by the teacher. After attempting the test, the student can view the marks scored by him as soon as he submits the test.

LMS can be considered as the future of the education system as it has provided a solution for the conversion of the conventional educational system to the smart educational system. Particularly in the pandemic situation, all the schools and colleges that are conducting their classes remotely have been using the software for the welfare of students.

10.5 VIRTUAL LABORATORIES

As the step was taken to shut down schools and colleges as a part of the safety measure, the laboratories present in them also got closed. Laboratory sessions are the sessions where the students implement the teachings learned in the theory classes, and it helps them understand the concept practically so they become more sound in the subject. The laboratory is also important as the concepts learned in the classes are critically evaluated and the difference between how the experiments are performed and how they are taught can be observed. It also required the introduction of smart education for the continuation of laboratory sessions, but the difficulty was that in laboratory sessions students perform the experiment themselves, and this cannot be achieved when the students are sitting at their homes. A method that incorporated an online education system and that can also give virtual hands-on experience to the students was required.

Virtual laboratories were hence introduced, which are software or applications where the students can get a virtual hands-on experience. The setup consists of virtual parts or kits that need to be arranged for the experiment to be performed. Once all the required components are placed, after running it, the output can be seen and further calculations or proceedings can be carried out. The teacher acts as the creator of the experiment and he chooses the components required for the experiment and also decides the time limit for experimenting. Besides, the software also monitors whether the student has performed the experiment or not and provides the generated report to the teacher. There are also some tutorial videos that are uploaded for the reference of students, and sometimes the teacher may also perform the experiment to depict the correct procedure of the experiment. At the end of the experiment, the student needs to submit the observations and results of the experiment on the website or application. Most of the virtual laboratory software consists of a self-evaluation test that the student needs to attempt after the submission of results. The whole procedure of virtual laboratories has been described in pictorial steps in Figure 10.4.

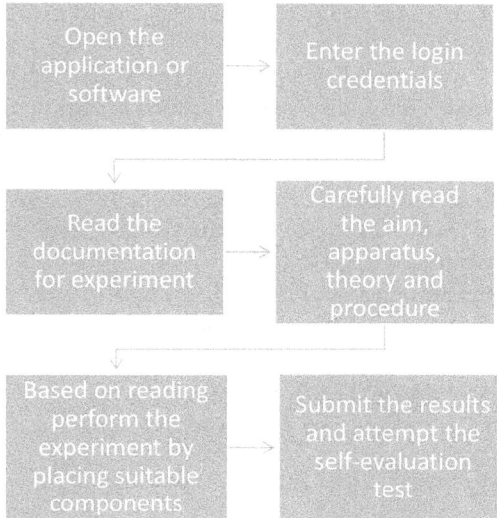

FIGURE 10.4 Procedure for experimenting with a virtual laboratory software or website.

Other than that, there are also other methods incorporated for the laboratory sessions. Simulation software is available in a large number of forms, which is being increasingly used for performing the experiments through simulation and to generate the results. This software consists of a graphical user interface (GUI) where readily available components are present that need to be arranged properly, and then the actual experiment is simulated. The advantage of this simulation software is that many of them are freely available and can be used for free while the virtual lab software is mostly subscription-based. Further, the laboratory sessions are also being conducted in a smart way where the instructor is performing the actual experiment with the actual apparatus in the laboratory live, like that of the online classroom session. This method is mostly employed for the subjects with laboratory experiments that cannot be performed through either virtual laboratory software or simulation software. However, in these sessions students cannot get proper hands-on experience as the instructor is performing the experiment, but they can at least watch the whole session going live in front of them.

The internet connection has been an issue in India, which may again prove to be an obstacle in the learning process due to poor connectivity. The laboratory session going live may get disconnected and the student may not be able to see the further processes. It can also happen that the student gets connected after a certain amount of time, because of which the student will miss a part of the session. For the virtual laboratory sessions, the connectivity issues may force the student to perform the same experiment again even though it was performed by him earlier and the work carried out was not saved. The simulation software varies in sizes, and some of them may be so large it may require a continuous high-speed network connection for a large amount of time. Also, the specification of the computer/laptop should be such that it can handle the software; otherwise, it may not open or the performance of the system will be affected. There is also a chance that the software may crash during usage.

Big Data analytics is something that is not only used to manage the collection and processing of some kind of data, but also shows flexibility to handle data and to manage a variety of data sets. This feature is being used and tested when online laboratories are concerned. The laboratory software consists of various types of equipment that are loaded into it and are made to work accordingly. So a lot of variety of data collection is observed, which includes the data of laboratory session, the components, the virtual apparatus, data of students or institutes using it, the time when the students perform the experiment, and finally the data of student and the experiment after it has been performed for further evaluation [11,12].

10.6   THE ORGANIZATION OF WEBINARS DURING THE COVID-19 SITUATION

Seminars are the sessions where either the participants try to obtain a solution to any problem via discussion made by the panel or a talk is organized to provide information or advancements in a topic of any particular field. The participants may attend them as a part of their academics or due to their interest in that field. COVID-19 affected the seminars; also, the prescheduled webinars were also canceled. The major importance of seminars lies in the fact that critical information is provided through these sessions on any field, making the participants technically sound in that field as the seminar session is delivered by an expert. Hence, the learning cycle of the students was affected badly. Moving towards a smart education system, the solution to this was the use of online mode for the seminar, and it is known as a webinar. Webinars were being organized even before the COVID-19 period, but it gained much popularity and acceptance during the pandemic situation.

The question arises on how to decide the various aspects regarding the webinar. Figure 10.5 depicts the answer to this question about how to think of a webinar and execute it in a step by step manner. First, the format is to be decided, followed by the topic of the webinar. The team for organizing is then to be finalized along with other important aspects such as the date and time for the webinar and the software or application. The material should be prepared and should be publicized in all the available formats. The decision of topic, speaker, and preparation of material are the most important aspects of organizing the webinar. Failing to present any of them properly may lead to many problems, including the failure of the event.

The webinar uses online available software or applications as its platform. These may also be accessed through smartphones. Some of the applications and software used for this purpose are the same ones as being used for online lectures. The invite link is generated by the coordinator or the host of the webinar and is shared among the interested participants who have pre-registered for the event. On the scheduled date and time, the host starts the webinar where, first, the participants are welcomed, and then the key speaker is introduced. The control is handed over to the speaker then who drives the session based on the topic of the webinar and explains every aspect of it with the help of detailed visuals. As it takes place in seminars, in the same

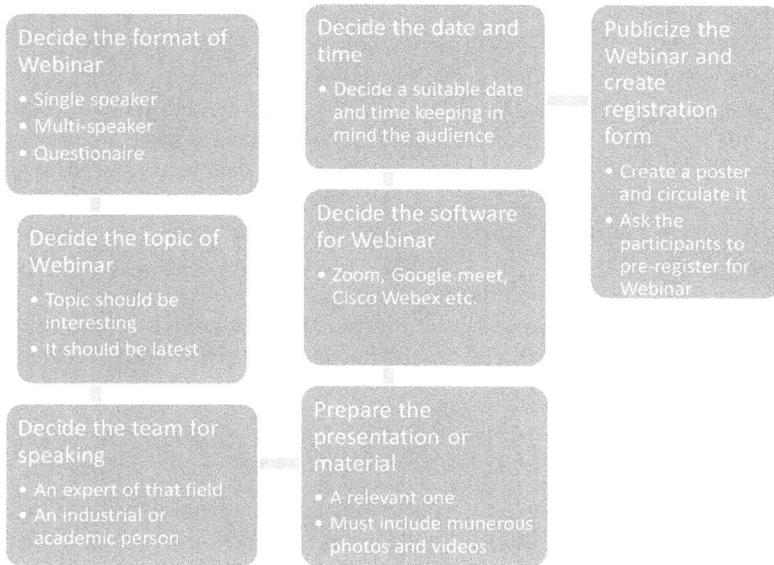

FIGURE 10.5   Process of organizing a webinar.

way the topic is explained using a presentation. All the participants are allowed to raise their doubts using the "raise hand" feature. Queries may be asked to the speaker by typing it in the chatbox or by using the microphone. Gradually, all the queries are addressed by the speaker, one at a time. All these software or applications employed for webinars have a certain limit on the total number of participants and, in many cases, the limit is crossed due to the presence of an extra number of students. In that case, other applications or software such as YouTube or Instagram are employed which are social video sharing and networking sites, respectively, but have a feature of live video feature that can incorporate thousands of participants. After the session ends, all the participants are given their certificate of participation through their respective email ids.

Again the internet connection may act as a barrier, particularly in rural areas of India. Even in the urban areas due to the absence of a stable internet connection, the session may get interrupted and may result in the disconnection from the entire session. There may be some mistake from the hosts where the pre-registered partici-pants may not receive the invite link, preventing them from attending the session. Also, when the number of participants exceeds the maximum, many times it has been observed that the extra students are not even allowed to enter into the session. Besides, switching over from the online software to other software may become a confusing process for the participants. It has also been seen that some participants do not receive any participation certificate from the hosts after the completion of the webinar.

Big Data analytics plays a huge role when the organization of the webinar is con-cerned. The simple reason behind it is that even though some of the applications are used for the webinar are also being used for the online lecture, the amount of data to be processed in the case of the webinar is much more than that of the online lectures. It can be seen as a huge challenge for the software or the application, but with the help of proper Big Data analytics algorithms, this large amount of data can be handled with high speed and accuracy. A large amount of data is first categorized, after this the smooth operation is ensured. As the amount of data increases, the algorithm ensures that the server does not crash and that the presentation being delivered by the organizer reaches the audience properly. It also makes sure that if due to any reason a person is disconnected from the internet, then his data is saved temporarily, and if he can come back within few minutes, he can start from the point he left. After the webinar is over, Big Data is responsible for providing the correct information about the audience who actually attended the session and then generating e-certificates for them [13,14].

10.7   ORGANIZATION OF E-CONFERENCES

Conferences play an important part in publishing research articles, and many research scholars and academic professors engaged in research look forward to it. When any conference is organized, it includes gathering of a large number of people; those attending the conference hail from all around the country. If it is an international conference, then the participants are from all around the world. Keeping these factors in consideration, the organizing institutes have either postponed the conferences or have canceled it. These aspects have created a lot of problems, particularly for the research scholars or academic professors. The only solution for this problem was a shift towards the smart education system where the conferences could be organized online. But there were a large number of problems encountered before the conference is organized. These problems are generated because there lies a huge difference between the organization of a lecture online and a conference online. There is a large number of events that take place in a conference, and many times some of the events are clubbed together, in a parallel sequence. But this is not possible in the online organization. Gradually, these problems were also addressed and a legitimate solution was proposed for each of them.

Like any offline conference, all the research articles are finalized after receiving the final draft of the article, including all the necessary changes asked by the editor of that section. The form is then filled in by the participants with their details; if more than one person has written that article, then it is also asked that who will be presenting the paper and who all will be attending the conference as an audience. After the collection of data, all the participants are sent the necessary details of the conference, which includes the date, time, etc. On the scheduled date and time, all the participants join the session through an application or software which may be the same as used for online lectures. All the participants are then welcomed and the opening ceremony takes place. The guest speaker talks, and he is from the organizing institute. After the completion of this ceremony, all the presentations are run in a parallel manner where the participants need to join the session with the separate link provided to them based on the category of their article. The candidates wait for their turn, and each of them is given 10 minutes for the presentation of their article through a presentation. At the end of the presentation, the panel asks a few questions to the presenter and, after that, the audience is also allowed to ask a few questions. Once all the presentations of all the candidates are finished, then the valedictory session takes place. Usually, the conferences are organized for two days; the morning session of the second day is when the presentation of all the candidates finishes. After that, the valedictory session takes place where the keynote speaker gives his speech. The various prizes for the various sessions are then distributed to the winning participants and the closing ceremony then takes place. Once the session is completed, all the participants are provided e-certificates.

The above paragraph describes how the conference takes place on the conference day. But the preparation for the conference requires many other steps, which are described in Figure 10.6.

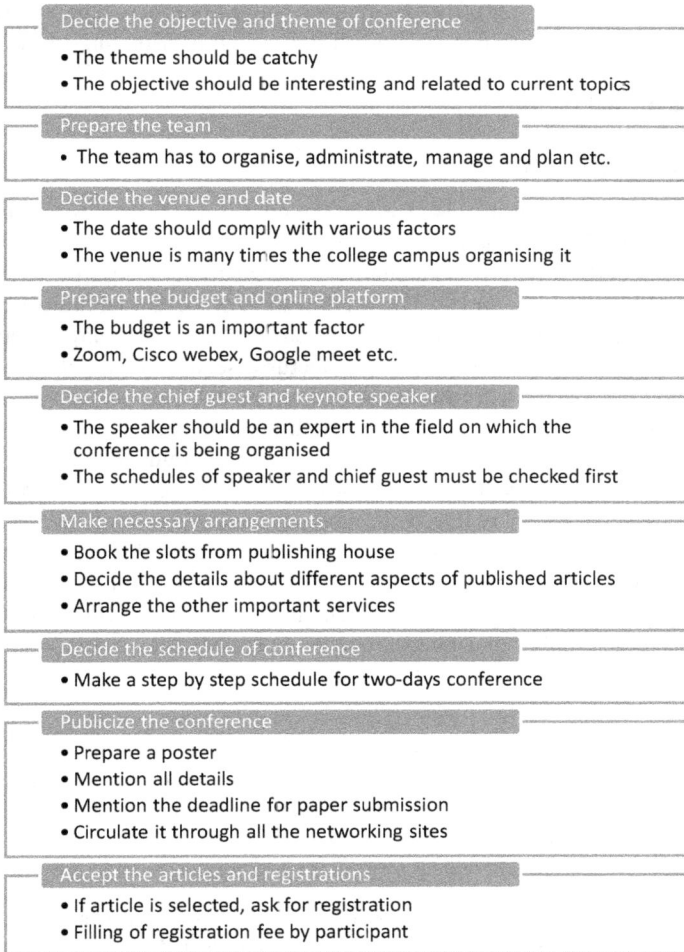

Decide the objective and theme of conference
- The theme should be catchy
- The objective should be interesting and related to current topics

Prepare the team
- The team has to organise, administrate, manage and plan etc.

Decide the venue and date
- The date should comply with various factors
- The venue is many times the college campus organising it

Prepare the budget and online platform
- The budget is an important factor
- Zoom, Cisco webex, Google meet etc.

Decide the chief guest and keynote speaker
- The speaker should be an expert in the field on which the conference is being organised
- The schedules of speaker and chief guest must be checked first

Make necessary arrangements
- Book the slots from publishing house
- Decide the details about different aspects of published articles
- Arrange the other important services

Decide the schedule of conference
- Make a step by step schedule for two-days conference

Publicize the conference
- Prepare a poster
- Mention all details
- Mention the deadline for paper submission
- Circulate it through all the networking sites

Accept the articles and registrations
- If article is selected, ask for registration
- Filling of registration fee by participant

FIGURE 10.6   Step-by-step process used for organizing an e-conference.

The network connectivity issue can prove a problem due to which the person may get completely disconnected from the session. The complexity of the problem would be apparent if any research scholar gets disconnected while he was giving the presentation. There may be some students or research scholars who did not get the link for attending the conference even after registering, which may also prove to be a problem. Sometimes it also happens that the person is not able to give a presentation due to some of the issues in the software. There may be some other minor problems such as the improper working of the microphone.

Big Data first of all helps in the collection of data of all the participants. Then it segregates them and stores them accordingly. The Big Data algorithm of the software used has to collect and organize a large volume of data as hundreds of participants may register for the conference. During the e-conference, a Big Data algorithm helps in a way that the online resources of the conference can be accessed by the participants from their home in an effective manner and without lag. The algorithm also ensures that the application does not crash due to the overloading of data. All the participants present their ideas and their research articles, and Big Data management helps in that as well [15,16].

10.8 CONCLUSION

This chapter focused on problems faced by the education sector due to the outbreak of the global COVID-19 pandemic. The chapter then described the process of switching over to smart education, which included the conduction of lectures, webinars, and conferences through online mode. The organization of virtual laboratories was also described. The chapter also presented the various problems that may be faced by the attendees during the processes. The online lectures seem to be an effective tool for learning, and the procedure is quite smooth. The students may face some difficulties due to switching over from conventional methods. Virtual laboratories or simulation software are proving helpful in making the students understand their theoretical subjects in a better way. The quality of laboratories may improve in the future. Webinars have quite nicely replaced the seminars, and this may prove very helpful in the coming years as people from different states or even different countries will be able to attend them. There are some areas of improvement in the organization of e-conferences. Some lack from the management side has been observed, which may be because a large number of aspects are to be managed at the same time, but this is getting better day by day.

REFERENCES

1. Patgiri, R. and Ahmed, A., 2016, December. Big data: The v's of the game changer paradigm. In 2016 *IEEE 18th International Conference on High Performance Computing and Communications; IEEE 14th International Conference on Smart City; IEEE 2nd International Conference on Data Science and Systems (HPCC/SmartCity/DSS)* (pp. 17–24). IEEE, Sydney, NSW.
2. Terzi, R., Sagiroglu, S. and Demirezen, M.U., 2020. Big data perspective for driver/driving behavior. *IEEE Intelligent Transportation Systems Magazine*, *12*(2), pp. 20–35.
3. Russom, P., 2011. Big data analytics. *TDWI Best Practices Report, Fourth Quarter*, 19(-4), pp. 1–34.

4. Najafabadi, M.M., Villanustre, F., Khoshgoftaar, T.M., Seliya, N., Wald, R. and Muharemagic, E., 2015. Deep learning applications and challenges in big data analytics. *Journal of Big Data*, 2(1), p. 1.
5. Viner, R.M., Russell, S.J., Croker, H., Packer, J., Ward, J., Stansfield, C., Mytton, O., Bonell, C. and Booy, R., 2020. School closure and management practices during coronavirus outbreaks including COVID-19: A rapid systematic review. *The Lancet Child & Adolescent Health*, 4, pp. 397–404.
6. Lee, J., 2020. Mental health effects of school closures during COVID-19. *The Lancet Child & Adolescent Health*, 4(6), p. 421.
7. Sahu, P., 2020. Closure of universities due to coronavirus disease 2019 (COVID-19): Impact on education and mental health of students and academic staff. *Cureus*, 12(4), e7541 (1–6).
8. Qinghua, Z., Bo, D., Buyue, Q., Feng, T., Bifan, W., Weizhan, Z. and Jun, L., 2019. The state of the art and future tendency of smart education. *Journal of Computer Research and Development*, 56(1), p.209.
9. Sharma, M., Ali, M.S. and Husain, S., 2017. Implementation of big data analytics in education industry. *Journal of Electrical and Computer Engineering*, 19(6), pp.36–39.
10. Titimus, P.M., 2019. Big data analytics in the higher education: Need of the future. In *Advances in Communication, Cloud, and Big Data* (pp. 23–28). Springer, Singapore.
11. Demchenko, Y., Belloum, A., de Laat, C., Loomis, C., Wiktorski, T. and Spekschoor, E., 2017, December. Customisable data science educational environment: From competences management and curriculum design to virtual labs on-demand. In *2017 IEEE International Conference on Cloud Computing Technology and Science (CloudCom)* (pp. 363–368). IEEE, Hong Kong.
12. Ray, S. and Saeed, M., 2018. Applications of educational data mining and learning analytics tools in handling big data in higher education. In *Applications of Big Data Analytics* (pp. 135–160). Springer, Cham.
13. Mikroyannidis, A., Domingue, J., Maleshkova, M., Norton, B. and Simperl, E., 2016. Teaching linked open data using open educational resources. In *Open Data for Education* (pp. 135–152). Springer, Cham.
14. Geri, N., Winer, A., Eshet-Alkalai, Y., Blau, I., Caspi, A., Geri, N. and Silber-Varod, V., 2015. Patterns of online video lectures use and impact on student achievement. In *Proceedings of the 10th Chais Conference for Innovation and Learning Technologies: Learning in the Technological Era* (pp. E9–E15), Israel.
15. Elia, G., Solazzo, G., Lorenzo, G. and Passiante, G., 2019. Assessing learners' satisfaction in collaborative online courses through a big data approach. *Computers in Human Behavior*, 92, pp. 589–599.
16. Ahuja, R., Jha, A., Maurya, R. and Srivastava, R., 2019. Analysis of educational data mining. In *Harmony Search and Nature Inspired Optimization Algorithms* (pp. 897–907). Springer, Singapore.

11 Role of IoT, Machine Learning, and Big Data in Smart Building

K. Manimala
Dr. Sivanthi Aditanar College of Engineering

CONTENTS

11.1 INTRODUCTION

One of the biggest challenges that needs immediate as well as long-term consistent action is the climate change [1]. The worst environmental effects and health issues brought out by this climate change are not a specific problem to any specific country but to the entire world. It is necessary to reduce greenhouse gas emissions and energy consumed by buildings to find sustainable solutions to save our environment. Indian

Green New Buildings Rating system (IGBC) advocates the necessity to use available resources in a sustainable manner. The Energy Statistics Report 2020, Government of India (GoI), has mentioned that almost 32% of total energy is consumed by domestic and commercial consumers, which is mainly used in lighting and in Heating, Ventilation, and Air Conditioning (HVAC) system. It is time the domestic and commercial services start saving energy; hence, smart buildings with improved energy conservation properties in addition to user's comfort assumes significance [2]. This is one way to alleviate the ever-increasing demand for additional energy supply.

The number of smart buildings (SBs) is growing today in the name of Smart Office, Smart Hospital, Smart educational facilities, etc. Today's computational and communications infrastructure [3] aids the development of smart buildings. With the latest sensors and technologies like Machine Learning (ML), Internet of Things (IoT), and big data, it is easy to convert any building to smart building with little modifications. SB can include services like smart thermostat that controls the temperature automatically based on time of the day or year and occupant's preference without human intervention. The crucial part of SB is to learn the occupant's preference before implementing technology for appropriate action. There comes the role of ML, which creates analytical models automatically for learning with the existing data. In general, the new generation of buildings that are provided with sensors, actuating elements, embedded chips, etc. for collecting information and control actions on appliances for occupant's comfort is called as smart building. They are actually designed to think of the user's comfort [3].

The smart buildings need to be equipped with IoT devices such as sensors and actuators. IoT devices along with ML algorithm helps to adjust the temperature and optimize the use of HVAC systems [1]. Forecasting temperature and tuning devices for optimized energy consumption based on sensors output is the task of the ML system. The impact of surrounding factors is essential for configuring the suitable HVAC system parameters, and this is required for designing the ML model. The large amount of data collected from sensors need to be analyzed to obtain information for knowledge extraction related to human activities and comfort preferences. ML and Big Data analytics will surely play a crucial role for such knowledge extraction and decision-making for smart services in smart building.

Several researchers have worked on this area and have employed various sensors and extracted various features as input to ML algorithms. This chapter summarizes the various literatures that focused on smart buildings, the sensors used for collecting data, features normally extracted, and the various ML algorithms employed for smart building.

A multi-agent system using sensors and ML can significantly reduce energy consumption and increase occupant comfort by limiting energy use to zones where only occupants are present. The occupant behavior is learned using ML algorithms and is optimized by observing patterns of occupancy behavior. Such techniques, if applied to commercial and residential buildings, could help in significant energy saving, thereby reducing emission effects [1].

Stochastic systems are used to model such a ML system due to the stochastic nature of the occupants of buildings. The challenges of such a model are simplification of occupant's actions, lack of proper standard for validation, and the necessity to deal actions of multiple occupants. Some works combined the impact of individual

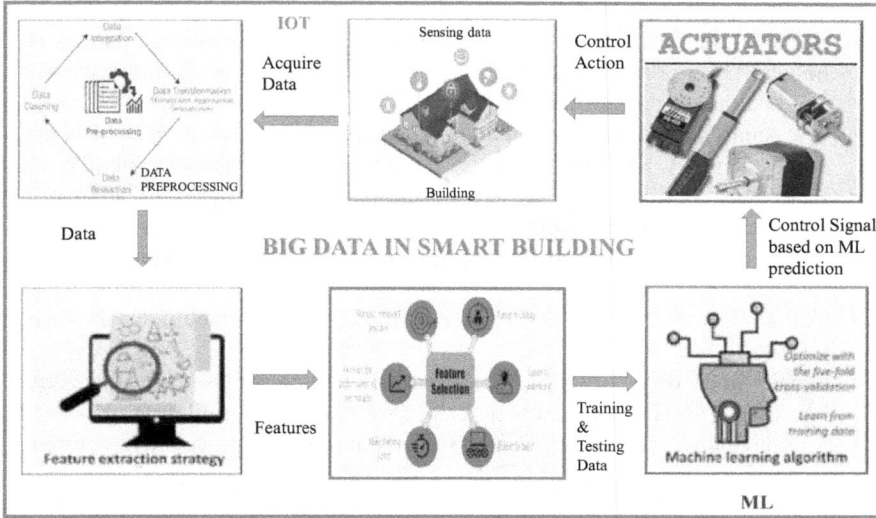

FIGURE 11.1   Framework of big data in smart building.

actions to address issues of multiple occupants. If the occupants are cooperative, the action can be based on a mutually agreed state of comfort. If not, conflict may arise while choosing comfort level. Literatures reported such problems as bargaining problems, culminating in disagreement. Several researchers have used different optimization-based algorithms for addressing this issue, but so far a universally accepted solution has not yet been reached due to its challenging nature [4]. The research gap is due to drawbacks of optimization algorithms and technological advancements. The latest technology and new optimization algorithms that have been tested for several other problems have not been tried for smart building. Therefore, there is a need for further research, and this chapter points out the outline of work carried out so far and the areas to strengthen to take this research to the next step.

There are three major steps in the process of big data analytics of SB, namely, detection, recognition, and prediction. The term detection represents the detection of events like human activity recognition, occupancy of room based on sensor readings, etc. Recognition is the process of categorizing an event into predefined categories like a human performing pattern recognition tasks intelligently. Prediction predicts the occurrence of future events based on past data and initiates control action accordingly. Figure 11.1 shows the structure of big data concept applied to smart building.

11.2   LITERATURE SURVEY

Several papers on SBs have been published, focusing on the role of IoT, ML, Optimization algorithms, and big data in the context of SBs.

- An overview of smart home research [5] was presented by Chan et al. in 2008 along with the details of assistive robots and wearable devices.

- Alam et al. [6] presented the details of sensors, actuating devices, algorithms, and the associated communication protocols utilized in smart homes. The work focused on the research goals, namely, comfort, security, and healthcare.
- The smart home features, smart grid technologies, the associated challenges, benefits, and future trends were detailed by Lobaccaro et al. [7].
- Pan et al. [8] studied the energy consumption in microgrids. The survey detailed the latest improvements in SBs and microgrids.
- Perera et al. [9] discussed IoT applications for SB from the view point of self-learning.
- The applications of data mining technologies in IoT was detailed by Tsai et al. [10].
- Sensor node deployment for occupant's detection and sensing coverage based on optimal method was described by Abdelraouf et al. [11].
- Mahdavinejad et al. [12] presented some commonly employed ML methods that can also be applied for IoT data analytics in SB.
- There is huge literature on sensors and actuating devices and the control actions initiated to design smart building [13–22].
- Data preprocessing like data cleaning, data integration, feature extraction, and feature selection are detailed in [23–37].
- Various supervised ML approaches like Support Vector Machine (SVM), Artificial Neural Networks (ANN), Decision tree, and unsupervised learning methods like clustering and other reinforced learning methods are briefed in [38–87].
- There are many recent literatures on smart building energy consumption using IoT, ML, and big data analytics [88–96].

11.3 CONTRIBUTION OF THE WORK

This chapter details SBs from the view point of IoT, ML, and big data, and the contributions of this work are as follows:

- Providing extensive literature review on the usage of IoT, ML, and big data analytics for SBs.
- Identifying the IoT-based sensor data to be collected for efficient implementation of SBs.
- Ascertaining the need for big data analytics in the successful implantation of SBs.
- Recognizing the research challenges in the area of SBs and the efficiency of ML algorithms in resolving those challenges.
- The implementation of ML algorithms to control low-level devices for energy consumption in a building.
- The contribution of SB in achieving users' comfort, energy efficiency, security, and convenience, and the feasibility of implementation of such smart feature in existing residential and commercial buildings.

11.4   LAYERS OF SMART BUILDING IMPLEMENTATION

The architecture of SB implementation is categorized into six layers beginning from the sensing layer, which incorporates different types of sensors installed in the building premises to collect information from SBs [13]. The network layer is responsible for providing data stream support and data flow control to ensure that messages are transported safely through transport medium such as Wi-Fi, Ethernet, Bluetooth, etc. [13]. As data are collected from various sources, Data Acquisition layer is responsible for assembling the data in a manner suitable for further processing. Context and semantic discovery layer is used to configure and store the relevant context and semantic information. Context processing and reasoning layer processes the vast information and extract the knowledge and takes decisions based on the situation [13]. The last layer that humans use to interact is application layer, which is responsible for comfort, health assistance, entertainment, security, energy efficiency, etc.

11.5   IoT-BASED SENSORS

Sensors are mainly used to collect information from the SB environment and initiate the intelligent control action based on the range of values obtained. Sensing is generally done by several IoT sensors in every part of the home like IR sensor, CO_2 sensor, camera, etc. Such sensing could provide important data for knowledge discovery process using ML algorithm. For example, motion sensors like IR sensors are utilized for detecting the presence of human in a room. Some of the commonly used sensors and their purpose are listed below [13]: Pressure sensors are attached to the objects in buildings to track the movement of the occupants. Contact switches placed on the doors of fridges or rooms detect the actions of the building occupant [14]. Table 11.1 lists the sensors and their purpose in SBs.

The environmental parameters, namely, temperature, CO_2 and sound sensed by the sensor agents are recorded in Timeline database. The number of occupants and the time of occupancy are determined by Occupancy Estimation Agents (OEA) based on the sensor data. The occupancy Prediction Agents (OPAs) forecast the occupancy of the room in the next hour based on this estimation. The Policy Agent (PA) or the control devices use these data for reactive actions when necessary. For example, the PA uses the prediction to activate the HVAC actuators, which in turn warm or cool the room before the arrival of users or shutdown when the room is expected to be empty in the next instant of time.

11.6   CONTROL DEVICES AND ACTUATORS

Big data processing collects data from sensors that are then transformed to a format suitable for further processing by ML algorithms. The prediction from ML algorithm initiates necessary control signal from control devices to actuators for proper action. The control devices need to communicate among themselves for taking proper control action. Actuator takes a decision based on sensor data and converts the electrical control signal into a physical action, enabling automated interaction with the environment. Table 11.2 lists out the control devices used in SB [13,15–22].

TABLE 11.1
Sensors and Usage

No.	Sensors	Purpose
1	Pressure sensor	Track the movement
2	IR sensor	Identify the presence of occupants
3	Contact switches	Track movement
4	Light sensor	Measure light intensity
5	Humidity sensor	Measure air humidity
6	Temperature sensor	Measure temperature
7	Power sensor	Power usage of electric devices
8	Wearable sensor (accelerometers, gyroscopes, magnetic sensor)	Occupant's motion and posture
9	Bio sensors	Monitor health condition
10	Electroencephalography sensors	Monitor brain activity
11	Electrooculography sensor	Monitor eye movement
12	Electromyography sensors	Observe muscle activity
13	Electrocardiography sensors	Observing cardiac activity
14	Thermal sensors	Observing body temperature
15	Sound sensor	Measure the sound energy
16	CO_2 sensor	Measure CO_2 content in a room

TABLE 11.2
Control Devices

No	Device	Purpose
1	WeMo- Wi-Fi enabled switch	ON/OFF electronic devices
2	Nest	Control temperature
3	Lockitron	Control door
4	SmartThings	Temperature and Motion control
5	Philips Hue	Controls lighting system
6	Blufitbottle	Controls drinking habits of user
7	Canary	All-in-all home security system
8	Amazon Echo	Controls home appliances
9	Honeywell Total Connect Remote Services	Security App

11.7 FEATURE SET

The selection of a suitable feature set is very important as it impacts the accuracy of the classifier used for SB activities. A set of features needs to be identified and extracted from the raw sensor data. One of the novel aspects of SB design is the identification of the feature set [23]. The feature set is comprised of building characteristics, energy use behavior of occupants, time of occupancy, and outdoor features like dew point temperature, solar radiation, wind speed and direction, cloudiness, rainfall, etc. Several building characteristics like compactness, surface and wall area, type of roof, height of building, glazing area, heat transfer coefficient of walls, window wall ratio, etc. are also considered as features [24]. Time features extracted are type of day (holiday or working day) and time of the day (day or night). Occupancy features include usage schedule of building, heat content due to people and lighting load, number of permanent occupants, and number of temporary occupants or visiting people.

The following are the list of features commonly extracted and are proved to have excellent capability for ML-based classifier. Depending on the situation, all or a subset of the features presented in Table 11.3 form the feature set. The procedure of selecting pertinent features that improve the accuracy of classifiers, thereby enhancing the performance of SB, is called Feature selection. Normally, optimization algorithms are employed for this purpose. But there are only few literatures which concentrate on this aspect of feature selection.

Some of the researches have utilized both indoor sensor data and outdoor weather condition as weather condition impacts the comfort of room occupants [1]. The features should be carefully chosen so that these parameters help to reduce the energy demand of HVAC system.

TABLE 11.3
Feature Set

No.	Feature	Description
1	Time	The total time of data collection in a day
2	Room occupying time	The total time of room occupation per day
3	Sound	Sound energy sensed in a room per minute
4	CO_2 during working hours	CO_2 in the presence of occupants
5	CO_2 during nonworking hours	CO_2 without occupants
6	Wide field motion	Huge movement of people within room
7	Narrow field motion	Less movement of people in a room
8	Average occupancy count	Average number of occupants
9	Temperature	Temperature of the room
10	Humidity	Humidity of the room
11	Light intensity	Brightness of the room
12	Number of door open	Number of people entering room
13	Number of door closing	Number of people leaving room
14	Solar Radiation	Level of solar radiation
15	Outdoor temperature	Temperature outside the building
16	Outdoor humidity	Humidity outside the building
17	Previous indoor temperature	Indoor temperature before HVAC system is tuned
18	Air conditioner (AC) status	ON/OFF condition of AC
19	Heater status	ON/OFF condition of heater
20	Air conditioner temperature	AC set temperature
21	Air conditioner humidity	Percentage of AC humidity

11.8   ROLE OF IoT IN DATA COLLECTION

11.8.1   Type and Size of Data

The types of data collected fall into three categories, namely, simulated, real, and public benchmark data. About 67% of the researchers have used real data taken in buildings, 19% of the literatures used simulated data, and 14% have used public benchmark data [24]. Real data represents the data collected by IoT sensors installed in the building in addition to electricity bills, energy surveys, statistics, and reports. When the quantity of real data is limited, data can be generated by modeling a building using building energy simulation software like EnergyPlus, DeST, DOE2, or Ecotect [24]. Such data can nearly represent a prototype of a real building. EUNITE data set and ASHRAE's Great Building Energy Predictor Shootout are the commonly used public data sets. Public data sets can serve as benchmark for performance comparison of different models [24–26].

Sensor data may be contaminated with noise, missing values, or outliers and, hence, require data processing to avoid the performance degradation of the prediction models [27]. Totally, 81% of the works have reported the results for nonresidential buildings, and the remaining for residential buildings.

A smaller data size could not be considered as a representative of sample data and hence not preferred. If the data size is large, the computational complexity of processing it increases. The data size so far reported varies from 2 weeks [28] to 4 years [29,30]. About 9% of the reported studies have utilized less than 1-month data, whereas 56% used 1-month to 1-year data and 31% used more than 1-year data.

11.8.2   Data Preprocessing

Any missing, inconsistent, or incorrect data will produce errors in the ML analysis, and hence a preprocessing step is needed [31]. The major steps of preprocessing are data integration, data cleaning, data transformation, and data reduction. Data integration is the process of combining the data collected from various sources. Data cleaning is used to cleanse the data by correcting or deleting inaccurate and incorrect information [32]. Data transformation converts the data into a format suitable for ML algorithm. Data transformation is the process of converting data from one domain to another domain suitable for further processing. From the transformed data, significant features are extracted to form the feature set. The feature set may contain several features. If the number of features is more, it is called as high-dimensional feature set.

11.8.2.1   Feature Extraction

There is a necessity to project the data collected from one domain to another domain to obtain useful information. In such a case, feature extraction step is important. Feature extraction method normally works in time, discrete, and frequency domains [33]. Wavelet transform, Fourier transform, principal component analysis, and S-transform are the commonly used feature extraction tools. They mathematically process a signal to convert from one domain to another domain. Euclidean-based distances and dynamic time warping are the vital approaches of discrete domain methods applied for classifying human activities and behavioral patterns.

11.8.2.2 Feature Selection

The dimensionality of data set is reduced by removing irrelevant features to enhance the performance of ML algorithm. As huge volumes of data are collected from various sensors from various sources, there is a possibility of redundancy, and they should be brought down to a less number of features using feature selection techniques without losing important information [34]. Feature selection methods identify the most significant features, and there are several literatures in smart building energy management that focuses on this crucial step. For a smart home environment, Fang et al. [35] have identified that different sets of features produce different accuracy rates for human activity; recognition and inclusion of inappropriate features deteriorated the accuracy of recognition. Two commonly followed approaches of feature selection are filter and wrapper approaches. Literatures suggest wrapper approach provides better performance [36]. The authors of [37] utilized different feature selection methods for recognizing human activities obtained from sensors and reported that the utilized features have a strong bond with the accuracy rate.

11.9 MACHINE LEARNING ALGORITHMS

IoT devices like sensors collect massive data and provide valuable information that should be mined with ML algorithms to extract useful information for the design of SB management. As the volume of data is very high, the traditional approaches-based predictive models are not enough, thereby necessitating big data-based tools. Big Data is now making a big impact in SB. ML can facilitate the process of mining the massive amount of data collected by the sensors and is now emerging as a powerful tool for the design of SB. ML algorithms learn a model from the portion of data set collected called training data and make predictions for new data based on that model automatically.

Several ML algorithms were employed to extract knowledge from the feature set formed using sensor data to predict future occupation of rooms to set the HVAC system accordingly to satisfy the occupants and to reduce the energy consumption. Sadi et al. [1] used 36 ML algorithms to mine the vast sensor data collected from real-time data to predict the room temperature of the test building and finally concluded that extra trees is the best among them. The most commonly used ML algorithms for model training are SVM, decision trees, and Artificial Neural Network. SVM is generally preferred as it is good in solving even nonlinear problems with less amount of data. It usually projects the problem into high-dimensional space and then applies kernel to convert into a linear problem. SVM is reported as one of the topmost influential data mining algorithm [38]. Decision trees is also a powerful algorithm, and it grows as the data size increases [39]. Smart building researchers have used several decision trees like classification and regression trees (CART), boosting trees (BT), chi-squared automatic interaction detector (CHAID), and random forest (RF) for prediction.

Some studies also compared the effectiveness of different algorithms in energy consumption prediction. The performance of SVM and BPNN was compared by Li et al. [40]; the performance of SVM and AR was compared by Borges et al. [41];

LS-SVM and BPNN was compared by Xuemei et al. [42]. There are more works reported on SVM, kNN, MLP, Bayesian Network, and Feed Forward Neural Network [43–52].

ML techniques are widely used to design smart systems that can sense and respond according to the activities of occupants in the SB [53]. Generally, ML is categorized into four groups handling different types of learning tasks as follows: Supervised learning, unsupervised learning, semi-supervised learning, and reinforcement learning algorithms. ML model uses input training data and goes through a training process that continues until the model reaches the desired level of accuracy [54,55]. Examples of some common supervised ML algorithms are decision tree, Bayesian classifier, support vector machines (SVMs), hidden Markov models (HMMs), artificial neural networks (ANNs), probabilistic neural network, genetic algorithm, etc. [56,57]. Supervised learning approaches are widely used to solve the problems in smart buildings.

A supervised learning system for hand washing using Markov decision processes was proposed by Boger et al. [58] to help people affected with dementia. Altun et al. [59] performed the supervised human activity classification using small sized IoT sensors. Mozer [60] designed an algorithm using artificial neural networks and reinforcement learning for controlling HVAC appliances in smart home environment. Bourobou et al. [61] developed an ML-based approach using ANN and clustering methods to predict occupant's activities in smart home environments. Supervised learning is further grouped mainly into classification and regression.

The task of classification algorithms is to classify an input data belonging to a specific class. The data set is classified into labeled and unlabeled data sets; the training done on labeled data set, and evaluation of classifier done on unlabeled data set. The number of data that are correctly classified defines the accuracy rate (AR) of ML algorithm. The classification result is mathematically described as [13]

$$\text{Accuracy Rate AR} = \frac{N_C}{N_t} \tag{11.1}$$

here N_C represents the number of data sets correctly classified and N_t represents the total number of data sets. The precision (P) and recall (R) parameters are used to measure the accuracy of classification results. Four measures, namely, true positive (TP), true negative (TN), false negative (FN), and false positive (FP), and the precision (P) and recall (R) are defined as follows [13]:

$$P = \frac{\text{TP}}{\text{TP+FP}} \tag{11.2}$$

$$R = \frac{\text{TP}}{\text{TP+FN}} \tag{11.3}$$

Commonly used classification techniques for SB are SVM, Decision trees, ANNs, and Bayesian networks. ML models do not need any information for building knowledge discovery systems other than providing input features and output targets (energy

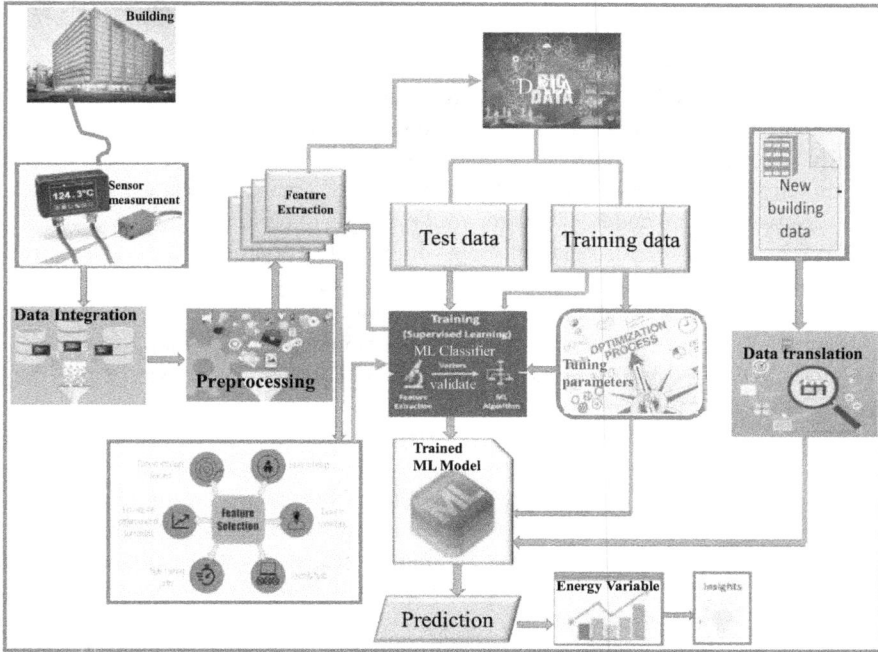

FIGURE 11.2 Smart building ML learning system.

performance). In other words, it functions as a black box to discover the relationship between input and output on its own [62]. ML needs sufficient amount of training data to predict the output for unseen input data. If the training data is insufficient, the accuracy of the ML classifier is affected, and thus the predictions made are inaccurate. The general scheme of ML supervised learning for smart building system is illustrated in Figure 11.2.

11.9.1 Support Vector Machine

SVM is a supervised ML algorithm used for classification tasks in several problems [63]. It is a popular technique utilized for several computer vision problems like character recognition, face recognition, compression, and other pattern recognition tasks [64]. The main concept is maximizing the margin that separates the hyperplane between two classes' closest data points. The points lying on the boundaries are called Support vectors and the optimal separating hyperplane is in the middle of the margin, thereby providing greater separation between data points belonging to different classes [65]. Fu et al. [66] proposed an SVM-based ML method to predict the electricity loads of public buildings that are provided with electricity sub-metering systems. Nguyen et al. proposed a human tracker system [67] using SVM for recognizing any abnormal human movement in the cameras installed in smart homes.

11.9.2 Artificial Neural Networks

SB researchers are widely employing ANNs for building energy prediction, and ANNs are considered as the chief of ML techniques. ANNs have the capacity to learn input–output relationship even in the presence of noise and fault [68]. ANN is actually a mathematical model of a biological neuron. The common ANNs used in this area are Radial Basis Function Network, Feed Forward Network, Probabilistic Neural Network, etc. Feed Forward network is simple to build as it does not have any feedback paths and the information moves in one direction only. ANN has a minimum of three layers, namely, input, hidden and output layer. Complicated problems may include few more hidden layers. Neurons are connected to other neurons with a connection weight. Every neuron at each layer sums up the input multiplied by the weight it receives from the previous layer neurons. Then applies activation functions like linear or sigmoidal function on the sum to produce final output [69].

Petersen et al. [70] developed an ANN-based classifier to provide context-aware services.

Refer [71] for a detailed survey focusing on the role of ANNs for smart homes. Ermes et al. [72] designed a hybrid classifier using a tree structure and neural networks for recognizing various actions of human. Ciabattoni et al. [73] designed an ANN-based ML model to forecast the output of solar power system and the energy utilized during specific times. ANN was used by Wong et al. [74] and Buratti et al. [75] for analyzing the energy consumption pattern of a commercial building in day time. They investigated different combination of input parameters to reduce the dimensionality of input features. A partial swarm optimization (PSO)-based ANN for prediction of energy utilized was proposed by Li et al. [76]. Khayatian et al. [77] and Deb et al. [78] also utilized ANN model for energy consumption prediction in buildings. An ANN-based ML model for determination of energy consumption and occupant's thermal comfort was designed by Ascinoe et al. [79]. They performed energy assessment and proposed a simulation-based analysis for the improvement of network parameters.

11.9.3 Decision Trees

Decision tree is one of the important ML modeling methods. It is used for both classification and regression tasks. The leaves of the decision tree represent class labels, while the branches represent the parameters attributed to those labels [80]. The accuracy and speed of Decision tree makes it a powerful choice for ML. The commonly used decision trees are Classification tree, regression tree, C 4.5, C5.0, ID3, Chi-squared, and conditional decision trees [81]. Degado et al. [82] proposed a decision tree to extract significant activities of human behavior in a smart environment. Patient monitoring and health services were designed using C4.5 algorithm [83]. The major drawback of this decision tree is it can overfit easily.

Hidden Markov Model, Bayesian algorithm, and deep learning algorithm are other commonly used supervised algorithms. The ML algorithm can be chosen depending upon the type of data and the classification task. Different algorithms can be tried, and the best one for the particular ML problem can be chosen based upon

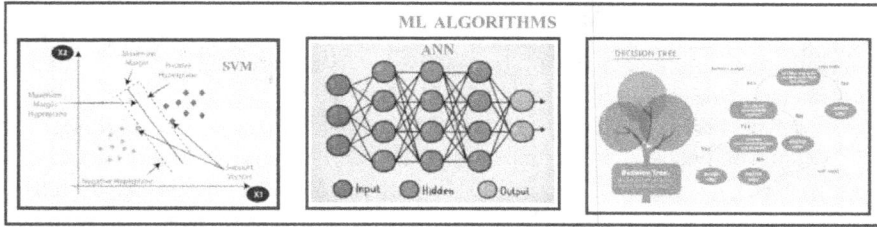

FIGURE 11.3 Supervised ML algorithms.

the accuracy of the results. Figure 11.3 shows the commonly used supervised ML algorithms.

11.9.4 UNSUPERVISED LEARNING

Unsupervised learning is a type of classification algorithm that clusters the input data into different groups based on the similarity between samples. It is an algorithm developed without any target labels to analyze the behavior of the system [84]. This unsupervised learning is applied in smart buildings to recognize different activities when the labels for input data are not available [85]. The commonly used unsupervised algorithms are A priori algorithm, K means, and fuzzy C means clustering algorithms. The direction for learning is absent, which is the major drawback of this algorithm.

11.9.5 REINFORCEMENT LEARNING

Reinforcement learning concept is designed to take actions for maximizing the concept of performance measure, which is a long-term objective. This type of learning is designed to learn control policies in the presence of massive training data. The commonly used approaches are Monte Carlo, Brute force, value function, temporal difference methods, and Q-learning. A reinforcement learning-based Q-learning algorithm was proposed in [86,87] for maximizing the benefits of smart building.

11.10 AVAILABLE ML PLATFORM AND TOOLS FOR SB DESIGN

There are so many available ML platforms and tools to design a learning algorithm for any area. The designers of SB using ML could utilize these tools and platforms. But the process of selecting an apt tool or platform is not easy as it is difficult to find a matching tool for our application. The designer has to weigh the merits and demerits of each tool and choose the appropriate one for their algorithm. The available ML platforms are H2O, MLlib (Spark), TensorFlow, Torch, Deeplearning4j, Massive Online Analysis (MOA), Caffe, Azure, etc. [13]. The list of existing ML tools include Apache Storm, Apache Kafka, Oracle, Amazon Kinesis Streams, Apache Spark Streaming, Apache Flume, and Apache SAMOA [13].

11.11 RESEARCH CHALLENGES

- Implementation of smart building concept in commercial buildings is complicated compared to residential building, as achieving consensus of users' comfort is difficult. More studies and knowledge discovery using ML algorithms are needed for the success of Smart Commercial buildings.
- To check whether the experiments carried out in a particular HVAC system are applicable to all other HVACs.
- There is a possibility that the experimental settings made for SB research may affect the results, which necessitates a common standard for experimental setup.
- The suggestion of various features set by various literature creates confusion to identify the best discriminatory feature set for designing SB, and hence further research should clearly assess the efficacy of features suggested in literatures and standardize the feature set model.
- As real-world data collected by sensors are prone to noise, the ML algorithms should be designed properly to predict outputs even in the presence of noise.
- The parameter settings of ML algorithms may impact the performance of the SB design, and hence a standard to tune the parameters of the ML classifier needs to be set up.
- The experiments need to be repeated several times during different weather conditions or during different times of year to validate the results obtained.
- The concept of maintaining security and privacy while designing SB is difficult as there is massive data collection, which necessitates the usage of cloud environments. As sensitive data is distributed through internet, there is a possibility of unauthorized users taking control over the devices of other's building, which should be strictly avoided.
- The services to be considered in SB should include all the automized control a house needs like water management, waste management, parking, security, power and electricity, lighting, HVAC, elderly and child care, etc.
- Analyzing the massive amounts of SB data necessitates the design of new technologies to store, organize, and process Big Data efficiently. Also, the requirement of high-performance processors to bring out the insights in Big Data is to be dealt with.
- The reliability concern of the occupants of SB is to be addressed properly as the control of many devices is now done automatically by ML algorithms.

11.12 CONCLUSION

The combination of IoT, ML, and big data Analytics has brought out a novel intelligent design of building called smart building. This chapter has defined the concept of smart building clearly and summarized the various Big Data research going in this area for energy conservation and occupant comfort. A detailed review of the application of IoT-based sensors and the processing techniques for occupant detection and human action recognition using ML algorithm for reducing energy consumption was done. Even though the advancements in recent technologies make the concept of SBs realistic, there are still various issues and challenges that limit the full-scale implementation of real-world SBs. Addressing these challenges is a powerful driving force for technical advancements in both industrial and academic areas of SB research.

REFERENCES

1. S. Alawadi, D. Mera, M. Fernández-Delgado, F. Alkhabbas, C. Magnus Olsson, and P. Davidsson, "*A Comparison of Machine Learning Algorithms for Forecasting Indoor Temperature in Smart Buildings.*" Springer, Energy Systems, January 2020, doi:10.1007/s12667-020-00376-x

2. S. Mamidi, Y.-H. Chang, and R. Maheswaran, "Improving building energy efficiency with a network of sensing, learning and prediction agents." *Proceedings of the 11th International Conference on Autonomous Agents and Multiagent Systems Innovative Applications Track (AAMAS 2012)*, V. Conitzer, M. Winikoff, L. Padgham, and W. van der Hoek Eds., 4–8 June 2012, Valencia, Spain.

3. A.S. Shah, H. Nasir, M. Fayaz, A. Lajis, and A. Shah, "A review on energy consumption optimization techniques in IoT based smart building environment." *Information*, vol. 10, no. 3, pp. 1–36, 2019.

4. M. Pal, and S. Bandyopadhyay, "Consensus of subjective preferences of multiple occupants for building energy management." *IEEE, Symposium Series on Computational Intelligence SSCI*, 2018. Bangalore, 1815–1822.

5. M. Chan, D. Estve, C. Escriba, and E. Campo, "A review of smart homesPresent state and future challenges." *Computer Methods and Programs in Biomedicine*, vol. 91, no. 1, pp. 55–81, Jul. 2008. [Online]. Available: http://www.sciencedirect.com/science/article/pii/ S0169260708000436.

6. M.R. Alam, M.B.I. Reaz, and M.A.M. Ali, "A review of smart homes - past, present, and future." *IEEE Transactions on Systems, Man, and Cybernetics, Part C (Applications and Reviews)*, vol. 42, no. 6, pp. 1190–1203, Nov. 2012.

7. G. Lobaccaro, S. Carlucci, E. Lfstrm, G. Lobaccaro, S. Carlucci, and E. Lfstrm, "A Review of systems and technologies for smart homes and smart grids." *Energies*, vol. 9, no. 5, p. 348, May 2016. [Online]. Available: https://www.*mdpi*.com/1996-1073/9/5/348.

8. J. Pan, R. Jain, and S. Paul, "A survey of energy efficiency in buildings and microgrids using networking technologies." *IEEE Communications Surveys Tutorials*, vol. 16, no. 3, pp. 1709–1731, 2014.

9. C. Perera, A. Zaslavsky, P. Christen, and D. Georgakopoulos, "Context aware computing for the Internet of Things: A survey." *IEEE Communications Surveys Tutorials*, vol. 16, no. 1, pp. 414–454, 2014.

10. C.W. Tsai, C.F. Lai, M.C. Chiang, and L.T. Yang, "Data mining for Internet of Things: A survey." *IEEE Communications Surveys Tutorials*, vol. 16, no. 1, pp. 77–97, 2014.

11. A. Ouadjaout, N. Lasla, D. Djenouri, and C. Zizoua. 2016. "On the effect of sensing-holes in PIR-based occupancy detection systems." In *SENSORNETS 2016- Proceedings of the 5th International Conference on Sensor Networks*, February 19–21, 2016. Rome, Italy, 175–180.

12. M.S. Mahdavinejad, M. Rezvan, M. Barekatain, P. Adibi, P. Barnaghi, and A.P. Sheth, "Machine learning for internet of things data analysis: A survey." *Digital Communications and Networks*, vol. 4, no. 3, pp. 161–175, Aug. 2018. [Online]. Available: http://www.sciencedirect.com/science/article/pii/S235286481730247X.

13. B. Qolomany, A. Al-Fuqaha, A. Gupta, D. Benhaddou, S. Alwajidi, J. Qadir, and A.C. Fong, "Leveraging machine learning and big data for smart buildings: A comprehensive survey." *EEEAccess*, vol. 7, pp. 90316–90356, 2019. doi: 10.1109/ACCESS.2019.2926642.

14. D. Ding, R.A. Cooper, P.F. Pasquina, and L. Fici-Pasquina, "Sensor technology for smart homes." *Maturitas*, vol. 69, no. 2, pp. 131–136, Jun. 2011. [Online]. Available: http://www.sciencedirect.com/science/ article/pii/S0378512211000983.

15. "WEMO," Dec. 2017. [Online]. Available: http://www.wemo.com/.
16. "Nest," Dec. 2017. [Online]. Available: https://www.nest.com/.
17. "Lockitron," Dec. 2017. [Online]. Available: https://lockitron.com/
18. "Philips Hue," Dec. 2017. [Online]. Available: http: //www2.meethue.com/.
19. "Oleo Apps Inc., BluFit Bottle," Dec. 2017. [Online]. Available: http://www.indigodomo. com/.
20. "Canary - A complete security system in a single device." Dec. 2017. [Online]. Available: https://canary.is/.
21. R. Crist and D. Carnoy, "Amazon Echo review," Dec. 2017. [Online]. Available: https:// www.*cnet*.com/products/amazon-echo-review/.
22. Home - Honeywell Security Group," Dec. 2017. [Online]. Available: http://www.security.honeywell.com/index.html.
23. S. Mamidi, Y.-H. Chang, and R. Maheswaran (2012). "Adaptive learning agents for sustainable building energy management." *AAAI Spring Symposium - Technical Report.* pp. 46–53.
24. K. Amasyali, and N.M. El-Gohary, "A review of data-driven building energy consumption prediction studies." *Renewable and Sustainable Energy Reviews* vol. 81, pp. 1192–1205, 2018.
25. R.E. Edwards, J. New, and L.E. Parker, "Predicting future hourly residential electrical consumption: a machine learning case study." *Energy Build* vol. 49, pp. 591–603, 2012. doi: 10.1016/j.enbuild.2012.03.010.
26. S. Karatasou, M. Santamouris, and V. Geros, "Modeling and predicting building's energy use with artificial neural networks: methods and results." *Energy Build* vol. 38, pp. 949–58, 2006. doi: 10.1016/j.enbuild.2005.11.005.
27. P.A. González, and J.M. Zamarreño, "Prediction of hourly energy consumption in buildings based on a feedback artificial neural network." *Energy Build* vol. 37, pp. 595–601, 2005. doi: 10.1016/j.enbuild.2004.09.006.
28. D. Liu and Q. Chen, "Prediction of building lighting energy consumption based on support vector regression." *2013 9th Asian Control Conf*erence, pp. 1–5, 2013. doi:10.1109/ASCC.2013.6606376.
29. P. Dagnely, T. Ruette, T. Tourwé, E. Tsiporkova, and C. Verhelst, "Predicting hourly energy consumption. Can you beat an autoregressive model." *Proceeding 24th Annual Review of Machine Learning. Conference Belgium Netherlands*, Benelearn, Delft, Netherlands, vol. 19, 2015.
30. B. Dong, C. Cao, and S.E. Lee, "Applying support vector machines to predict building energy consumption in tropical region." *Energy Build* vol. 37, pp. 545–53, 2005. doi: 10.1016/j.enbuild.2004.09.009.
31. J.M. Hellerstein, Quantitative data cleaning for large databases. White Paper, United Nations Economic Commission for Europe. http://db.cs.berkeley.edu/ jmh/papers/-cleaning-unece.pdf, 2008.
32. K. Pattipati, A. Kodali, J. Luo, K. Choi, S. Singh, C. Sankavaram et al. "An integrated diagnostic process for automotive systems." In: Prokhorov D, editor. *Computational Intelligence in Automotive Applications*, Berlin, Heidelberg: Springer Berlin Heidelberg, 2008. pp. 191–218. doi: 10.1007/978-3-540-79257-4_11.
33. F. Alas, J. Socor, X. Sevillano, F. Alas, J.C. Socor, and X. Sevillano, "A review of physical and perceptual feature extraction techniques for speech, music and environmental sounds." *Applied Sciences*, vol. 6, no. 5, p. 143, May 2016. [Online]. Available: http:// www.mdpi.com/2076-3417/6/5/143
34. A. Ghodsi, *"Dimensionality Reduction A Short Tutorial."* Department of Statistics and Actuarial Science, Univ. of Waterloo: Ontario, Canada, vol. 37, 2006. [Online]. Available: https://www.*math.uw*aterloo.ca/~aghodsib/courses/f06stat890/readings/tutorialstat890. pdf

35. H. Fang, L. He, H. Si, P. Liu, and X. Xie, "Human activity recognition based on feature selection in smart home using back-propagation algorithm." *ISA Transactions*, vol. 53, no. 5, pp. 1629–1638, Sep. 2014. [Online]. Available: http://www.sciencedirect.com/science/ article/pii/S0019057814001281

36. I. Guyon and A. Elisseeff, "An introduction to variable and feature selection." *Journal of Machine Learning Research*, vol. 3, pp. 1157–1182, Mar. 2003. [Online]. Available: http://dl.acm.org/citation.cfm?id=944919.944968

37. H. Fang, R. Srinivasan, and D.J. Cook, "Feature selections for human activity recognition in smart home environments." *International Journal of Innovative Computing, Information and Control*, vol. 8, pp. 3525–3535, May 2012.

38. X. Wu, V. Kumar, J. Ross Quinlan, J. Ghosh, Q. Yang H. Motoda et al. "Top 10 algorithms in data mining." *Knowledge and Information Systems,* vol. 14, pp. 1–37, 2008. doi: 10.1007/s10115-007-0114-2.

39. P. Domingos, "A few useful things to know about machine learning." *Communications of the ACM,* vol. 55, pp. 78–87, 2012. doi: 10.1145/2347736.2347755.

40. Q. Li, Q. Meng, J. Cai, H. Yoshino, and A. Mochida "Applying support vector machine to predict hourly cooling load in the building." *Applied Energy* vol. 86, pp. 2249–56, 2009. doi: 10.1016/j.apenergy.2008.11.035.

41. C.E. Borges, Y.K. Penya, I. Fernández, J. Prieto, and O. Bretos, "Assessing tolerance-based robust short-term load forecasting in buildings." *Energies* vol. 6, p. 2110, 2013. doi: 10.3390/en6042110.

42. L. Xuemei, L. Jin-hu, D. Lixing, X. Gang, and L. Jibin, "Building cooling load forecasting model based on LS-SVM." *2009 Asia-Pacific Conference on Information Processing* vol. 1, pp. 55–8, 2009. doi: 10.1109/APCIP.2009.22.

43. R. Platon, V.R. Dehkordi, and J. Martel, "Hourly prediction of a building's electricity consumption using case-based reasoning, artificial neural networks and principal component analysis." *Energy Build* vol. 92, pp. 10–8, 2015. doi: 10.1016/j.enbuild.2015.01.047.

44. R.K. Jain, T. Damoulas, and C.E. Kontokosta, "Towards data-driven energy consumption forecasting of multi-family residential buildings: feature selection via The Lasso." *Computing in Civil and Building Engineering* 2016. doi: 10.1061/9780784413616.208.

45. Z. Hou, Z. Lian, Y. Yao, and X. Yuan, "Cooling-load prediction by the combination of rough set theory and an artificial neural-network based on data-fusion technique." *Applied Energy* vol. 83, pp. 1033–46, 2006. doi: 10.1016/j.apenergy.2005.08.006.

46. Y.K. Penya, C.E. Borges, and I. Fernández, "Short-term load forecasting in non-residential buildings." *AFRICON* vol. 2011, pp. 1–6, 2011. doi: 10.1109/AFRCON.2011.6072062.

47. C. Fan, F. Xiao, and S. Wang "Development of prediction models for next-day building energy consumption and peak power demand using data mining techniques." *Applied Energy* vol. 127, pp. 1–10, 2014. doi: 10.1016/j.apenergy.2014.04.016.

48. J.-S. Chou and D.-K. Bui, "Modeling heating and cooling loads by artificial intelligence for energy-efficient building design." *Energy Build* vol. 82, pp. 437–46, 2014. doi: 10.1016/j.enbuild.2014.07.036.

49. Q. Li, Q. Meng, J. Cai, H. Yoshino, and A. Mochida, "Predicting hourly cooling load in the building: a comparison of support vector machine and different artificial neural networks." *Energy Conversion and Management* vol. 50, pp. 90–6, 2009. doi: 10.1016/j.enconman.2008.08.033.

50. Q. Li, P. Ren, and Q. Meng, "Prediction model of annual energy consumption of residential buildings." *2010 International Conference on Advances in Energy Engineering*, pp. 223–6 2010. doi: 10.1109/ICAEE.2010.5557576.

51. J. Massana, C. Pous, L. Burgas, J. Melendez, and J. Colomer, "Short-term load forecasting in a non-residential building contrasting models and attributes." *Energy Build* vol. 92, pp. 322–30, 2015. doi: 10.1016/j.enbuild.2015.02.007.

52. I. Fernández, C.E. Borges, and Y.K. Penya, "Efficient building load forecasting." *ETFA*, pp. 1–8, 2011. doi: 10.1109/ETFA.2011.6059103.

53. S. Belaidouni and M. Miraoui, "Machine learning technologies in smart spaces." pp. 52–55, 2016.

54. T. Hastie, R. Tibshirani, and J. Friedman, "Unsupervised learning," in *The Elements of Statistical Learning: Data Mining, Inference, and Prediction*, Second Edition, 2nd ed. New York, NY: Springer, 2016.

55. R.O. Duda, P.E. Hart, and D.G. Stork, *Pattern Classification*, 2nd ed. New York: Wiley-Interscience, Nov. 2000.

56. J. Lapalu, K. Bouchard, A. Bouzouane, B. Bouchard, and S. Giroux, "Unsupervised mining of activities for smart home prediction," *Procedia Computer Science*, vol. 19, pp. 503–510, Jan. 2013. 57. [Online]. Available: http://www.sciencedirect.com/science/article/pii/S1877050913006753.

57. A.N. Aicha, G. Englebienne, and B. Krse, "Modeling visit behaviour in smart homes using unsupervised learning," in *Proceedings of the 2014 ACM International Joint Conference on Pervasive and Ubiquitous Computing: Adjunct Publication*, ser. UbiComp '14 Adjunct. New York, NY, ACM, pp. 1193–1200, 2014 [Online].

58. J. Boger, J. Hoey, P. Poupart, C. Boutilier, G. Fernie, and A. Mihailidis, "A planning system based on Markov decision processes to guide people with dementia through activities of daily living." *IEEE Transactions on Information Technology in Biomedicine*, vol. 10, no. 2, pp. 323–333, Apr. 2006.

59. K. Altun, B. Barshan, and O. Tunel, "Comparative study on classifying human activities with miniature inertial and magnetic sensors." *Pattern Recognition*, vol. 43, no. 10, pp. 3605–3620, Oct. 2010. [Online]. Available: http://linkinghub.elsevier.com/retrieve/pii/S0031320310001950.

60. M.C. Mozer, "The neural network house: An environment that adapts to its inhabitants," Proceedings of AAAI Spring Symposium Intelligent Environments, 1998, p. 5.

61. S.T.M. Bourobou and Y. Yoo, "User activity recognition in smart homes using pattern clustering applied to temporal ANN algorithm," *Sensors (Basel, Switzerland)*, vol. 15, no. 5, pp. 11953–11971, May 2015. [Online]. Available: https://www.*ncbi*.nlm.nih.gov/pmc/articles/PMC4481973/

62. S. Seyedzadeh, F.P. Rahimian, I. Glesk, and M. Roper, "Machine learning for estimation of building energy consumption and performance: A review", *Visualization in Engineering*, vol. 6, p. 5, 2018. doi: 10.1186/s40327-018-0064-7.

63. A. Desarkar and A. Das, "Big-data analytics, machine learning algorithms and scalable/parallel/distributed algorithms," in *Internet of Things and Big Data Technologies for Next Generation Healthcare, ser. Studies in Big Data*, C. Bhatt, N. Dey, and A.S. Ashour, Eds. Cham: Springer International Publishing, pp. 159–197, 2017 [Online]. doi: 10.1007/978-3-319-49736-5.

64. R. Burbidge and B. Buxton, "An introduction to support vector machines for data mining." in *Keynote Papers, Young or12,* University of Nottingham, Operational Research Society, pp. 3–15, 2001.

65. D. Meyer, "Support vector machines." *The Interface to libsvm in package e1071*, e e1071, Technikum Wien, 2015.

66. Y. Fu, Z. Li, H. Zhang, and P. Xu, "Using support vector machine to predict next day electricity load of public buildings with sub- metering devices," *Procedia Engineering*, vol. 121, pp. 1016–1022, Jan. 2015. [Online]. Available: http://www.sciencedirect.com/science/article/pii/S1877705815029252

67. Q.C. Nguyen, D. Shin, D. Shin, and J. Kim, "Real-time human tracker based on location and motion recognition of user for smart home," in *2009 Third International Conference on Multimedia and Ubiquitous Engineering*, pp. 243–250, Jun. 2009. China. doi: 10.1109/MUE.2009.51.

68. G.K.F. Tso and K.K.W. Yau, "Predicting electricity energy consumption: A comparison of regression analysis, decision tree and neural networks." *Energy*, vol. 32, no. 9, pp. 1761–1768, 2007. doi: 10.1016/j.energy.2006.11.010.

69. Y.-S. Park and S. Lek, "Artificial neural networks: Multilayer perceptron for ecological modeling". In *Developments in Environmental Modelling*, Elsevier, Vol. 28, pp. 123–140, 2016. doi: 10.1016/B978-0-444-63623-2.00007-4.

70. J. Petersen, N. Larimer, J.A. Kaye, M. Pavel, and T.L. Hayes, "SVM to detect the presence of visitors in a smart home environment," *Conference proceedings: Annual International Conference of the IEEE Engineering in Medicine and Biology Society. IEEE Engineering in Medicine and Biology Society. Annual Conference*, pp. 5850–5853, 2012. San Diego, CA.

71. R. Begg and R. Hassan, "Artificial neural networks in smart homes," in *Designing Smart Homes: The Role of Artificial Intelligence, ser. Lecture Notes in Computer Science*, J.C. Augusto and C.D. Nugent, Eds. Berlin, Heidelberg: Springer Berlin Heidelberg, 2006, pp. 146–164. [Online]. doi: 10.1007/117884859

72. M. Ermes, J. Prkk, J. Mntyjrvi, and I. Korhonen, "Detection of daily activities and sports with wearable sensors in controlled and uncontrolled conditions," *IEEE Transactions on Information Technology in Biomedicine*, vol. 12, no. 1, pp. 20–26, Jan. 2008.

73. L. Ciabattoni, G. Ippoliti, A. Benini, S. Longhi, and M. Pirro, "Design of a home energy management system by online neural networks," *IFAC Proceedings Volumes*, vol. 46, no. 11, pp. 677–682, Jan. 2013. [Online]. Available: http://www.sciencedirect.com/science/article/pii/S1474667016330221.

74. S. Wong, K.K. Wan, and T.N. Lam, "Artificial neural networks for energy analysis of office buildings with daylighting." *Applied Energy*, vol. 87, no. 2, pp. 551–557, 2010. doi: 10.1016/j.apenergy.2009.06.028.

75. C. Buratti, M. Barbanera, and D. Palladino, "An original tool for checking energy performance and certification of buildings by means of Artificial Neural Networks. *Applied Energy*, vol. 120, pp. 125–132, 2014. doi: 10.1016/j.apenergy.2014.01.053.

76. K. Li, C. Hu, G. Liu, and W. Xue, "Building's electricity consumption prediction using optimized artificial neural networks and principal component analysis." *Energy and Buildings*, vol. 108, pp. 106–113, 2015. doi: 10.1016/j.enbuild.2015.09.002.

77. F. Khayatian, L. Sarto, and G. Dall'O', "Application of neural networks for evaluating energy performance certificates of residential buildings." *Energy and Buildings*, vol. 125, pp. 45–54, 2016. doi: 10.1016/j.enbuild.2016.04.067.

78. C. Deb, L.S. Eang, J. Yang, and M. Santamouris, "Forecasting diurnal cooling energy load for institutional buildings using Artificial Neural Networks." *Energy and Buildings*, vol. 121, pp. 284–297, 2016. doi: 10.1016/j.enbuild.2015.12.050.

79. F. Ascione, N. Bianco, C. De Stasio, G.M. Mauro, and G.P. Vanoli, "Artificial neural networks to predict energy performance and retrofit scenarios for any member of a building category: A novel approach." *Energy*, vol. 118, pp. 999–1017, 2017. doi: 10.1016/j.energy.2016.10.126.

80. Y.-y. Song and Y. Lu, "Decision tree methods: applications for classification and prediction." *Shanghai Archives of Psychiatry*, vol. 27, no. 2, pp. 130–135, Apr. 2015. [Online]. Available: https://www.ncbi.nlm.nih.gov/pmc/articles/PMC4466856/

81. L. Rokach and O. Maimon, "Decision trees," in *Data Mining and Knowledge Discovery Handbook*, O. Maimon and L. Rokach, Eds. Boston, MA: Springer US, 2005, pp. 165–192. [Online]. doi: 10.1007/0-387-25465–X 9

82. M. Delgado, M. Ros, and M. Amparo Vila, "Correct behavior identifi- cation system in a tagged world," *Expert Systems with Applications*, vol. 36, no. 6, pp. 9899–9906, Aug. 2009. [Online]. Available: http://www.sciencedirect.com/science/article/pii/S0957417409 00116X

83. M. Viswanathan, "Distributed Data Mining in a Ubiquitous Healthcare Framework," in *Advances in Artificial Intelligence*, Canadian AI 2007, Z. Kobti and D. Wu, Eds. Lecture Notes in Computer Science, vol. 4509. Springer, Berlin, Heidelberg, 2007. doi: 10.1007/978-3-540-72665-4_23

84. M. Usama, J. Qadir, A. Raza, H. Arif, K.-L.A. Yau, Y. Elkhatib, A. Hussain, and A. Al-Fuqaha, "Unsupervised machine learning for networking: Techniques, applications and research challenges," *arXiv preprint arXiv:1709.06599*, 2017.

85. P. Rashidi, D.J. Cook, L.B. Holder, and M. Schmitter-Edgecombe, "Discovering activities to recognize and track in a smart environment," *IEEE Transactions on Knowledge and Data Engineering*, vol. 23, no. 4, pp. 527–539, Apr. 2011.

86. M.C. Mozer, "Lessons from an adaptive home," in *Smart Environments*. Wiley-Blackwell, 2005, pp. 271–294. [Online]. Available: https://onlinelibrary.wiley.com/doi/abs/10.1002/047168659X.ch12

87. D. Li and S.K. Jayaweera, "Reinforcement learning aided smart-home decision-making in an interactive smart grid," in 2014 *IEEE Green Energy and Systems Conference (IGESC)*, Long Beach, CA, Nov. 2014, pp. 1–6.

88. J.R. Santana, L. Sánchez, P. Sotres, J. Lanza, T. Llorente, and L. Muñoz, "A privacy-aware crowd management system for smart cities and smart buildings." *IEEE Access*, vol. 8, pp. 135394–135405, 2020. doi: 10.1109/ACCESS.2020.3010609.

89. M. Razmara, G.R. Bharati, M. Shahbakhti, S. Paudyal, and R.D. Robinett, "Bi-level optimization framework for smart building-to-grid systems." *IEEE Transactions on Smart Grid*, vol. 9, no. 2, pp. 582–593, Mar. 2018. doi: 10.1109/TSG.2016.2557334.

90. M. Masera, E.F. Bompard, F. Profumo, and N. Hadjsaid, "Smart (Electricity) grids for smart cities: Assessing roles and societal impacts." *Proceedings of the IEEE*, vol. 106, no. 4, pp. 613–625, Apr. 2018. doi: 10.1109/JPROC.2018.2812212.

91. B.S. Çiftler, S. Dikmese, İ. Güvenç, K. Akkaya, and A. Kadri, "Occupancy counting with burst and intermittent signals in smart buildings." *IEEE Internet of Things Journal*, vol. 5, no. 2, pp. 724–735, Apr. 2018. doi: 10.1109/JIOT.2017.2756689.

92. S. Abrol, A. Mehmani, M. Kerman, C.J. Meinrenken, and P.J. Culligan, "Data-enabled building energy savings (D-E BES)." *Proceedings of the IEEE*, vol. 106, no. 4, pp. 661–679, April 2018. doi: 10.1109/JPROC.2018.2791405.

93. Xu Weitao, Zhang Jin, Kim Jun Young, Huang Walter, S. Kanhere Salil, K. Jha Sanjay, Hu Wen, "The design, implementation, and deployment of a smart lighting system for smart buildings." *IEEE Internet of Things Journal*, vol. 6, no. 4, pp. 7266–7281, Aug. 2019. doi: 10.1109/JIOT.2019.2915952.

94. H. Dagdougui, A. Ouammi, and L.A. Dessaint, "Peak load reduction in a smart building integrating microgrid and V2B-based demand response scheme." *IEEE Systems Journal*, vol. 13, no. 3, pp. 3274–3282, Sept. 2019. doi: 10.1109/JSYST.2018.2880864.

95. W. Li, "Application of economical building management system for singapore commercial building." *IEEE Transactions on Industrial Electronics*, vol. 67, no. 5, pp. 4235–4243, May 2020. doi: 10.1109/TIE.2019.2922946.

96. C.K. Metallidou, K.E. Psannis, and E.A. Egyptiadou, "Energy efficiency in smart buildings: IoT approaches." *IEEE Access*, vol. 8, pp. 63679–63699, 2020. doi: 10.1109/ACCESS.2020.2984461.

12 Design of Futuristic Trolley System with Comparative Analysis of Previous Models

Balla Adi Narayana Raju, Deepika Ghai, and Kirti Rawal
Lovely Professional University Phagwara

CONTENTS

12.1 INTRODUCTION

Over the past decade, internet and the economy have flourished simultaneously leading to the birth of an era of hassle-free shopping from the convenience of one's living room. Internet technologies have always had the edge in tracking the purchase habits of the customers and exploited it [1]. However, the traditional retail stores have not had the similar level of targeted advertisement and curated shopping experience. The retail stores have remained relatively untouched by the advancements in automation, image processing, and Artificial Intelligence (AI). Considering that the supermarkets are where customers get their constant supply of daily-use items like food products, garments, electronics devices, and household appliances [2], any amount of advancement based around the customer's preferences can cause a direct spike in the sales and perception of product. Supermarkets face constant challenges like managing store floors and bill payments at counters, which is a time-consuming process [3], as illustrated in Figure 12.1. Self-services with shopping trolleys or bins decreases labor cost,

FIGURE 12.1   Problem of long queue for billing in supermarket [3].

and numerous store chains are endeavoring further decrease by shifting to self-service check-outs [4]. So stores are working on new innovative ideas to promote self-service procedure in order to decrease the labor work charges . The chapter focuses on the trolley that gives the customer a self-serving service with less time utilization and makes a store with less labor work prerequisites. This chapter provides an inaugural and elementary comprehensive survey of trolley-based system and a potential research idea for autonomous-based system for smart shopping in the mall, i.e. Big Bazaar, Reliance Trends, Best Price, etc. In summary, this literature review can serve as a good reference for researchers in the area of designing trolley-based systems.

12.1.1   APPLICATIONS OF THE AUTONOMOUS TROLLEY SYSTEM

The applications of the autonomous nature of the trolley system are as follows:

  i. A smart trolley which can follow the customer in the store and can transport the things from one spot to the next. It can also track the items that a customer picks and can monitor the surrounding of the store [5–7].
 ii. This can form the basis of a luggage trolley that follows the user while maintaining a certain distance [8].
iii. This can also be used in wheelchairs for people to reduce the strain on the supporting individual. The wheelchair can follow the person piloting the same from one place to another without any major hassle.
 iv. This can also be used by parents of infants to carry them around in baby strollers, without the associated strain of dragging the stroller [9].
  v. There are a lot more uses of this approach in the fields of emergency clinics, eateries, houses, armed forces, and many more.

12.1.2   CHALLENGES

The major challenges faced by the trolley system in the supermarket setting for automation, detection, recognition, and analyzing can be roughly categorized into seven types:

i. *Exertion of Heavy Loads to the Self-Propelled System*: A heavy load hauling usually takes up high torque, and overhauling could damage the motors.
Also, as the load in each cart is variable (this can also be accomplished with
Pulse Width Modulation (PWM)), managing the speed of the trolley is a
major concern for automatic movement in the system.

ii. *Multiple Tag Readings*: The cart in a practical scenario has many items
present, and reading each tag in the cart accurately grows harder with each
added object. The reader used to scan barcodes should be able to read the
number of barcodes in a particular time with maximum range of products.

iii. *Facial Analysis*: The detection of customer is also a major challenge.
Detecting the correct client and getting the data from the database to provide a better shopping experience accurately is a concerning error. The
face data set of the customers should be properly taken to remove the error
caused by the misinterpretation of faces [10].

iv. *Background Complexity*: The backgrounds in real-life scenarios are more
complex compared with the technical implementation data sets. Counters
with items can be of uneven form, thus making it hard to detect and analyze
if a large number of items are placed.

v. *Data Interference*: Various factors like data mismatch and data manipula-
tion can occur due to criminal activities like shoplifting. Noise in the system
can also lead to the misinterpretation of data.

vi. *Logistics Inventory Management*: Managing and supervising the supply
chain management from warehouse to the sale point is a difficult task with
large data manipulations and manual intervention of barcode with physical
records, whereas Radio-Frequency Identification(RFID)tags scanning can
be alternatively fast.

vii. *Developing a Unified System*: Building up this kind of new framework and
using it in the store will require a huge number of assets and information at
the underlying stages to prepare the model. Many testing stages need to be
done to improve the procedure stream of the trolley from facial recognition,
human following, and tag detection to monitoring the surroundings of the
store. Another challenge is making users take up these changes in a positive
light, creating an initial data set that would help in future recommendations
following principles of data security and privacy.

12.1.3 MOTIVATION

In this chapter, we are concentrating on comparative analysis of previous trolley
models and also giving a proposed idea for futuristic trolley system. Though there
are a number of research articles available on the topic of trolley-based systems, none
of the research papers surveyed this particular work considering the comparative
comprehensive surveys of trolley-based system for smart shopping. Till now, many
researchers have put emphasis only on a specific application, such as human following [5], huge system for product identification [6], RFID-based smart shopping [7],
supervised instance segmentation of supermarket products [11], and secure smart
shopping system [12]. To the best of our knowledge, this is the first survey chapter

which carries out a detailed study about state-of-the-art techniques of trolley-based models along with comparative analysis of different systems with respect to the strengths and weaknesses. This chapter sheds light on the design of the proposed system that should be considered while addressing these problems. The main objectives of the chapter are as follows:

    i. To study and analyze various trolley-based models that can be used in the supermarkets to provide the customers with a better shopping experience and a user-friendly assistance.

    ii. To propose an efficient and effective trolley-based model for monitoring the stores as well as the customers buying habits for a prediction on the items for better profitability.

The rest of the chapter is organized as follows: In Section 12.2, we provide a detailed review in the field of trolley-based system along with strengths and weaknesses. Then, the system architecture of the futuristic trolley system is discussed in Section 12.3. The idea on the autonomous trolley system is presented in Section 12.4. Finally, conclusion remarks and promising research directions in the future are given in Section 12.5.

12.2   RELATED WORK

Researchers have proposed numerous methods for trolley-based systems. Nowadays, shopping includes standing in lines before cash counters for buying specific items. The technology used in stores for identifying the items is barcode scanning, which was created in the 1970s [13]. Today, barcode identifications are found in each item; it has become common. The disadvantage of this technology is that the item should be in the line of sight of the scanner, which is a tedious procedure. Public awareness of utilizing RFID technology was featured in recent years, and the retail giants like Wal-Mart and U.S. Division of Defense (DOD) asked their suppliers to utilize this technology. In 2006, Wal-Mart Chief Information Officer (CIO) stated that utilizing RFID brought about 26% decrease in out of stock in stores, and the items are filled many times faster than those items which are physically scanned [14]. The University of Arkansas Information Technology Research Institute in 2009 studied the business benefit of utilizing RFID innovation at significant dimensions. The aftereffect of using RFID labels in the chain management board showed overall precision improved by 27%, under stocks decreased by 21%, and overstock decreased by 6%. This study likewise compared the time required to count the products in the store, utilizing the current strategy and RFID. With RFID, it is seen that it took 2 hours to check 10,000 products; through the standardized barcode identification, it took 53 hours to scan similar products, which makes an average of 4,767 things for every hour utilizing RFID; for barcode, it was 209 products for every hour, a 96% decrease in counting time [3].

    It is seen that almost 10 billion fashion items and 16 billion shoe items are dispatched from producers annually. Utilizing the traditional barcode scanning, it is hard to trace the items precisely and within the time limit. Inventory management utilizing this technology is making it hard to take quick business choices for creating in-store sales lift [15].

In 2011, Agarwal, Sultania, Jaiswal, and Jain [16] built an RFID-based automatic shopping cart that allows the customer with a superior approach for shopping in-store environment. This shopping trolley comprises a touch screen with a graphical interface attached to the cart, a client card, and an item reader for scanning the labels of the items placed in the cart, which makes it customer intuitive.

In 2013, McBeath suggested that the retailers need to adapt to another technology to remove the complexity in items happening on daily basis . They require exact, real-time, and item-level pictures of stock. RFID has improved the store deals by 20%–30% [17]. J.C. Penney improved stock accuracy from 75% to 99% in the gathering of things utilizing RFID [18].

Gupta, Kaur, Garg, Verma, Bansal, and Singh [19] stated that initially all the stocks from various suppliers arrive at the store; then, these items are registered by using the registration form, which consist of the detailed information of each product including Identification Number (ID). Then when the customer enters the store, the items will be moved to the registration counter where he/she will be given an RFID rechargeable loyalty card. The customer will then take the smart cart, which is locked, and start shopping. The items placed on the cart are scanned, and then, the cart is moved to the payment counter. At the billing counter, the cart is connected to the server via a serial port and authentication is done for opening the locked cart. The data of the purchased RFID tagged product is automatically transferred to the server. The bill payment is done with the loyalty card. Although the time of long queues is removed, it still has some drawbacks such as wired connection between the counter and the cart, time for connecting billing counter to cart, and time for analysis of each item on the cart. Items can be only removed at the time of billing. Payment is only done by RFID loyalty card, and it has to be recharged each time at the billing counter.

Kumar, Gopalakrishna, and Ramesha [7] proposed an intelligent shopping cart system. In this system, RFID Technology, ZigBee, and microcontrollers are used. This system has a start button for making the trolley active. When an item with RFID tag is placed in the trolley, the tag ID is read by RFID Reader; if the tag number is matched with database, then it displays product details tagged to that ID on the Liquid Crystal Display (LCD) screen. The display also has an end shopping button; when this button is pressed, the total list of items present in the trolley are then send to the master PC via ZigBee. There is also an option in the LCD screen to remove a particular item from the cart. The bill will be updated according to customer choice. The limitation of this system is that payments at the end are done on the billing counter for which user should again waste some time in billing of items and the trolley is not smart enough to interact with the surrounding of the store.

Gangwal, Roy, and Bapat [20] proposed a smart shopping cart for automated billing purpose using Wireless Sensor Networks (WSN), as shown in Figure 12.2, in which the cart is attached with monitor, camera, and a scanner for tag identification. The camera attached to the trolley is used for decision-making process. Initially, the product is scanned; simultaneously, the picture of the product is taken. Then, an image comparison algorithm is run on the images that are acquired by the cart

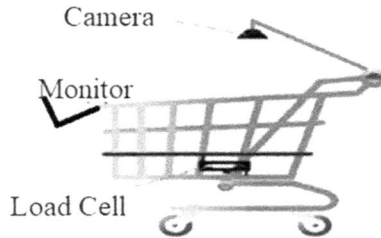

FIGURE 12.2   Smart shopping trolley.

to ensure that the items placed in the cart are matched with the scanned item. The major drawback of this system is the images are compared each time when the item is placed in the cart, which requires large memory space.

In 2014, Kamble, Meshram, Thokal, and Gakre [21] proposed a multitasking shopping cart system using the RFID technology. The RFID reader is used to read the product information which is retrieved at the time of billing. This system consists of three modules: (i) Server Communication Component (SCC), (ii) User Interface and Display Component (UIDC), and (iii) Automatic Billing Component (ABC), which are coordinated together to provide smart trolley that reduces billing time. This chapter consists of certain issues such as the stock maintenance is not mentioned. Further, this system is not able to correspond about the missing products dropped in the trolley. .

Jayshree, Gholap, and Yadav [22] proposed RFID-based billing trolley. This trolley is having the feature of headsets attached to the trolley wherein the scanned item name and cost are repeated by the headphone. They have also attached a LCD screen to the trolley to display the total amount. The drawback of this system is paying the bill in the counter at the end of the purchase, which is time consuming.

Gupta and Garg [6] proposed a system as an analytical model for automating purchases using RFID-enabled shelf and cart. In this system, they have used weight-sensing mat integrated with RFID reader, ZigBee, and microcontroller for implementation. The shelf of the stores is layered with these weight-sensing mats which tell where the product is placed on the shelf depending on the extent of pressure expended on these sensors. Similarly, the cart also consists of these sensing mats. When the item is taken from the shelf and placed on the trolley, the product is detected and displayed on the LCD screen. Then, the product cost is added to the total bill amount and sent to the base station by wireless network for payment.

In 2015, Bhatwal and Chamoli [23] proposed an automated self-billing system using RFID technology which consists of display panel, RFID reader, and master computer. Initially, the passive tags are scanned by the reader as shown in Figure 12.3, which is then shown on the touch display panel. If the user wants to remove the item, then he/she can remove the item using the cancellation button in the display panel. The display panel also prints the bill at the end, which the customer should pay using cash or card. The drawback of this is the display panel that the author has described is a master computer, which, in reality, makes the billing process complex.

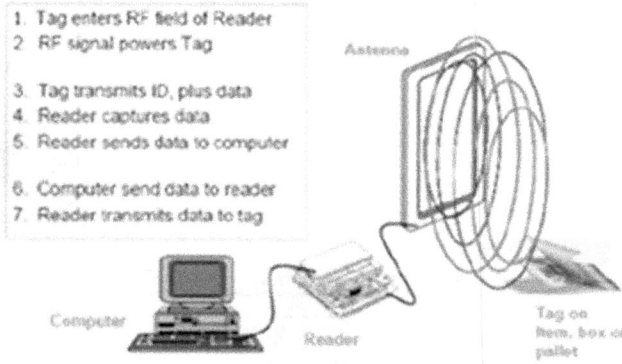

1. Tag enters RF field of Reader
2. RF signal powers Tag

3. Tag transmits ID, plus data
4. Reader captures data
5. Reader sends data to computer

6. Computer send data to reader
7. Reader transmits data to tag

Antenna

Computer

Reader

Tag on Item, box or pallet

FIGURE 12.3 Working of RFID tag.

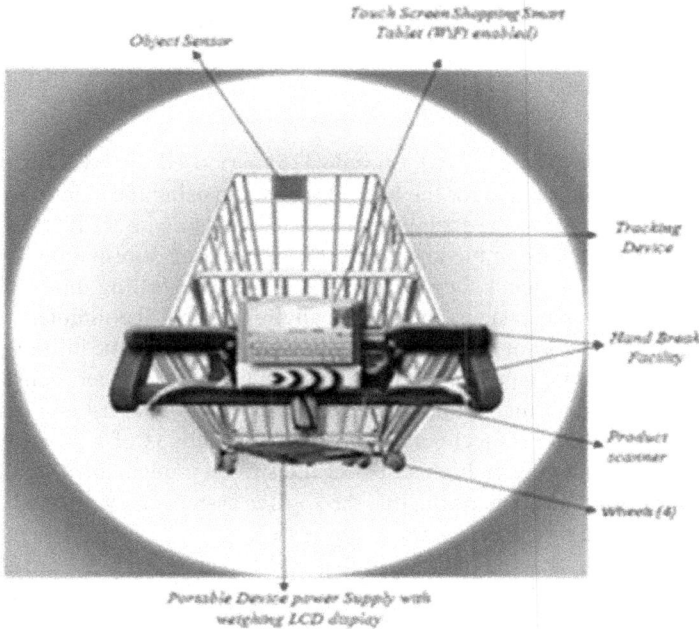

Object Sensor

Touch Screen Shopping Smart Tablet (WiFi enabled)

Tracking Device

Hand Break Facility

Product scanner

Wheels (4)

Portable Device power Supply with weighing LCD display

FIGURE 12.4 Architecture of smart trolley.

Ali and Riaz [24] proposed a smart trolley (Figure 12.4) which consists of a hand-brake LCD screen, product scanner, and object sensor. The product scanner and object sensors are used to detect the items in the store, whereas the LCD screen is a touchscreen user interface where the customer can search for the specific item in the store. The user interface (Figure 12.5) will show the items placed in the store with the shelves map of the product for better user-friendly shopping.

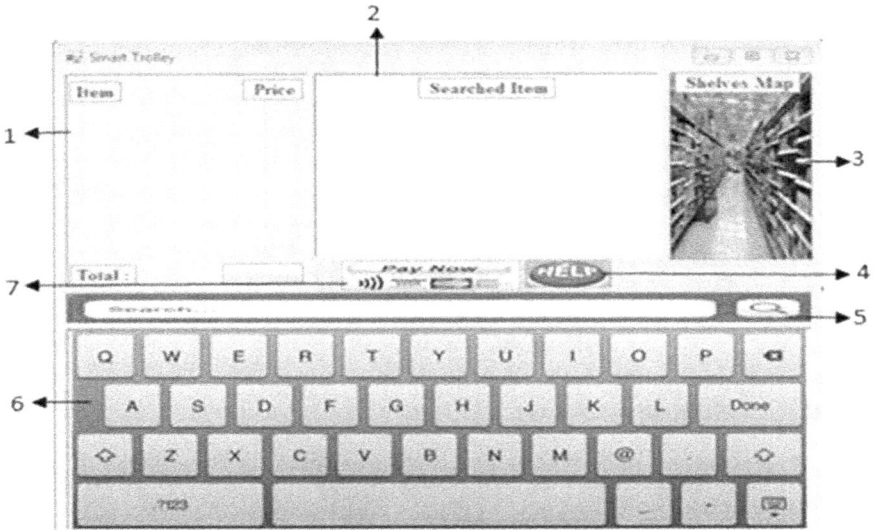

FIGURE 12.5   User interface design.

Sawant, Krishnan, Bhokre, and Bhosale [25] proposed an RFID-based smart shopping in which alternative for the barcodes is made using the smart labels called the RFID tags. The cart in the shopping malls are enabled with RFID technology to recognize the products using the tags and send the product information to the shopping malls database, which is then retrieved at the time of billing thus reducing the billing time, but this paper does not discuss anything about the maintenance of the product list; also, it does not provide any mechanism for scaling the billing process in malls when two or more customers bill at the same time. This paper focuses majorly on reducing the billing time, but no management of the product list is provided. The cost of implementing RFID technology in shopping malls is of a greater consideration because the cost may be high when RFID is to be implemented in all the trolleys in the shopping mall. The traffic that the ZigBee system can withstand also plays a major role in deciding the efficiency of the system.

In 2016, Suganya, Swarnavalli, Vismitha, and Rajathi [26] proposed automated smart trolley with smart billing using Arduino. In this paper, the author developed a mobile application in which the customer in store is asked to enter the items that he wants to purchase; then the items list are shown in a sorted order, with the cost and location displayed according to the list for ease of finding the item in the store.

Iyappan, Jana, Anitha, Sasirega, and Venkatesan [27] proposed an enhanced shopping model for improving smartness in the supermarket using Smart Arduino-Based Intelligent System (SABIS) architecture in which trolley and racks of the stores are mounted with the microcontrollers. The cart also consists of Infrared (IR) module. The information of items is transmitted through rack to the trolley by wired connection using jumper wires. When the item is brought from the rack and placed in the trolley, the IR rays emitted are blocked so to know about the particular item that

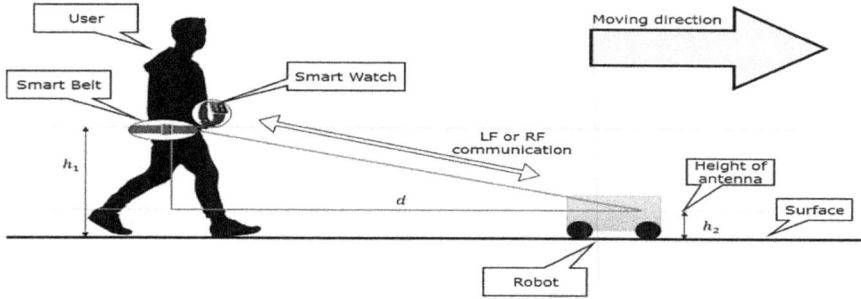

FIGURE 12.6 Architecture of smart mobile robot.

is placed and the information of that item is shown on the LCD screen attached to the trolley. The drawback of this model is that for purchasing each item the controllers of the trolley and rack should be connected manually each time, which is a time-consuming process.

Lee, Kim, Hwang, and Kang [5] proposed a mobile robot which can follow and lead humans by detecting user location and behavior with wearable devices, as shown in Figure 12.6. This paper suggested the prototype of small robot which can follow the user by using the smart watch and belt to track the person using Radio Frequency (RF) communication with Low-Frequency (LF) transmitter. The drawback of this idea was tracking or finding person was difficult and also was not so accurate.

Chiang, You, Lin, Shih, Liao, Lee, and Chen [28] designed a smart shopping cart along with face recognition feature to the customer. The customer is assisted with the recommendation of products based on the history of purchase. In this manner, the purchasing efficiency can further be enhanced and the shopping time can be reduced. However, the algorithms used for navigation of the cart and face recognition are complex.

Wu, Yao, Hou, Chang, Tseng, Huang, Chen, and Yang [29] proposed an Intelligent Shopping Assistant System (ISAS) with steady recognition and navigation of the trolley (Figure 12.7). This model has an additional app feature where the customers can see the shopping list and direct the shopping trolley to get products. The app also has product recommendation and information on the discount details. The advantages of the model include hands-free shopping experience with least time spent at the shopping mall.

Swamy, Seshachalam, and Shariff [30] proposed a smart RFID-based interactive kiosk cart using wireless sensor node. The model consists of RFID reader, kiosk, and ZigBee module. The authors developed their own database for product details. The kiosk attached to the cart shows the scanned items with their maximum retail price and selling price after discount. For food items, the expiry date is been displayed; then, the product details can be seen in the web browser.

In 2017, Chavan, Gadgay, and Pujari [31] proposed a smart shopping cart using RFID which shows the expired details of the items of the store on the LCD screen. There is also a mode button to add and remove the items from the cart. The buzzer

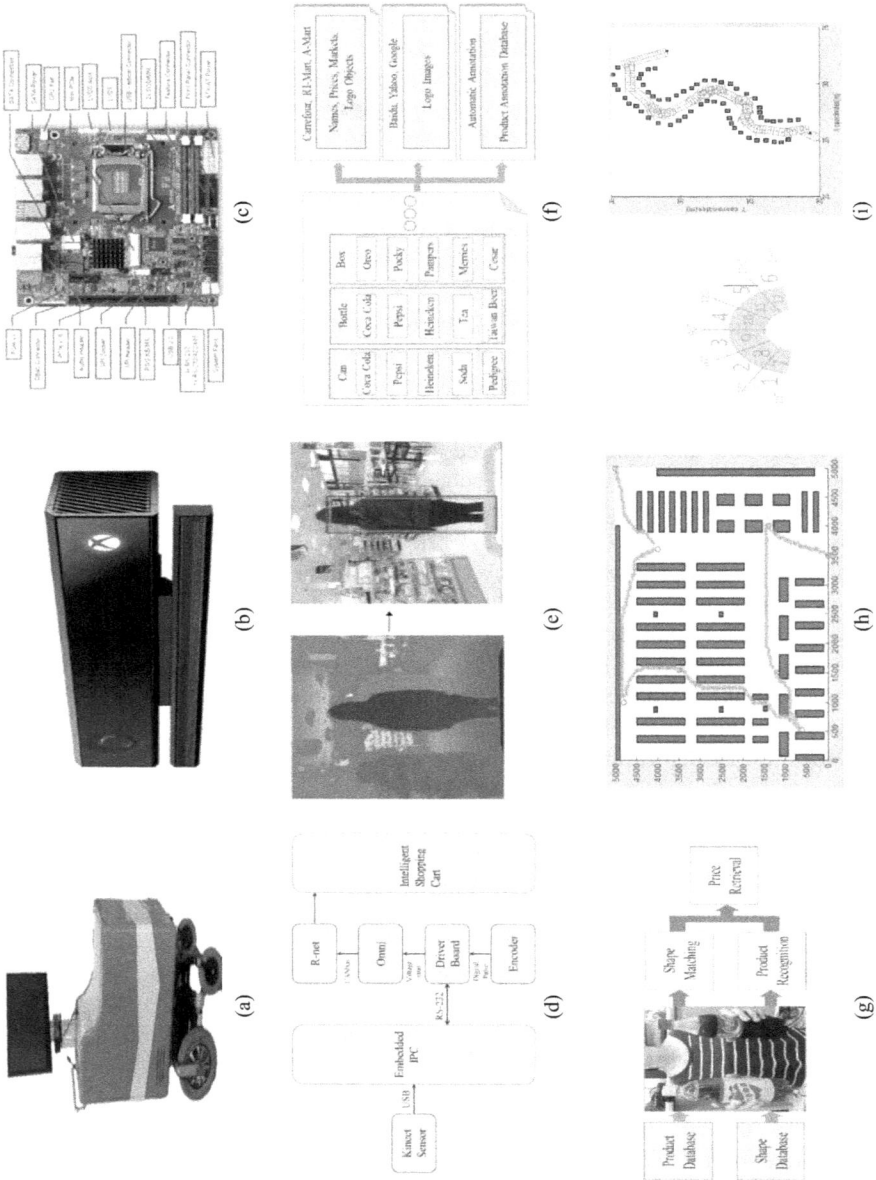

FIGURE 12.7 Intelligent shopping assistant system [29]: (a) robot platform, (b) Microsoft XBOX one Kinectv2, (c) IPC: AmITX-HL-G-Q87 mini-ITX MB, (d) hardware architecture of robot on the wheelchair, (e) human detection, (f) product database, (g) product and price detection, (h) path planning result, and (i) Q-learning obstacle avoidance.

attached to the cart is used when the item is expired. The item details of the cart are communicated to the administration PC through Digi XBee wireless network.

Li, Song, Capurso, Yu, Couture, and Cheng [32] proposed Internet of Things (IoT) applications on secure smart shopping system, as shown in Figure 12.8. This model is the first to use Ultra-High-Frequency (UHF) RFID technology to achieve automatic reading of the items with a proper range. It was designed for the secure protocol for the communication. The design also consists of RFID reader, ZigBee adaptor, and LCD touchscreen with weight scanner. The model is highly secure and accurate because the metal outside the cart blocks the signals to a pretty high extent such that, when the reader is inside the cart, no item outside the cart can be read.

In 2018, Suryanto, Siagian, Angin, Sashanti, and Yogen [33] designed an automatic mobile trolley using ultrasonic sensors (Figure 12.9) in which the authors used three ultrasonic sensors connected to the trolley in a line at a particular distance. These sensors are used for distance calculation from the customer to product of the store so as to follow in a particular path.

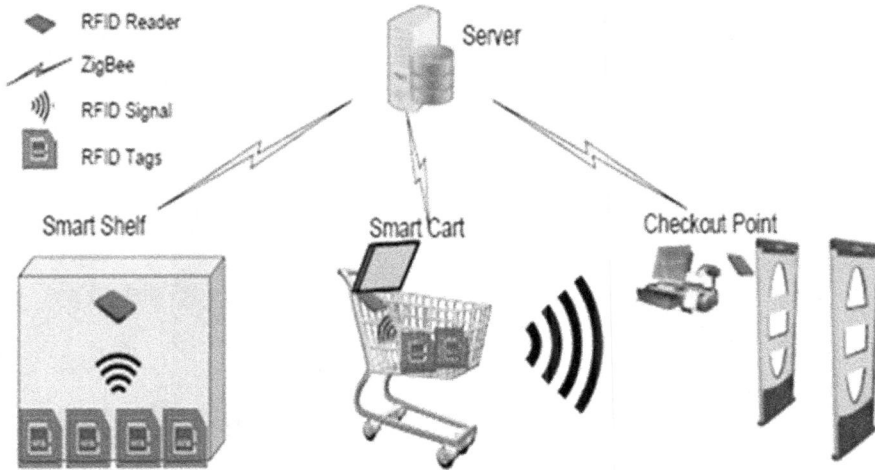

FIGURE 12.8 System model for secure smart shopping.

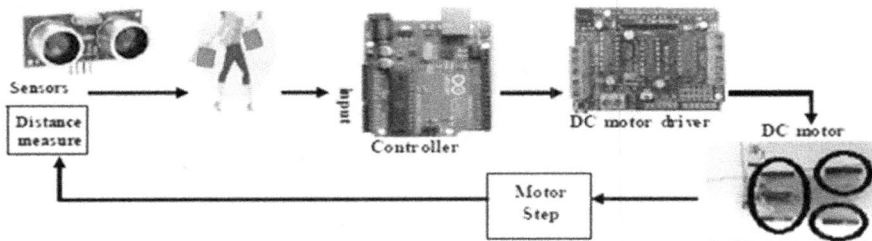

FIGURE 12.9 Block diagram of automatic trolley using ultrasonic sensors.

Chithra, Sunil, Sneha, Shruthi, and Sowmya [34] proposed an IoT-based futuristic trolley for intelligent billing with the amalgamation of RFID and ARMLPC2148. In this framework, they have used a microcontroller; the RFID reader reads the RFID tag number and compares it with stored tag numbers; if it is present, then the product cost is added to the total bill amount, and product details (like product cost, manufacturing year, brand name, etc) are displayed on the LCD screen. If it is not found in the database, it will display product is not found or not present, and this process will continue until the end of shopping. This total bill information will be sent to master PC using ZigBee at the receiver and the receiver-side the bill is generated. The output of this proposed framework is better than the existing technology, but it has certain limitations as it cannot be used to keep all sales on track and assess item availability at the shopping markets.

Athauda, Marin, Lee, and Karmakar [12] proposed robust low-cost passive UHF RFID-based smart shopping trolley, as shown in Figure 12.10. The technology used

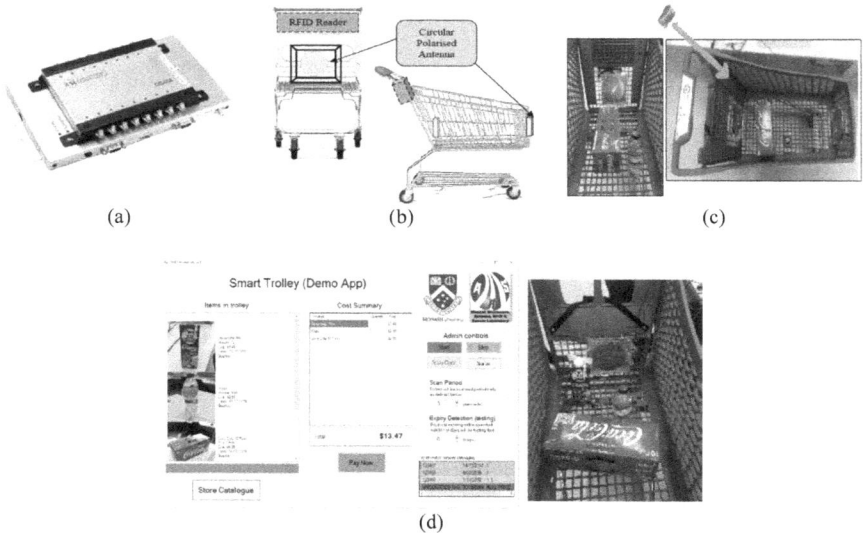

(a)   (b)   (c)

(d)

FIGURE 12.10 Robust low-cost passive UHF RFID-based smart shopping trolley [12]: (a) convergence systems limited- CSL468 RFID reader, (b) front and side view with circular polarized antenna, (c) circular polarized antenna attached to the shopping trolley for product testing, and (d) shopping trolley user interface.

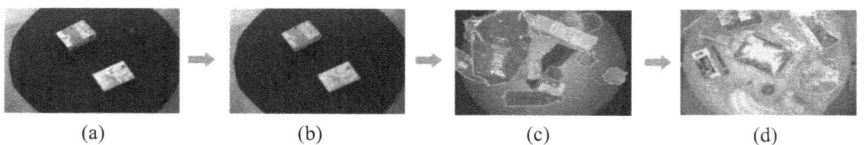

(a)   (b)   (c)   (d)

FIGURE 12.11 Instance segmentation of supermarket products [11]: (a) image acquisition of store items, (b) automatic labeling, (c) augmentation and training of the images with respect to D2S data set, and (d) output image in a segmented format.

was RFID with Circular Polarized (CP) antenna, which can detect items irrespective of orientation, size, and shape. The RFID reader (CSL468) used in this model can detect 300 tags per second with a range of 12 m. The disadvantage of this model is it is not autonomous and that no processing techniques was used for prediction.

Follmann, Drost, and Böttger [11] proposed a fast segmentation system for the products of supermarket. The process and steps they used are acquiring product with automatic labeling and then augmentation with training the labeled images. At last, segmented image is seen. Figure 12.11 shows the instance segmentation in the product of supermarket. The images of the grocery items are acquired. Images are then labeled by using automatic labeling techniques. Thereafter, the images are augmented and trained with respect to the densely segmented supermarket (D2S) data set. At last the segmentation of the images is done to find the various grocery items present in that image. The disadvantage of the system is that it is not a portable and the model is huge.

The comparative analysis of various existing trolley-based models is shown in Table 12.1.

Even though many researchers have proposed a large number of systems, a single unified trolley-based model needs to be proposed that fits for all applications. In this chapter, a generic trolley-based framework is proposed that uniformly describes the smart way of shopping in stores with a trolley interactive mechanism for showing the path and predicting the customer purchasing patterns.

12.3 METHODOLOGY

The trolley methodology is classified into two types: (i) system architecture and (ii) trolley architecture.

12.3.1 SYSTEM ARCHITECTURE

The system architecture encompasses a network ecosystem of the different technologies that need to work in unison for creating the backbone of the technology as shown in Figure 12.12. The proposed idea at the most talks about the inside working of shopping trolley and its functionalities. Some of the technologies like cloud computing and server-side services are shown to give a brief idea about how the shopping trolley will be connected to push and pull the data which is obtained by the trolley from store surroundings.

  i. *Management Interface System (MIS)*: It is a Graphical User Interface (GUI) for the product that forms the basis of the software, such as an Enterprise Resource Planning (ERP) tool. The inventory management, sales insight, product information, billing solutions, trolley management solutions, etc. are provided in this MIS, subject to user authorization.
  ii. *Cloud*: The cloud server is a virtual server where storage and high-speed processing occur with high-fidelity security options for protection of information. This is shared by the supermarket chains centrally. These cloud servers are brains of the management interface system, centralized computer, mobile interface, and trolleys.

TABLE 12.1
Comparison of Existing Trolley-Based Models

Author (Year)	Technology	Description	Strength	Weakness
Gupta, Kaur, Garg, Verma, Bansal, and Singh [19] (2013)	RFID, Loyalty smart card, micro-controller	First, passive tags present in the cart are detected by the RFID reader, and then the cart is sent to the billing counter where the wired connections are made between the counter interface and the cart for product details transfer. At last, the payment is done using rechargeable loyalty cards.	Less time for shopping compared to the existing technology	Payments are done using loyalty card, need to be recharged every time from the counter The details of product are transmitted using wired connection which is a time consuming payment process Items can be removed only at the billing counters
Kumar, Gopalakrishna, and Ramesha [7] (2013)	RFID, LCD display, ZigBee module	It explains how to access real-time information about the diverse products inside the shopping cart. Tags are detected using the RFID reader. Thereafter, the product details are shown on the LCD screen which is attached to the cart. Items can also be removed by using the LCD touchscreen. Finally, the details of the items are transferred to the master PC using ZigBee.	User friendly No manual help needed for item details Less time consumption in purchasing the items	Due to ZigBee, the high data rate of product details cannot be handled Payments of products can be given at counter only
Gangwal, Roy, and Bapat [20] (2013)	WSN (smart display monitor, camera)	Initially items are scanned by the reader; simultaneously camera takes the images of the items present on the cart. Finally, image comparison is done to find that the scanned items are same as the items seen by the camera.	Accurate for detecting items User friendly with smart display Less purchasing time	Images are compared each time when the items are placed in cart, which requires a large memory space

(Continued)

TABLE 12.1 (*Continued*)
Comparison of Existing Trolley-Based Models

Author (Year)	Technology	Description	Strength	Weakness
Kamble, Meshram, Thokal, and Gakre [21] (2014)	RFID, LCD display	Items are read by the RFID reader and shown on LCD. Then, the data is send to the central billing unit for automatic billing process.	Less shopping time / Smart billing	Stock maintenance is not noticed / No remedial action mentioned when the item is not detected
Jayshree, Gholap, and Yadav [22] (2014)	RFID, headsets, LCD display	The headsets and LCD display are attached to the trolley; when a product is scanned by the RFID reader, the items details are shown in display with a voice in headsets.	Smart item identification / Smart headsets which tell about the product details / Customer friendly system	No stock information / No online payment options
Gupta, and Garg [6] (2014)	RFID, weight-sensing mats, LCD display	This paper provides an idea of LCD use for offers, discount, and total bill. Weight-sensing mats are connected with the microcontroller on the racks of the store. These sensing mats are also placed in trolleys of the store. When items are taken from the rack and placed in the trolley, the sensing mats detects the items and shows the details in the display attached to the cart.	Accurate item identification / Smart display / Less time required for item purchase	More hardware and costly
Bhatwal, and Chamoli [23] (2015)	RFID, display panel	Passive tags are scanned and details are shown on the display panel. According to the users interest, the item remove button is present in the display panel.	Less time consuming as compared to the traditional method of shopping / Bill of items scanned is printed by the display panel	Bill payments can only be done on the counters with no online payment options

(*Continued*)

TABLE 12.1 (*Continued*)
Comparison of Existing Trolley-Based Models

Author (Year)	Technology	Description	Strength	Weakness
Ali, and Riaz [24] (2015)	Hand breaks, LCD display with touch sensing	User can search for an item on the LCD screen attached to the cart. The cart shows the shelf maps of the particular item in store.	Smart store view and item tracing User friendly display	As store changes their items, each time tracking of store patterns is difficult
Sawant, Krishnan, Bhokre, and Bhosale [25] (2015)	RFID, ZigBee	The cart in the shopping malls are enabled with RFID technology to recognize the products using the tags and send the product information to the shopping malls database, and the data are retrieved at the time of billing, thus reducing the billing time	Reduced billing time Quick object identification	Since ZigBee module is used for transferring information, the high data rate of product details cannot be handled
Suganya, Swarnavalli, Vismitha, and Rajathi [26] (2016)	Mobile application, Arduino	Developed a mobile application in which the customer in store is asked to enter the items that he wants to purchase. Then, the items list is shown in a sorted order with the cost and location is displayed in the mobile.	Friendly mobile application Item storing according to the product arranged in stores	
Iyappan Jana, Anitha, Sasirega, and Venkatesan [27] (2016)	LCD display, IR module, micro-controller	Trolley and racks of the stores are mounted with the microcontrollers. The cart also consists of IR module. The information of items is transmitted through rack to the trolley by wired connection using jumper wires. When the item is brought from the rack and placed in the trolley, the IR rays emitted are blocked so as to know about the particular item.	Accurate product identification using IR rays	Connecting wires from store racks to trolley, each time for knowing item details is a time consuming process No online payment options Requires huge hardware for each racks making it costly Can't place multiple items in trolley at same time

(*Continued*)

TABLE 12.1 (*Continued*)
Comparison of Existing Trolley-Based Models

Author (Year)	Technology	Description	Strength	Weakness
Lee, Kim, Hwang, and Kang [5] (2016)	Smart watch, smart belt, RF communication	The robot follows the user by using the smart watch and belt to track the person using RF communication with LF transmitter.	Human following	Not accurate enough for human tracing Problems while turning the system from one direction to another
Chiang, You, Lin, Shih, Liao, Lee, and Chen [28] (2016)	IoT, RFID, embedded computer	Proposed a smart shopping cart along with face recognition feature to the user. The customer is also assisted with the recommendation of products based on the history of purchase. Therefore, the purchasing efficiency can further be enhanced.	Reduced shopping time	Complex algorithms for navigation of the cart
Wu, Yao, Hou, Chang, Tseng, Huang, Chen, and Yang [29] (2016)	Image detection, RFID, Touch screen, LASER Rangefinder	ISAS with steady recognition and navigation of the cart. This model has an additional app feature where the customers can view the shopping list and direct the shopping cart to pick up goods. The app also has product recommendation and information about the discount details.	The model include a hand free shopping experience with minimum time spent at the shopping mall	
Swamy, Seshachalam, and Shariff [30] (2016)	RFID, Kiosk, ZigBee	Kiosk is attached to the trolley which shows the scanned items with their maximum retail price, selling price after discount in item. For food items, the expiry date is displayed. Thereafter, the product details can be seen in the web browser.	Smart kiosk Expiry information of item with buzzer User-friendly kiosk interface	Cannot handle high throughput of data

(Continued)

TABLE 12.1 (*Continued*)
Comparison of Existing Trolley-Based Models

Author (Year)	Technology	Description	Strength	Weakness
Chavan, Gadgay, and Pujari [31] (2017)	Digi XBee wireless module, LCD, RFID	A mode button is used to add and remove the items from the cart. Buzzer attached to the cart is used when the item is expired. The item details of the cart are communicated to administration PC through Digi XBee wireless network.	User-friendly interface Smart alarm system when item is expired	For paying bill, the customer has to stand in front of counters
Li, Song, Capurso, Yu, Couture, and Cheng [32] (2017)	UHF RFID reader, micro controller, LCD touchscreen, ZigBee adapter, and weight sensor	This model is highly secure and accurate as the metal outside the cart blocks the signals to a pretty high extent. When the reader is inside the cart, no item outside the cart can be read. Lastly, the details of the items are seen in the LCD display.	Most secure and accurate User-friendly display.	
Suryanto, Siagian, Angin, Sashanti, and Yogen [33] (2018)	Ultrasonic sensors, micro-controllers	Sensors are used to calculate distance between the customer and trolley. Sensor helps the cart to follow the customers in a particular path.	Human following using three ultrasonic sensors	Not accurate enough to follow human in store surroundings
Chithra, Sunil, Sneha, Shruthi, and Sowmya [34] (2018)	RFID, ARM LPC2148, LCD, ZigBee	The time required for billing in the shopping malls is cut down in self-scanning. Initially, RFID reader reads the RFID tag number and compares it with stored tag numbers. If it is present, then the product cost is added to the total bill amount, and product details are displayed on LCD screen (such as product cost, manufacturing year, brand name, etc.)	Product details extraction in low time.	Since ZigBee module is used for transferring information, the high data rate of product details cannot be handled

(Continued)

TABLE 12.1 (*Continued*)
Comparison of Existing Trolley-Based Models

Author (Year)	Technology	Description	Strength	Weakness
Athauda, Marin, Lee, and Karmakar [12] (2018)	Circular polarized antenna, UHF RFID	Circular polarized antenna is mounted in the cart which can detect items irrespective of item orientation, size and shape.	can identify a large number of tags in a single go, i.e. 300 tags/sec	-
Follmann, Drost, and Böttger [11] (2018)	Digital camera, controller	Acquisition of store items is done with automatic labeling of the items. Thereafter, augmentation and training of the images using D2S data set is shown. Finally, output of the image is shown in a segmented format.	Image identification with trained D2S data set	Huge system Not a portable model

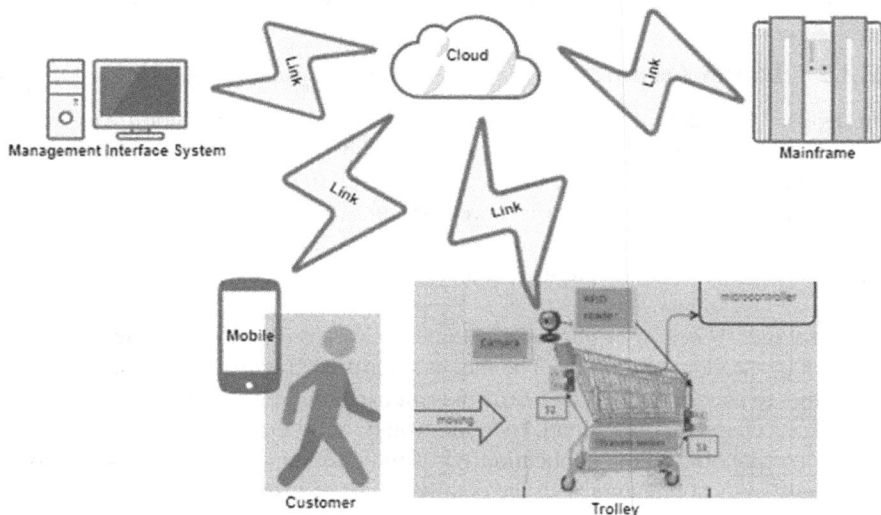

FIGURE 12.12 System architecture of proposed futuristic trolley.

iii. *Mainframe*: These are the extensive Beowulf clusters of PCs that work simultaneously to process the colossal information, aid in data exchange handling, undertake data statistics, help with transaction processing, etc. Mainframe servers can run various working frameworks at the same time. In this job, a solitary centralized computer can supplant higher-working equipment administrations accessible to conventional servers. Mainframe servers are intended to deal with the high volume of information and with fast, highly accurate throughput. The data rate of sending the data varies from gigabyte to terabyte in a short span of time. Compared with a regular Personal Computer (PC), mainframe is hundred times much faster in getting data from the web.

iv. *Mobile*: It is a gadget at the client level utilized for making payments and for getting product details of the items in the store. An interactive User Interface (UI) with options of adding the requirements, removing products, paying bills, etc. is used. The common components found in mobile are: a CPU to manipulate the data, a battery to provide the power source for the mobile capacities, and an information mechanism to enable the user to connect with the server.

v. *Trolley*: The trolley comprises sensors, camera, and motor controller for making the framework self-sufficient. The sensors are utilized for following the client. Camera is used for facial recognition, classifying, and for monitoring the surrounding of the stores. RFID reader is used for getting the tag details of the items which the customer has added into the trolley.

12.3.2 TROLLEY ARCHITECTURE

The trolley architecture comprises four modules: (i) face recognition, (ii) trolley movement, (iii) items detection, and (iv) object and item classification, as shown in Figure 12.13. Each module has its own advantages from getting customer details, product details, and classifying the store surrounding to customer following. Image classification techniques are used to monitor the surroundings for product identification and customer purchase pattern prediction for marketing strategies.

i. *Module 1-Face Recognition*: Customer faces are identified by the trolley, using the camera module. The face is coordinated with the data set of the store to check whether the face is present in the database or not. If matching facial characters are found, then the trolley will be activated and the data corresponding to the application ID of the customer will be retrieved into the client's application with the detailed chart of his/her interest and previous transactions. The client can also log in manually using his front facing camera, receiving a Quick Response (QR) code that can be scanned at the trolley. In this event, if the face isn't found in the data set, at that point the face is added to the database (DB) and another ID is generated and activated with the new customer ID. The identification of the facial features is done using highly accurate face recognition algorithms to avoid the error associated with the detection of faces.

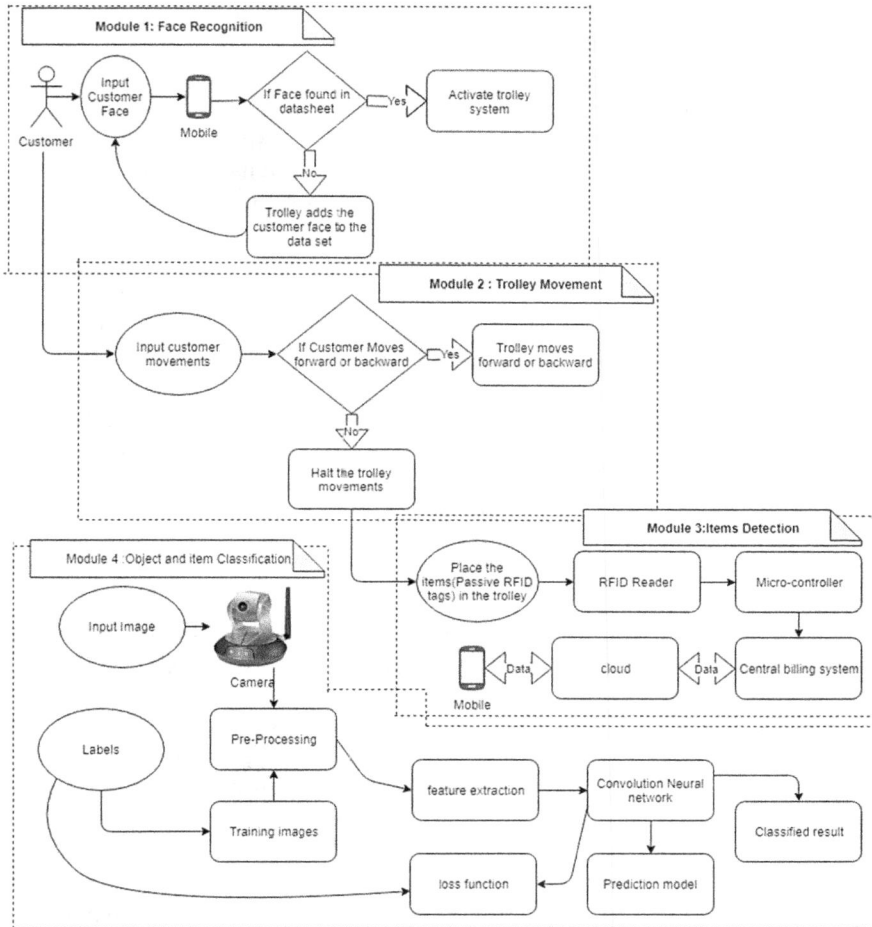

FIGURE 12.13 Architecture of proposed futuristic trolley model.

ii. *Module 2-Trolley Movement:* This module comprises sensors for making the trolley an independently moving user-governed vehicle. The sensors are set on either sides of the trolley to act accordingly. The trolley movements are governed by the client movements; if the client is moving forward, then the trolley follows while maintaining a certain distance between them automatically. Thus, hauling the check-in products around the store makes the system simple to utilize, and the client can have better shopping experience.

iii. *Module 3-Items Detection:* The client needs to pick the commodity which he/she is keen on and place it in the trolley. When the items are placed in the trolley, the RFID reader channels and extracts the label ID of each item and adds it with the item details. The details of the items are then sent to the central billing system of the market and subsequently to the cloud and to the user mobile. The aggregate sum will be paid using the client application in the mobile, which is linked to the server. These passive tags are scanned more than once in order to remove and know any label which is not scanned. The client can incorporate the item using the application. In the end, all items are shown in the application as a summary to the client; then, the payment can be initiated online without a fuss created at the checkout counters.

iv. *Module 4-Object and Item Classification:* This module contains a camera which is built into the trolley with numerous assignments and processes attached to it, from face acknowledgment to picture arrangement. The propelled grouping procedures and calculations are utilized to distinguish different items in the stores for their better arrangement. Further, it extracts the data and utilizes different sorts of classifier to anticipate the user and customer in a more friendly way. . The process flow of object classification is as follows: (i) initially, the various store images are labeled and trained for analyzing the store surroundings; (ii) real-time images are acquired by the camera that are then preprocessed; (iii) feature extraction of the images is done and passed to the Convolutional Neural Network (MaskR-CNN) [30]; and (iv) classified result of output image is seen by customers present in the particular aisle. Second, it can also anticipate the product details which are held by the customers in the particular aisle.

Initially the trolley is picked by the customer. Then, the trolley snaps the customer image to extract the facial features for accessing the data of his/her previous purchase and also for login authentication. The customer is compared with the data set to check whether he/she is a regular user or new user. If the customer is new user, then his/her data is updated to the data set as a new user. The trolley is also a moving governed vehicle in which sensors are attached to the trolley to act accordingly. Considering customer movements, the trolley moves forward or backward. The front sensor attached to the trolley is used for avoiding obstacles. An RFID reader is also attached to the trolley where the customer can scan the product they want to buy by placing the products in the trolley. The products placed in trolley are finally added and billing amount is shown in the mobile with payment option. The user is connected to server with mobile and can initiate the payment for purchase. The object classification algorithms (MaskR-CNN) used in the trolley can also identify the customer and products present near it for branding and product purchasing patterns, even in brick and mortar stores.

12.4 CONCLUSION

Convenience forms the backbone of major technological resolution. What intrigues people is an immersive and hassle-free experience. This includes prospects like guiding customer around the store recommending products which suits the customers taste, building up buying habits by leveraging customer purchase data, and using the same to offer discounts to him. The incentive behind the development of this smart trolley system is that it creates this immersive experience by catering to the customer like a concierge while simultaneously creating a huge amount of curated data. This helps in increasing efficiency in exploiting purchase patterns, easy accessibility, improving productivity, predicting interest, and saving administrative and labor cost in a plethora of domains such as shopping markets, restaurants, hospitals, offices, airways, etc.

The current work in the field has made advancements such as RFID-based object detection using circular polarized antennas and weight sensing mats on the racks and on trolley to sense a product being picked up from the self and placed in the cart, but some improvements still remain like an Automatic Ground Vehicle (AGV) to remove the manual hauling of the goods, face recognition of customer for product preference, smart web-based billing and convenience check-outs. A secondary benefit includes data such as the number of customers in a certain aisle, number of customer interacting with a certain product, its sales conversion ratio, etc. This chapter builds up on the previous work and provides a comparative analysis of all the trolley-based models, suggesting improvements that can create a better and more complete model. This chapter may help readers who are working in the field of autonomous trolley-based models to understand and proceed in their research work. It is concluded that till date there is no single unified development of trolley-based model that can perform suitably for all kinds of applications. Each model has its own strengths and weaknesses. The challenge for the future is to design an efficient and effective trolley-based model that is flexible, fast, low cost, and accurate with minimum hardware requirements.

REFERENCES

1. C.K. Sunitha, E. Gnanadhas, "Online shopping - an overview," *B-Digest*, vol. 6, 2011, pp. 16–22.
2. A. Thomas, R. Garland, "Supermarket shopping lists: Their effect on consumer expenditure," *International Journal of Retail & Distribution Management*, vol. 21, no. 2, Feb. 1993, pp. 8–14.
3. A. Sarac, N. Absi, S.D. Peres, "A literature review on the impact of RFID technologies on supply chain management," *International Journal Production Economics*, vol. 128, Nov. 2010, pp. 77–95.
4. T. Fernandes, R. Pedroso, "The effect of self-checkout quality on customer satisfaction and repatronage in a retail context," *Service Business*, vol. 11, Mar. 2017, pp. 69–92.
5. J.G. Lee, M.S. Kim, T.M. Hwang, S.J. Kang, "A mobile robot which can follow and lead human by detecting user location and behavior with wearable devices," in: *International Conference on Consumer Electronics*, IEEE, Las Vegas, NV, Jan. 7–11, 2016, pp. 209–210.
6. V. Gupta, N. Garg, "Analytical model for automating purchases using RFID-enabled shelf and cart," *International Journal of Information and Computation Technology*, vol. 4, no. 5, 2014, pp. 537–544.
7. R. Kumar, K. Gopalakrishna, K. Ramesha, "Intelligent shopping cart," *International Journal of Engineering Science and Innovative Technology*, vol. 2, Jul. 2013, pp. 499–507.
8. C. Nuñez, A. García, R. Onetto, D. Alonzo, S. Tosunoglu, "Electronic luggage follower," in: *Florida Conference on Recent Advances in Robotics*, Jacksonville, FL, May 20–21, 2010, pp. 1–7.
9. Hands-free smart stroller, https://www.dailymail.co.uk/sciencetech/article-3410783/Would-trust-baby-self-driving-stroller-2-750-buggy-allows-parents-run-it.html.
10. Efficient face recognition algorithms and techniques, https://www.rankred.com/face-recognition-algorithms-techniques/.

11. P. Follmann, B. Drost, T. Böttger, "Acquire, augment, segment & enjoy: Weakly supervised instance segmentation of supermarket products," arXiv:1807.02001v2 [cs.CV], July 2018, pp. 1–14.
12. T. Athauda, J.C.L. Marin, J. Lee, N.C. Karmakar, "Robust low-cost passive UHF RFID based smart shopping trolley," *IEEE Journal of Radio Frequency Identification*, vol. 2, Sep. 2018, pp. 134–143.
13. History of barcode scanning, https://www.smithsonianmag.com/innovation/history-barcode-180956704/.
14. R.B. Ferguson, "Wal-Marts CIO dishes on RFID," *NRFTech Conference*, California, Aug. 9, 2006.
15. M. Liard, *"RFID Item-Level Tagging in Fashion Apparel and Footwear,"* Oyster Bay, NY, ABI Research, 2009, pp. 1–8.
16. A.A. Agarwal, S.K. Sultania, G. Jaiswal, P. Jain, "RFID based automatic shopping cart," *Control Theory and Informatics*, vol. 1, no.1, 2011, pp. 39–44.
17. B. McBeath, *"The Explosion of Retail Item-Level RFID: A Foundation for the Retail Revolution,"* ChainLink Research, Inc. Newton, Apr. 2013, pp. 1–16.
18. J.C. Penney, RFID tagging, RFID Journal, https://www.rfidjournal.com/articles/view? 10368.
19. S. Gupta, A. Kaur, A. Garg, A. Verma, A. Bansal, A. Singh, "Arduino based smart cart," *International Journal of Advanced Research in Computer Engineering & Technology*, vol.2, Dec. 2013, pp. 3083–3090.
20. U. Gangwal, S. Roy, J. Bapat, "Smart shopping cart for automated billing purpose using wireless sensor networks," in: *Seventh International Conference on Sensor Technologies and Applications, Barcelona, Spain*, Aug. 25–31, 2013, pp. 168–172.
21. S. Kamble, S. Meshram, R. Thokal, R. Gakre, "Developing a multitasking shopping trolley based on RFID technology," *International Journal of Soft Computing and Engineering*, vol. 3, Jan. 2014, pp. 179–183.
22. G. Jayshree, R. Gholap, P. Yadav, "RFID based automatic billing trolley," *International Journal of Emerging Technology and Advanced Engineering*, vol. 4, 2014, pp. 136–139.
23. P. Bhatwal, O. Chamoli, "Automated self-billing system using RFID technology," *International Journal of Emerging Trends & Technology in Computer Science*, vol. 4, Mar. 2015, pp. 209–212.
24. S. Ali, M. Riaz, "Smart trolley," 2015, pp. 1–14. Doi: 10.13140/RG.2.1.2230.2567.
25. R. Sawant, K. Krishnan, S. Bhokre, P. Bhosale, "The RFID based smart shopping cart," *International Journal of Engineering Research and General Science*, vol. 3, Mar. 2015, pp. 275–280.
26. R. Suganya, N. Swarnavalli, S. Vismitha, G.M. Rajathi, "Automated smart trolley with smart billing using Arduino," *International Journal for Research in Applied Science & Engineering Technology*, vol. 4, Mar. 2016, pp. 897–902.
27. P. Iyappan, S.S. Jana, S. Anitha, T. Sasirega, V.P. Venkatesan, "An enhanced shopping model for improving smartness in markets using SABIS architecture," in: *2016 International Conference on Wireless Communications, Signal Processing and Networking*, IEEE, Chennai, India, Mar. 23–25, 2016, pp. 140–145.
28. H.H. Chiang, W.T. You, S.H. Lin, W.C. Shih, Y.T. Liao, J.S. Lee, Y.L. Chen, "Development of smart shopping carts with customer-oriented service," in: *2016 International Conference on System Science and Engineering*, IEEE, Puli, Taiwan, Jul. 7–9, 2016, pp. 1–2.
29. B.F. Wu, S.J. Yao, L.W. Hou, P.J. Chang, W.J. Tseng, C.W. Huang, Y.S. Chen, P.Y. Yang, "Intelligent shopping assistant system," in: *2016 International Automatic Control Conference*, IEEE, Taichung, Taiwan, Nov. 9–11, 2016, pp. 236–241.

30 JC.N. Swamy, D. Seshachalam, S.U. Shariff, "Smart RFID based interactive Kiosk cart using wireless sensornode," in: *2016 International Conference on Computational Systems and Information Systems for Sustainable Solutions*, IEEE, Bengaluru, India, Oct. 6–8, 2016, pp. 459–464.

31. P. Chavan, B. Gadgay, V. Pujari, "Smart shopping cart using RFID," *International Journal for Research in Applied Science & Engineering Technology*, vol. 5, June 2017, pp. 697–701.

32. R. Li, T. Song, N. Capurso, J. Yu, J. Couture, X. Cheng, "IoT applications on secure smart shopping system," *IEEE Internet of Things*, vol. 4, Dec. 2017, pp. 1945–1954.

33. E.D. Suryanto, H. Siagian, D.P. Angin, R. Sashanti, S. Yogen, "Design of automatic mobile trolley using ultrasonic sensor," *Journal of Physics: Conference Series*, vol. 1007, 2018, pp. 1–6.

34. G. Chithra, P.V. Sunil, M. Sneha, R. Shruthi, L.N. Sowmya, "IoT based futuristic trolley for intelligent billing with amalgamation of RFID and ArmLPC2148," *International Journal of Electrical, Electronics and Computer Systems*, vol. 6, 2018, pp. 104–106.

35. Mask R-CNN for object detection and instance segmentation on Keras and TensorFlow https://github.com/matterport/Mask_RCNN.

13 Big Data for Smart Health

Chetan S. Arage
Sanjay Ghodawat University

K. V. V. Satyanarayana
Koneru Lakshmaiah Education Foundation

Nikhil Karande
GH Raisoni Institute of Engineering and Technology

CONTENTS

13.1 INTRODUCTION

Day by day, data is bring out in a speed of various functions, gadgets, and topographical exploration actions for the reason of climate determining, climate indicator, calamity assessment, infraction identification, and the haleness organization, to name a some. In recent work, Big Data is related with automation, and different firms like Google, Facebook, and IBM distillate important information from the big amount of data gathered [1–3]. A stage of accessible data in health protection is currently at bottom in the work. Big Data is bring out briskly in all the areas in addition to health protection, corresponding to patient prevention, conformity, and different managerial needs but also we need to focus on how to improve doctors convey care too. Human services IoT can likewise help quiet commitment and fulfillment by enabling patients to invest more energy associating with their doctors.

The web of things has various applications in human services, from remote checking to keen sensors and restorative gadget reconciliation. It can possibly keep patients sheltered and secure, but human services IoT is not without its challenges. The quantity of associated gadgets and the huge measure of information they gather can be a test for clinics IT teams to oversee. Here comes the idea of Big Data, on the grounds that big data utilized for the most part for gathering and keeping up the enormous amount of information. As the worldwide populace keeps on expanding alongside the increasing human life expectancy, treatment conveyance models are developing rapidly, and a portion of the choices hidden in these quick changes must be found to produce information [4]. Human services investors are guaranteed new learning from enormous information, and the pair is known for its amount (volume) just as its multifaceted nature and speed. Medicine-industry specialists and investors have started to commonly break down enormous information to acquire understanding, yet these exercises are still in the beginning periods and must be facilitated to address social insurance conveyance issues and improve human services quality. Early frameworks for enormous information examination of human services informatics have been built up crosswise over numerous situations, e.g., the examination of patient qualities and assurance of analysis charge and conclusions to recognize the best and most economical medicines [4]. Well-being information processing is portrayed as the knowledge of social insurance education, registering education and data education in the investigation of medicinal services data. Well-being informatics includes information procurement, stockpiling, and recovery to give better outcomes by social insurance suppliers.

Data: The amounts, characters, or images on which tasks are performed by a PC, which might be put away and transmitted as electrical flags and recorded on attractive, optical, or mechanical account media.

13.1.1 WHAT IS BIG DATA?

Big Data is still data, but with a huge scale. Enormous Data is a concept used to represent a set of data that is enormous in scale, but then grows exponentially over time. In brief, such data is so immense and dynamic that none of the traditional information can be preserved or analyzed productively by the executive's computers. Today,

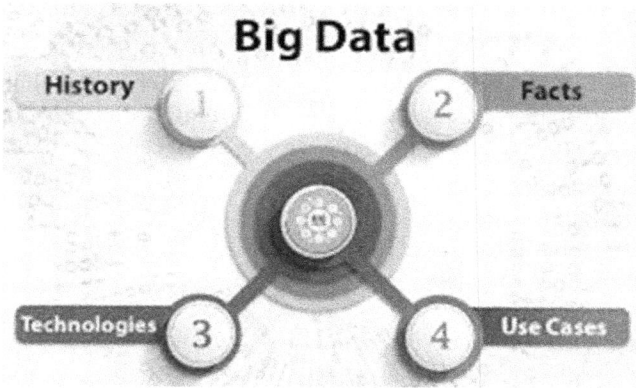

FIGURE 13.1   Big Data functions.

there are many drawbacks in the health system-related corporate medical care field and cloud information server, primarily about how to access, handle, and interpret healthcare or medical-related data from sources that are various and multiple systems that are stored in the cloud and available from anywhere (Figure 13.1).

As technology, Machine Learning helps to process a vast data volume, ease the role of data scientists in an environment which is automated, and gain equal recognition and acknowledgement everywhere at any time, the clod figure outing and Big Data are acquiring popularity for their advancements & sharing of data and accessible everywhere at any given time. The methods we used for data drilling were from several years earlier, but they were not successful in getting perfect outputs because they were unable to use an efficient output algorithm. Machine Learning is a methodology that focuses on the creation of computer programs that can find and use data and learn about themselves.

13.1.2   TYPES OF BIG DATA

Big Data could be found in three forms:

1. Structured
2. Unstructured
3. Semi-structured

13.1.2.1   Structured

Any information that can be packed aside, obtained, and viewed as a set structure is referred to as "ordered" data. Over the timeframe, software engineering capabilities have made more substantial strides in designing approaches to work with such knowledge (where the arrangement is visible ahead of time) in addition to inferring an opportunity from it. Be that as it may, these days, we are expecting problems of scale as such information grows to a monumental degree; run of the mill sizes are now in the territory of multiple zeta bytes.

13.1.2.2 Unstructured

It delegates unstructured information to any information with an obscure framework or layout. Despite the enormous amount, unorganized data poses multiple difficulties with regard to its management to derive an opportunity from it. An ordinary case of unstructured data is a heterogeneous source of information containing a mix of basic records, photographs, videos, and so on. Currently, associations have lots of information available to them, but frankly, they do not have the foggiest understanding of how to assess an opportunity from it because this information is in its crude framework or unstructured framework.

13.1.2.3 Semi-structured

Semi-organized data can contain all kinds of data. We may see semi-ordered data as an organized framework, but in social Database Management System (DBMS), for instance, it is often not defined by a table description. The semi-organized information case is information that is referred to in an XML text.

13.1.3 INTERNET OF THING

The web of element is a modeling point of expertized, social, and monetary criticalness. Customer things, tough products, trucks and rickshaw, modern and service parts, social insurance framework, sensors, and other daily news stories are being tied with web availability and ground-breaking information illustrative abilities that assurance to variation the manner in which we breathing, act and show. Extensions for the result of IoT on the web and economy are amazing, with few envisioning the highest of the 100 billion combined IoT applications and the international economic estimate of greater than $11 trillion by 2025. Simultaneously, nonetheless, the Internet of Things raises noteworthy difficulties that could obstruct understanding its potential advantages.

The web of things has various applications in human services, from remote observing to brilliant sensors and medicinal gadget joining. Prior to Internet of Things, patients' cooperation with specialists were restricted to visits, and telephone and content correspondences. There were no chances that specialists or medical clinics could screen patients' well-being consistently and make proposals as needs be (Figure 13.2).

> *Internet of Things (IoT)*: Empowered gadgets have rendered remote observation possible in the field of health care, realizing the potential to hold patients healthy and powerful, and empowering doctors to express superlative concern. As interactions with professionals have turned out to be easier and more efficient, patient devotion and fulfilment have also increased. In addition, patients benefit from remote well-being to decrease the dimension of clinic stay and counteract potential allegations. Similarly, IoT has a significant effect on reducing government benefit premiums and enhancing patient quality.
>
> Without a question, IoT is transforming social protection organizations by rethinking the reach of procedures and collaboration between persons

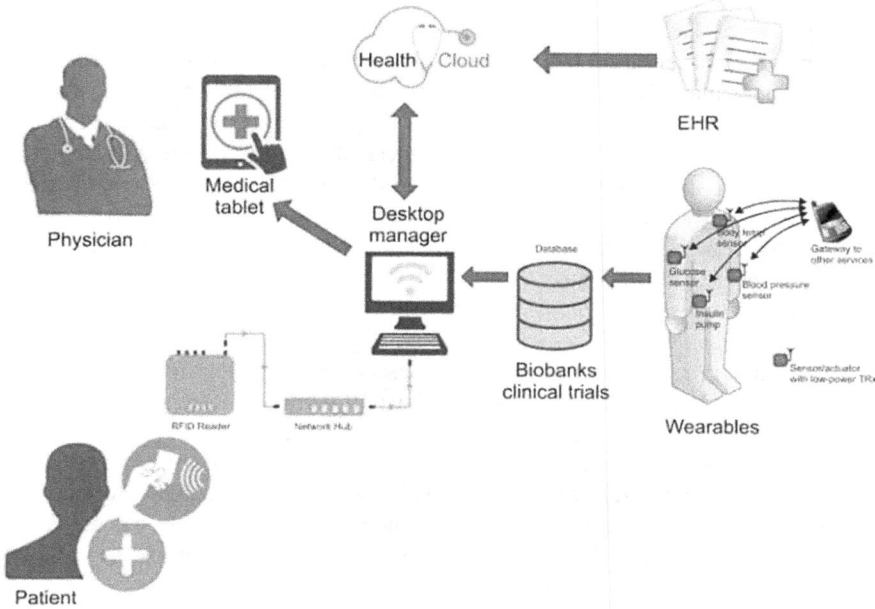

FIGURE 13.2 Use of IoT in healthcare.

in the implementation of medical service arrangements. IoT has human resources responsibilities that are useful to patients, families, physicians, medical centers, and insurance providers.

IoT for Patients: Wearable gadgets such as wellness communities and other remote-related gadgets such as circulatory strain and pulse monitor sleeves, glucometer, etc., offer tailored care to patients. To recall calorie tally, practice search, arrangements, circulatory tension varieties and considerably more, these devices can be modified.

By motivating regular follow-up of wellness symptoms, IoT has changed individuals, especially old patients. This mainly concerns people and their families living alone. Ready tool delivers signs to families and worried well-being suppliers about some disturbing impact or shifts in an individual's regular workouts.

IoT for Physicians: Doctors can track the well-being of patients all the more viably by using wearable and other home checking devices implanted with IoT. They will pursue the commitment of patients to medication schedules or some timely clinical intervention condition. IoT empowers specialists in human resources to be increasingly responsive and proactively engage with customers. Input obtained from IoT gadgets will make it easier for doctors to differentiate patients for the right clinical process and obtain good outcomes.

IoT for Hospitals: There are several different places where IoT technologies are useful in hospitals, in addition to monitoring the well-being of patients.

Sensor-labeled IoT gadgets are used for the constant tracking of clinical appliances such as wheelchairs, defibrillators, nebulizers, oxygen syphons, and other observation equipment. Sending restorative workers to separate locations may also be helpful to sort out the issue of broken-down gadgets on an ongoing basis. A notable problem for patients in emergency centers is the transmission of pollutants. Cleanliness observing gadgets operated by IoT helps to prevent patients from being infected. IoT systems also assist the board in resource management, such as grocery store inventory control, and atmospheric monitoring, such as icebox temperature control, dampness and atmosphere or heat control.

IoT for Health Insurance Industries: There are various open doors for well-being backup plans with IoT-associated savvy gadgets. Insurance agencies can use information caught through well-being observing applications for their endorsing and claims works. This information will empower them to distinguish extortion claims and recognize anticipation. IoT gadgets get straightforwardness among backup plans and clients the guaranteeing, evaluating, claims taking care of, and hazard appraisal forms. In the light of IoT-caught information driven choices in all activity forms, customers will have satisfactory possibility into invisible thought behind each choice made and process results.

Backup plans may offer motivating forces to their clients for utilizing and sharing well-being information produced by IoT gadgets. They can remunerate clients for utilizing IoT gadgets to monitor their standard exercises and adherence to treatment plans and prudent well-being measures. This will assist backup plans with reducing claims altogether. IoT applications can similarly allow insurance organizations to accept the information caught by these applications (Figure 13.3).

The significance points of interest of IoT in medicinal services include:

- *Cost Reduction*: IoT empowers patient checking continuously, thus essentially chopping down superfluous visits to specialists, medical clinic stays, and re-confirmations

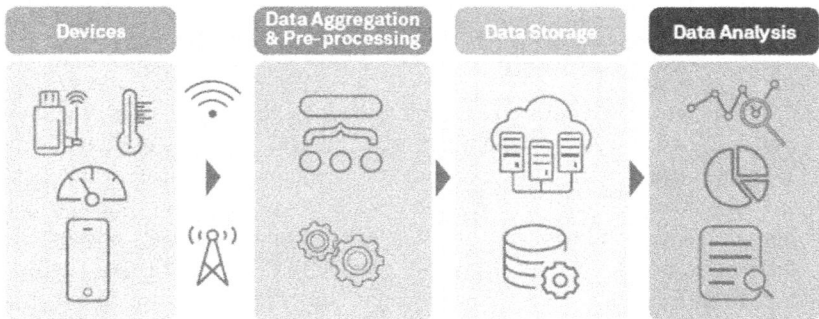

FIGURE 13.3 The four stages of IoT solutions.

- *Improved Treatment*: It empowers doctors to settle on proof-based educated choices and brings supreme straightforwardness
- *Faster Disease Diagnosis*: Continuous patient observing and ongoing information helps in diagnosing illnesses at the early stage even before the malady results in symptoms
- *Proactive Treatment*: Continuous well-being checking opens the door for giving proactive medicinal treatment
- *Drugs and Equipment Management*: Management of medications and restorative hardware is a noteworthy test in a social insurance industry. Through associated gadgets, these are overseen and used effectively with diminished expenses
- *Error Reduction*: Data developed by IoT gadgets contribute to strong basic leadership and ensure smooth operations of medical services with decreased blunders, duplication, and framework costs.

The IoT for medical facilities is not without complications. Related IoT-empowered gadgets grab immense data steps, including touchy data, leading to many information security issues.

It is important to actualize capable defense efforts. Via constant well-being observation and access to the well-being information of patients, IoT explores new aspects of patient care. This data is a gold mine for social security partners to enhance the well-being and encounters of patients while making revenue openings and enhancing the roles of medicinal services. Being set up to equip this advanced force in the inexorably connected universe will prove to be the differentiator.

13.2 CHALLENGES

13.2.1 Data Protection and Privacy

Information preservation and defense are the most important risks that IoT postures face. IoT gadgets steadily collect and relay information.

13.2.1.1 Incorporation: Different Gadgets and Conventions

The inclusion of numerous devices also blocks the introduction of the IoT in the social security sector. The reasoning behind this impediment is that gadget makers have not reached a consensus with regard to norms and guidelines for communications.

Along these lines, irrespective of whether the ranges of devices are related, the difference in their communication convention confuses and inhibits the mechanism of collection of knowledge. This inconsistency of the standards of the related gadget hinders the whole process and reduces the degree of IoT flexibility in social security.

13.2.2 Information Over-Burden and Exactness

Due to the usage of separate communication norms and benchmarks, as examined before, information conglomeration is problematic. IoT gadgets, in any case, still record a large amount of details. The information obtained by IoT devices is used to improve critical bits of knowledge.

13.2.2.1　Cost

Amazed to see cost factors in the research areas? I know the vast majority of you will be; but the primary issue is: IoT has not yet made human care fair for the simple man to promote. The explosion in healthcare costs is a stressful sign for all, particularly the nations that are developing. The scenario is to such a degree that it led to "Medical Tourism" in which patients with essential situations get to the generating countries' human resources departments and undertake surgery there, costing them as little as one-tenth. An fascinating and promising concept is, the IoT in social security is a theory.

13.3　FRAMEWORK ANALYSIS

With the appearance of the savvy City, keen social insurance administrations are developing to improve urban natives' personal satisfaction. Be that as it may, to associate and access shrewd medicinal services administrations, individuals, and sensors, flawlessly, any time, and any place, Mobile Cloud Computing (MCC) assumes an essential job. In MCC administrations, clients can offload (part of) an errand in cloudlet for quicker execution. The existing writing talks about a few systems to recognize when a customer ought to offload task in a cloudlet [4,5,6,7]. Be that as it may, little work has been done in the field of VM movement in MCC frameworks. A traffic product, cross-site VM relocation model is proposed.

In this model, when numerous VMs require movement, a self-assertive arrangement of VM relocation blocks the data transfer capacity of between-site joins along these lines, diminishing the quantity of effective VM movements. At that point, the VM movement issue is detailed as a Mixed Integer Linear Programming (MILP) issue and a heuristic calculation is utilized to get an estimated ideal outcome. Other than this, a portability-instigated administration relocation for MCC is proposed. In this work, a limit-based ideal administration movement approach is created, where the VM relocation issue is demonstrated as a Markov choice procedure (MDP). This proposed framework considered versatile clients to pursue a one-dimensional uneven irregular walk portability model. An administration is moved starting with one smaller scale cloud then onto the next when a client is in states limited by a lot of predefined edges. Furthermore, three lightweight undertaking movement models are created in [8]: Cloud-wide task relocation, where the assignment movement choice is made by a focal cloud, which boosts the targets of a cloud supplier. Server-driven undertaking relocation, where all movement choices are made by the server, where the assignment is as of now executing. Errand-based relocation, where movement is started by the undertaking itself. In this methodology, the movement choice is settled on after every choice age, in light of client's portability and remaining assignment execution time. This proposed strategy considers the expanding information volume move time during errand relocation starting with one cloud then onto the next. In the interim, in our past work, we proposed a portability and burden-mindful Genetic Algorithm-based VM movement approach, GAVMM, to limit task-execution time. Be that as it may, this methodology basically attempts to limit the provisioned errand execution time without considering over-provisioned assets in the cloudlets. Hence, the GAVMM neglects to limit asset over-provisioning in the cloudlets. Be that as it may, all best in class works utilize the single VM movement approach, where a cloudlet relocates a solitary VM to another cloudlet. This methodology loosens up the issue plan but neglects to adequately advance the entire framework destination. Rather, we here utilize a joint VM relocation, where many VMs together move to a lot of cloudlets, permitting powerful streamlining of asset utilization and errand execution time. In synopsis, most VM movement philosophies do not successfully consider client portability close by the burden state of cloudlet servers in a heterogeneous MCC framework. This builds administration personal time, particularly for those applications where the client regularly connects with the provisioned cloudlet. Likewise, when moving a VM, over-provisioned assets in the objective cloudlet must be additionally thought to be something else; also, the complete number of provisioned VMs in the objective cloudlet will significantly be decreased. As far as we could possibly know, this work is the first to proficiently use ACO framework to build up a VM movement approach for limiting the errand execution time and advancing the cloudlets asset use. In addition, we stretch out the VM movement model to mutually move a lot of VMs to a lot of cloudlets so as to limit task-execution time and to limit asset over-provisioning contrasted with single VM-based relocation draws near.

Enormous wellbeing is a promising industry, which is portrayed by human focus, dealing with an individual's wellbeing from birth to death, from counteractive action to recovery and including industry from government to advertise. The space of enormous well-being covers well-being items (counting the medications, therapeutic gadgets, and senior items), well-being administration (counting medicinal administrations, annuity administrations, and versatile social insurance), well-being land (counting benefits, human services), and well-being (counting health care coverage and other money-related items).

Human illnesses are of this kind: 33% may be fully prevented, 33% should be detected early, and 33% could be able to lead to personal well-being with forceful treatment. By reinforcing early detection, actual diseases may be limited. The transition from well-being to illness is witnessed. The well-being status is, as a rule, from well-being to okay status, to high-hazard status, to early sores status, to severe side effects status, and eventually to the status of illness.

13.4 BIG DATA ANALYTICS IN HEALTH INFORMATICS

The key gap between the middle customary well-being analysis and tremendous well-being analysis of knowledge is the introduction of PC programming. The social insurance agency relies on numerous companies for tremendous knowledge processing in the normal setting. In view of the substantial outcomes, many human services investors support technology creation, their working processes are realistic, and they will work on knowledge for institutionalized systems. Nowadays, with the test of taking care of increasingly increasing enormous social security records, the human services agency is investigated. The area of large information investigation is evolving and will potentially supply the social insurance system with useful bits of expertise. As stated above, a large proportion of the gigantic information measures provided by this system are spread in written copies, which should be digitized at a later stage [9]. Enormous data will increase the conveyance of social insurance and decrease its cost, while helping to encourage patient concern, advance medical outcomes, and reduce pointless expenses [6]. As of now, tremendous knowledge analysis has helped to predict the consequences of doctor decisions, such as whether to continue with the heart operation based on the age, medical health, and well-being perceptions of the patient. Basically, after looking into all this, we can infer that the role of tremendous data in the well-being field is to track knowledge indexes associated with medicinal facilities, which are confounding and difficult to supervise using existing facilities, programming, and software. Notwithstanding the increasing amount of knowledge on social security, recovery strategies are still evolving [10]. In this way, intentional usage and execution-dependent pay have evolved in the social security category as key elements. In 2011, public resources associations already generated more than 150 additional bytes of information [11] in order to be useful to the social care framework [12], which would all be productively investigated [12]. The requirement to include knowledge related to social security in EHRs exists in a number of systems. In the field of bioinformatics, where various terabytes of information are provided by genomic sequencing, an abrupt rise in information connected with medicinal services information processing has also been seen [12]. There is an assortment of scientific frameworks required for restorative deciphering, which could then be used for patient consideration [13]. The social insurance informatics network is testing the diverse origins and forms of tremendous details to establish information processing strategies. There is a huge interest in a process that incorporates diverse sources of information [7]. It is necessary to use multiple logical methodologies to perceive inconsistencies in enormous calculations of data from different data sets (Figure 13.4).

There is currently no system that holds all cloud records relating to medicine or healthcare, such as diagnostic testing, scans, or prescriptions from a patient between appointments that can be viewed with a safe device from anywhere. Many medical-related agencies today use computer devices and software to store data on the machine and implement a manual method that eliminates the effort a human has

FIGURE 13.4 Contributions of big data in healthcare system.

to put to access control, and decreases the effort and time to access those control. But it also takes more time for consumers who are unable to view data electronically from their own location to enter the location manually.

This can be a daunting challenge as there is no one source where complete data store relevant to medical/health the consumer can access from any point of location so that customer commitment and time can be minimized. Interoperability and multiple data specifications or formats are a big obstacle for data fusion activities. Many researchers have been working on the system for many years to store all data which are medical-related stores dispersed or consolidated and capable of being access them from any place, but they are unable to enforce them due to security issues. Furthermore, multiple cloud data storage is feasible, but it increases the sharing of records and creates uncertainty and time. These needless and risky examinations, though placing the patient at risk at the same time, are not used. A similar challenge is also posed by drug theft. From this point on, another daunting challenge is to keep data protected from unauthorized persons. Many organizations have been focusing on data protection over the last few years so that data access on the server can be shielded from unwanted persons. As a result, providers of health care are searching for other reliable ways to protect accessible data records that are personally stored on the device or network. The aim of forecasting a known output or goal begins with supervised learning. In Machine Learning contests, where human competitors are judged on their results on standard data sets, repetitive supervised learning challenges include handwriting identification (such as identifying handwritten digits), classifying photographs of items (e.g., is this a cat or a dog?), and text labelling (e.g., is this a clinical experiment for a financial report or a cardiac disease). Notably, all these are tasks that could be performed best by a skilled individual, and so the program also tries to imitate human output. Where there are no results to predict, as with unsupervised learning.

Instead, in the results, we attempt to identify naturally occurring trends or groupings. This is necessarily a more complex challenge to judge and also its success in subsequent supervised learning activities tests the importance of certain classes gained by unsupervised learning.

Generally, there are various basic barriers for dealing with them in the area of the healthcare sector. Some of the main problems involved in handling EHRs are as follows:

1. It is very difficult to exchange medical or treatment data with adequately covered consumers.
2. Patients have restricted access, for safety reasons, to their own data and to their administrators.
3. Any data not accessible on any decentralized authentication scheme.
4. Not allowed to receive health-related alerts from time to time.

13.4.1   THE FRAMEWORKS AVAILABLE FOR THE ANALYSIS OF HEALTHCARE DATA

A systematic methodology composed of six key skills that companies need to consider when building up a Big Data enterprise is the architecture of Big Data. The diagram below illustrates the Big Data Framework (Figure 13.5).

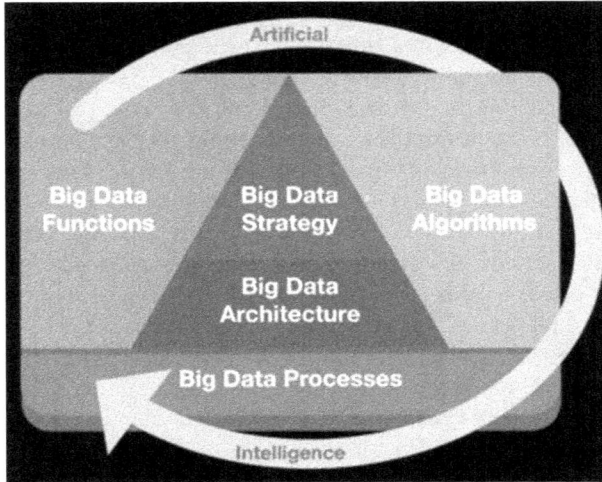

FIGURE 13.5   Big data process.

13.4.2   THE SIX KEY FEATURES OF THE ARCHITECTURE

13.4.2.1   Strategy of Big Data

For a number of organizations, data has become a competitive tool. The ability to analyses vast data sets and detect trends in the data will be a strategic edge for organizations. For example, in choosing what movies or series to make, Netflix looks at consumer behavior. By defining which providers to loan money and advice on their site, the Chinese sourcing site Alibaba has become one of the global giants. Large Data is Big Business today.

Enterprise companies need a sound Big Data plan to generate positive benefits from investments in Digital Data. How can investment gains be obtained, and where can Big Data research and analytics target efforts? There are practically infinite options for research, and companies may potentially get lost in the data bytes of zetta. The first phase to the sustainability of Big Data is a solid and organized Big Data approach.

13.4.2.2   Big Data Architecture

Organizations must have the caliber to store and handle vast volumes of data in order to deal with huge data sets. In order to achieve this, to facilitate Big Data, the enterprise should have the underlying IT infrastructure. Therefore, organizations should provide a robust infrastructure of Big Data to enable the study of Big Data. To promote Big Data, how do businesses plan and set up their architecture? And, from a storage and transmission standpoint, what are the requirements?

The technological capabilities of Big Data ecosystems are called by the Big Data Design aspect of the Big Data Platform. It addresses the different positions within a Big Data Framework that are present and looks at the best architectural practices. This section would suggest the reference Big Data architecture of the National Institute of Standards and Technology (NIST) in keeping with the vendor-independent nature of the system.

13.4.2.3   Big Data Algorithms

To have a detailed knowledge of statistics and algorithms is a profound capacity to deal with results. Therefore, to deduct insights from evidence, Big Data experts need to have a strong background in analytics and algorithms. Algorithms are simple descriptions of how a class of problems should be solved. Calculations, data analysis, and automatic logic functions may be done by algorithms. Valuable knowledge and observations can be gained by applying algorithms to vast quantities of data.

The system portion of Big Data algorithms relies on the skills of someone who has aspirations to work with Big Data. It seeks to create a stable base including fundamental operations of mathematics and offers an introduction to various algorithm groups.

13.4.2.4   Big Data Processes

It is a profound skill to work with outcomes and provide a clear understanding of algorithms and statistics. Therefore, Big Data experts are expected to have a deep background in algorithms and analytics to deduct lessons from evidence. Simple examples of ways to address a group of issues are known as Algorithms. Algorithms can perform data analysis, automatic logic functions, and calculations. Through application of algorithms to large data amounts, useful insights and information can be obtained.

Algorithms of Big Data rely on the expertise of someone who has an aspiration to work with Big Data in the machine section. It aims to establish a consistent basis that requires operations that are simple and mathematical and provides an introduction to algorithms of different classes.

13.4.2.5   Big Data Functions

The functions of Big Data involve the functional aspects of the application of Big Data in companies. This part of the Big Data system explores how organizations should coordinate themselves to define positions in Big Data and address duties and obligations in Big Data organizations. The effectiveness of Big Data programs is profoundly influenced by corporate culture, organizational frameworks, and work functions. Therefore, we will study some "best practices" in setting up Big Data firms.

The nontechnical elements of Big Data are addressed in the Big Data Roles portion of the Big Data System. You'll learn how to set up a Center of Excellence for Big Data (BDCoE). In addition, it also discusses crucial performance drivers for beginning the organization's Big Data mission.

13.4.2.6   Artificial Intelligence

Artificial Intelligence (AI) tackles the last part of the Big Data System. AI is one of today's big fields of concern and promises a whole world of opportunity. We discuss the relationship between Big Data and Artificial Intelligence in this section and describe core AI characteristics.

Many companies are willing to launch ventures in the area of AI, but others are not sure where to commence their journey. In the sense of delivering market opportunities to corporate organizations, the Big Data Architecture takes a practical overview of AI. Therefore, the last segment of the framework illustrates how AI is a sensible next move for companies who have set up the Big Data Platform's other capabilities. The last feature of the Big Data System has been interpreted for purposes as a lifecycle. In order to have long-term benefit, AI should continue to learn constantly from the Big Data in the enterprise (Figures 13.6 and 13.7).

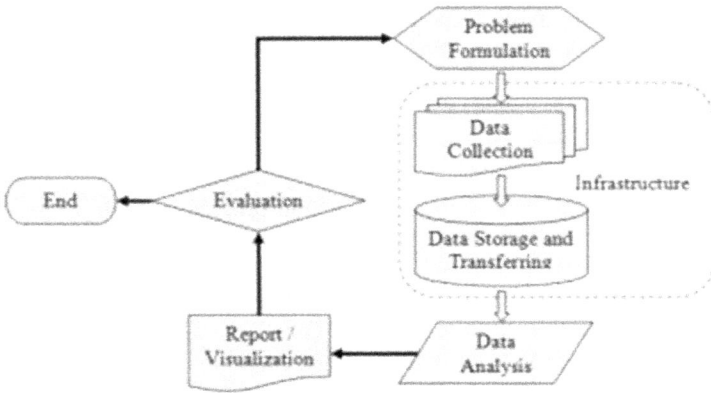

FIGURE 13.6   Generalized workflow of big data.

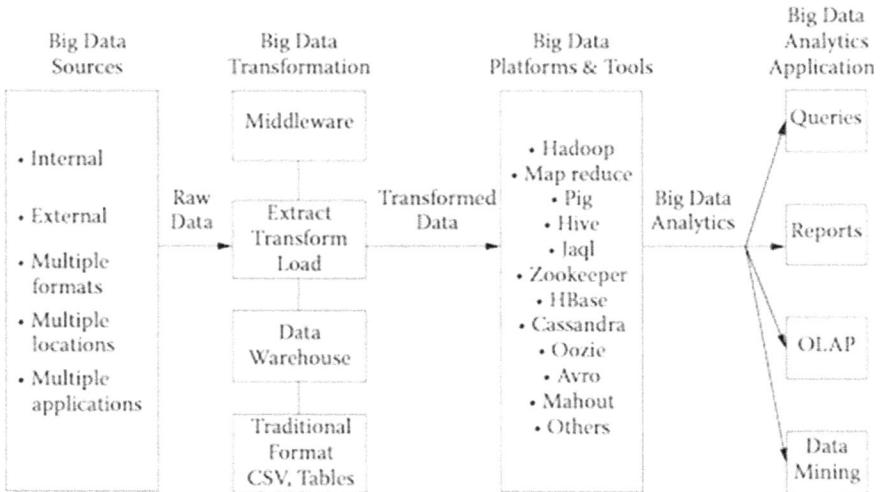

FIGURE 13.7   Big data process for data utilization.

13.5 IMPACT OF BIG DATA ON THE HEALTHCARE SYSTEM

In relation to the most suitable or accurate patient assessment and the quality evidence used in the well-being informatics framework [14], the capacity of tremendous knowledge could affect outcomes. In this regard, the review of enormous data measures would have a significant effect on the framework of clinical administrations in five respects, or "pathways." Improving conditions for patients on this path, as seen below, will be the focus of the system for medicinal treatment which will have a direct effect on the patient. Correct living: Correct living alludes to a greater and more beneficial life for the patient [14]. By careful living, patients might supervise themselves by making the right choices for themselves, making wise decisions, and enhancing their prosperity in the light of the use of data drilling. Patients should expect a working job of recognizing a balanced life by selecting the right way for their day-to-day well-being, with regard to their dietary schedule, preventive thought, fitness, and multiple real life exercises [15]. Right care: This path means that patients have access to the most acceptable research and that all vendors receive comparable knowledge and provide common priorities to prevent unnecessary structure and effort. In the time of tremendous knowledge, the angle has turned out to be increasingly rational. Right provider: In this manner, healthcare professionals will obtain a broad outlook on their patients by consolidating data from multiple outlets, such as clinical hardware, broad well-being perspectives, and financial information [14]. The existence of this data allows individual expert cooperatives to perform assessments and improve the skills to recognize and provide patients with stronger diagnostic alternatives. Correct innovation: This direction perceives the latest epidemic conditions, innovative drugs, and experimental medicinal therapies that will continue to advance [14]. Similarly, improvement in patient care structures, such as the revision of drugs and the success of creative work efforts, would empower stronger approaches to promoting development and patient well-being through the national social security framework. For stakeholders, the usability of previously tentative information is critical. This data will be used to examine high-potential targets and identify tactics for improving traditional methods of medical research. Right value: Suppliers must give their patients careful and continuing attention to enhance the efficiency and evaluation of well-being-related administration. The highest useful results accepted by their social security framework must be reached by patients. For example, separating and wrecking data deception, monitors, and squandering, and enhancing resources are allocations that should be taken to ensure the savvy use of information (Figure 13.8).

13.5.1 DATA SCIENCE OF HEALTHCARE DATA ANALYTICS

In the world of healthcare, there has been a data boom in Big Data. It took more than a decade for conventional methods implemented earlier to study genomics, DNA, and cancer with testing and techniques through the Human Genome Project to understand and evaluate the structure of DNA and data patterns. In order to evaluate chronic illnesses for treatment and recovery, Big Data Analytics has implemented groundbreaking instruments and techniques. In order to clarify the possible root

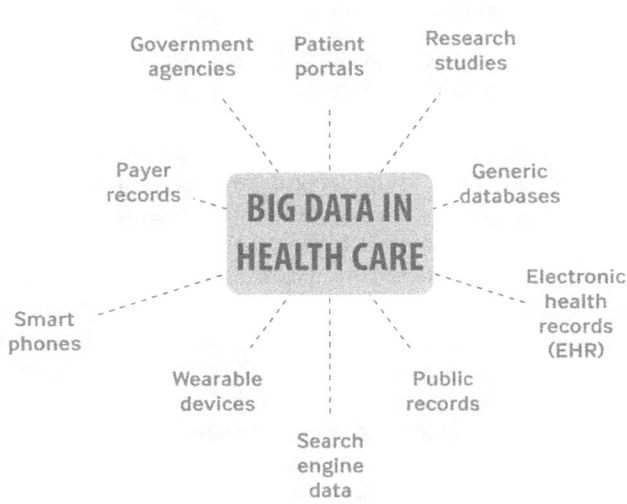

FIGURE 13.8   Healthcare system using IoT and concept of big data.

causes of tumor growth causing disease, gene sequencing has been used. From tera-bytes to exabytes, data has evolved exponentially. X-Ray, CT scan, and MRI health-care results have risen by leaps and bounds in terms of the amount of results. Via Big Data analytics, advanced medical technology allowed the diagnosis of the patient's records and a comparison of this with the global population to isolate the noises from the signal to recognize the dynamics of tumor growth that were not historically available to speed up the diagnosis and care. Though there are many hypotheses and methods that can be implemented for the diagnosis of the illnesses, this chapter briefly examines some of the main techniques (Figure 13.9).

It gets train data in less time span, one of the most relevant ones to merge health-care with computer learning or data analysis. Unmonitored learning is the primary purpose of learning from trained data, since unmonitored learning does not require training data. The knowledge used for unsupervised learning is not organized and processed correctly somewhere in the system. We are able to operate on health data and provide protection by using the various approaches. Mostly, by creating the clus-ter of data and correctly evaluating those data, we work on captured data. These details are then saved on the cloud so that we can view the information at any time and place. Many of the targets that can be accomplished by focusing on health data are as follows:

- Improve interoperability through the processing and preservation of distrib-uted cloud data.
- Design and eliminate effective data validation methods from the possibility of hacking.
- Save patients time and money. Creating an effective system for maintaining access to cloud-based electronic health information.

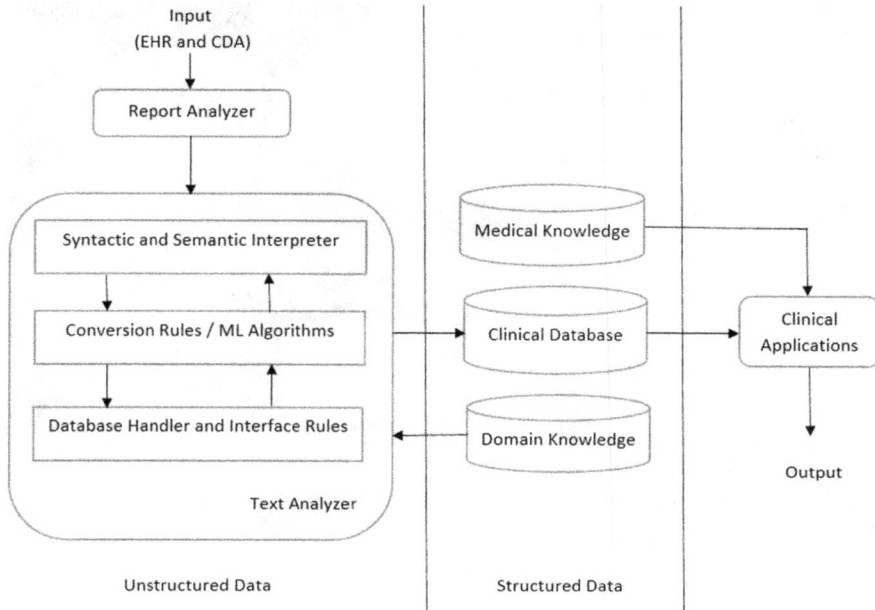

FIGURE 13.9 How healthcare-related data processed.

- Providing home health care facilities that encourage the importance of life for patients:
- Minimizing sickness, impairment, and disorders in patients.
- Maximizing the possible degree of freedom for patients.
- To maximize efficiency, security, results, and transparency.

13.6 APPLICATIONS OF IOT IN HEALTHCARE WITH BIG DATA

Due to its different variety of usage in various industries, the growth of the IoT is thrilling for everybody. It has many uses in healthcare. Healthcare IoT assists with (Figure 13.10):

- Reducing wait time in emergency departments
- Patient, employees, and product monitoring
- Strengthening drug treatment
- Ensuring vital hardware availability

IoT has also unveiled many wearables and accessories that have made patients' lives easy.

13.7 CONCLUSION

In view of the ever-growing number of computerized business processes and the vast volume of data available in the healthcare sector to be processed in parallel, it

FIGURE 13.10 Healthcare monitoring system.

is now unavoidable to accept and use IoT. Through applying various techniques of Big Data in the healthcare system and data made accessible to patients by IoT, it is possible to make reliable forecasts or projections about potential outcomes. In different grouping processes, this analysis used a community of people who are in the phase. Since the use of Big Data and IoT techniques in classification studies results in detailed results followed by substantial time and cost reductions, it is widely recommended that these techniques be used in data processing. This research is considered to be beneficial for healthcare organizations and individuals working in all fields of employment that have adapted to the computerization of their business processes and use large-scale data.

REFERENCES

1. Wenjin Yu, Tharam Dillon, Life Fellow, IEEE, Fahed Mostafa, Wenny Rahayu, Member, IEEE, and Yuehua Liu, "A global manufacturing big data ecosystem for fault detection in predictive maintenance", *IEEE Transactions on Industrial Informatics*, Vol. 16, No. 1, January 2020.
2. Pau Suan Mung, and Sabai Phyu," Effective analytics on healthcare big data using ensemble learning". 2020 IEEE Conference on Computer Applications (ICCA), Yangon, pp. 1–4, 2020. doi: 10.1109/ICCA49400.2020.9022853
3. Stefano Proto, Evelina Di Corso, Daniele Apiletti, Luca Cagliero, Tania Cerquitelli, Giovanni Malnati, and Davide Mazzucchi, "REDTag: A predictive maintenance framework for parcel delivery services", *IEEE*, January 6, 2020.
4. Aras Can Onal, Omer Berat Sezer, Murat Ozbayoglu, and Erdogan Dogdu, "Weather data analysis and sensor fault detection using an extended IoT framework with semantics, big data, and machine learning", *2017 IEEE International Conference on Big Data (BIGDATA)*. Boston, MA, 11–14 Dec. 2017.

5. Sunder Ali Khowaja, Aria Ghora Prabono, Feri Setiawan, Bernardo Nugroho Yahya, and Seok-Lyong Lee, *"Contextual Activity Based Healthcare Internet of Things, Services, and People (HIoTSP): An Architectural Framework for Healthcare Monitoring Using Wearable Sensors"*, 1389-1286/© 2018 Elsevier B.V. All rights reserved.

6. Lei Wang, Chen Yibo, Peng Li, and Lingxiao Zhao, *"Multimode Data Fusion Based Remote Healthcare Framework"*, BDET 2018, August 25–27, 2018, Chengdu, China © 2018 Association for Computing Machinery.

7. Bikash Kanti Sarkar, *"Big Data for Secure Healthcare System: A Conceptual Design"*, Received: 7 October 2016 / Accepted: 8 March 2017 © The Author(s) 2017.

8. Yueyao Wang, Qinmin Hu, Yang Song, and Liang He, *"Potentiality of Healthcare Big data: Improving Search by Automatic Query Reformulation"*, 978-1-5386-2715-0/17/$31.00 ©2017 IEEE.

9. Ali Al-Badia, Ali Tarhinia, and Asharul Islam Kha, *"Exploring Big Data Governance Frameworks"*, 1877-0509 © 2018 The Authors. Published by Elsevier Ltd.

10. Shiva Raj Pokhrel, Keshav Sood, Shui Yu, and Mohammad Reza Nosouhi, *"Policy-based Bigdata Security and QoS Framework for SDN/IoT: An Analytic Approach"*, 978-1-7281-1878-9/19/$31.00 ©2019 IEEE.

11. Qing Wang, Xiaodong Wang, and Ye Tao, *"A User Profile Analysis Framework Driven by Distributed Ma-chine Learning for Big Data"*, AICS 2019, July 12–13, 2019, Wuhan, Hubei, China © 2019 Association for Computing Machinery.

12. Ahmed E. Youssef, "A framework for secure healthcare systems based on big data analytics in mobile cloud computing environments", *International Journal of Ambient Systems and Applications (IJASA)* Vol.2, No.2, 1–11, June 2014.

13. G. Shwetha, P.R Visali Lakshmi, and N. Sri Madhava Raja, *"Analysis of Medical Image and Health Informatics Using Bigdata"*, 978-1-5090-4855-7/17/$31.00 ©2017 IEEE.

14. Haiping Huang, Tianhe Gong, Ning Ye, Ruchuan Wang, and Yi Dou, *"Private and Secured Medical Data Transmission and Analysis for Wireless Sensing Healthcare System"*.

15. Hai Tao, Md Zakirul Alam Bhuiyan, Ahmed N. Abdalla, Mohammad Mehedi Hassan, Jasni Mohamad Zain, and Thaier Hayajneh, *"Secured Data Collection with Hardware-based Ciphers for IoT-based Healthcare"*, 2327–4662 (c) 2018 IEEE.

Index

For Product Safety Concerns and Information please contact our EU
representative GPSR@taylorandfrancis.com
Taylor & Francis Verlag GmbH, Kaufingerstraße 24, 80331 München, Germany